T0258114

Advances in
Ceramic Armor VIII

Advances in Bulk Cryst

Advances in Bulk Crystal and Thin Film Formation

Edited by **Sharon Levine**

New York

Published by NY Research Press,
23 West, 55th Street, Suite 816,
New York, NY 10019, USA
www.nyresearchpress.com

Advances in Bulk Crystal and Thin Film Formation
Edited by Sharon Levine

International Standard Book Number: 978-1-63238-029-6 (Hardback)

Printed in the United States of America.

Contents

Preface

Every book is a source of knowledge and this one is no exception. The idea that led to the conceptualization of this book was the fact that the world is advancing rapidly; which makes it crucial to document the progress in every field. I am aware that a lot of data is already available, yet, there is a lot more to learn. Hence, I accepted the responsibility of editing this book and contributing my knowledge to the community.

In contemporary research and development, materials manufacturing crystal growth is referred to as a method to solve a broad spectrum of technological tasks in the fabrications of materials with stipulated properties. This all-inclusive profound book enables a reader to achieve insight into essential characteristics of the field. It includes important topics under the section bulk crystal growth.

While editing this book, I had multiple visions for it. Then I finally narrowed down to make every chapter a sole standing text explaining a particular topic, so that they can be used independently. However, the umbrella subject sinews them into a common theme. This makes the book a unique platform of knowledge.

I would like to give the major credit of this book to the experts from every corner of the world, who took the time to share their expertise with us. Also, I owe the completion of this book to the never-ending support of my family, who supported me throughout the project.

Editor

Bulk Crystal Growth

The Growth and Properties of Rare Earth-Doped NaY(WO₄)₂ Large Size Crystals

Chaoyang Tu*, ZhenYu You, Jianfu Li, Yan Wang and Zhaojie Zhu

Key Laboratory of Photoelectric Materials Chemistry and Physics of CAS,
Fujian Institute of Research on the Structure of Matter,
Chinese Academy of Sciences,
P.R. China

1. Introduction

Recently, strong attention has been focused on development of a new-advanced material for optoelectronics applications. $MRe(WO_4)_2$ [M=alkali metal, Re=rare earth] single crystals is noticed as an interesting self-frequency conversion solid-state laser host material because of stimulated Raman scattering[1]. $NaY(WO_4)_2$ crystal is classified among the disorder crystalline host for lasing rare-earth ions[2]. Because of the disorder structure, the optical features in the absorption and emission spectrum even at low temperature are broadened.

The lattice parameters of $NaY(WO_4)_2$ crystal are a=b=5.205 Å and c=11.251 Å respectively with the space group of $I4_1/a$ [3]. This crystal is a typical tetragonal scheelite-type crystal with a formula $MT(WO_4)_2$, where M is a monovalent alkali cation and T a trivalent cation . In these materials the M and T cations are randomly distributed in the 2b and 2d sites [4], which can be replaced by rare earth ions, such as Nd^{3+}, Yb^{3+}, Tm^{3+}, Ho^{3+} and Ce^{3+}. As a consequence, the optical absorption and emission lines of rare earth doping ions become broadened, which allow some laser tunability as well as a better match with the available diode laser emissions used for pumping. As it melts congruently, large size single crystal can be easily obtained by the Czochralski (CZ) method. Furthermore, the higher concentration of rare earth ions can be accepted in the crystal because of the higher covalent characteristic results in the lower luminescent quenching efficiency. Compared to the other laser host crystals such as YAG and YVO_4 crystal, $NaY(WO_4)_2$ crystal has lower melting point and its raw materials for crystal growth is in-nocuity. As a result, $NaY(WO_4)_2$ crystal can serve as an excellent laser host. In this chapter, the crystal growth, thermal characteristic, optical and spectrum and laser properties of rare earth doped- $NaY(WO_4)_2$ crystals are presented.

2. The growth of large size crystals

Rare earth-doped $NaY(WO_4)_2$ crystals were grown in air along <001> direction by using Czochralski method[1~3]. The chemicals used were analytical grade Na_2CO_3, WO_3, Y_2O_3 and spectral grade Re_2O_3 (Re=Yb, Tm, Ho,Ce, Nd, Er). The starting materials were prepared by mixing Y_2O_3, Na_2CO_3, WO_3 and Re_2O_3 powders according to reaction formula:

$$Na_2CO_3+(1-x)Y_2O_3+x\ Re_2O_3+4WO_3\rightarrow 2NaY_{1-x}Re_x(WO_4)_2+CO_2\uparrow$$

The weighed materials with doping 6 at% Re^{3+} were thoroughly mixed and pressed and put into a platinum crucible with $\Phi50\times50$ mm³, then heated to 750 °C and kept for 18 h to decompose the Na_2CO_3, and ground, mixed again, and then reheated to 800 °C, kept for 24 h. The obtained sample was very hard ceramics.

The synthesized material melted congruently at 1210 °C. The platinum crucible was heat by conventional RF-heating method. Crucible size is 50 mm in diameter and 50 mm in height. The pulling rate was 1-1.5 mm/h and the crystal rotation rate at 12-20 r.p.m. To release the stress produced in the temperature-lowering process, the crystals were annealed at 1200 °C for 5-6 h and then cooled down to room temperature at a rate of 20 K/h.

The earlier grown crystals as shown in Fig.2.1 occur screwy crack during anneal process. In order to avoid the cracking of the crystal, the designed after-heater should be used and the above crystal must be taken to anneal again in O_2 atmosphere according to a special temperature-controlled procedure.

As a result, high-quality $(Tm^{3+},Ho^{3+},Nd^{3+},Yb^{3+},Er^{3+}/Yb^{3+})$ rare earth doped-NYW cylinder crystal with dimension of $\Phi25$ mm×100 mm (shown in Fig.2.2~2.5). The result shows that its optical homogeneity is 4×10^{-5}, as shown in Fig.2.5. It shows that it has excellent quality.

In order to obtain the large-sized rare earth doped-NY(WO$_4$)$_2$ crystals with high optical homogeneity, the control of growing processes and conditions are very important firstly. Then the used raw materials must be highly pure. Furthermore, to get the defined composition of the melt, the preparation of chemicals was found to be important. Thirdly, the seed surface must be melted to remove the defect in the seed before growing. And the growth point temperature must be a little higher than the saturation point. The control of the pulling rate, rotating rate and annealing rate is also very important. Finally, a designed after-heater should be used to avoid the crack of large crystal.

The concentration of rare earth ions in the $NaY(WO_4)_2$ crystal has been measured to by the inductively coupled plasma-atomic emission spectrometry (ICP-AES) method. A sample for the experiment has been cut from the top to eh boule. The concentration of Yb^{3+} ions has been 1.73 wt%. The distribution coefficient (K_0) of Yb^{3+} ions in the $Yb:NaY(WO_4)_2$ crystal has been calculated using the following relation:

$$K_0=C_A\ /C_0\ ;$$

Where C_A is the Yb concentration at the top of the grown crystal and C_0 is the initial concentration of the admixture. The result indicates that the segregation coefficient of Yb^{3+} ions in Yb: $NaY(WO_4)_2$ crystal is approximately 1.02.

3. The thermal characteristic

The a and c axes were obtained by the YX-2 X-ray Crystal Oridentation Unit (produced by Dandong Radiative Instrument Co,Ltd). Two pieces of square samples with the size 5×5×5 mm³ having polished faces perpendicular to the a and c crystallophysical directions were used to carry out the measurements. The thermal expansion of as-grown $Yb^{3+}:NaY(WO_4)_2$ crystal was measured by using Diatometer 402 PC instrument from 300 K to 1273 K[1]. Because of the relatively lower reliability of the room temperature cell parameter arising out of presence of water in the sample chamber, only the data from 473 to 1273 K is considered in calculating the expansion coefficients. The thermal expansion pattern was obtained (shown in the Fig.3.1). The thermal expansion coefficients of the Yb^{3+}:NaY(WO$_4$)$_2$ crystal were calculated over different temperature ranges. In this case, the linear thermal expansion coefficients for different crystallographic direction c- and a-axes are , 1.83×10^{-5} K^{-1}, 0.85×10^{-5} K^{-1}, respectively.

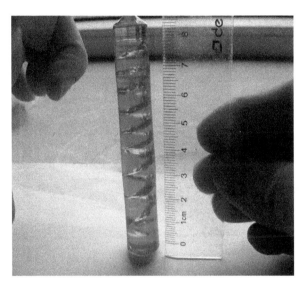

Fig. 2.1 The cracking Nd^{3+}:NaY(WO$_4$)$_2$ crystal.

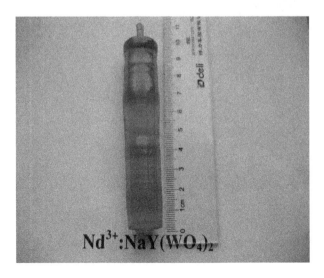

Fig. 2.2 The grown Nd^{3+}:NaY(WO$_4$)$_2$ crystal.

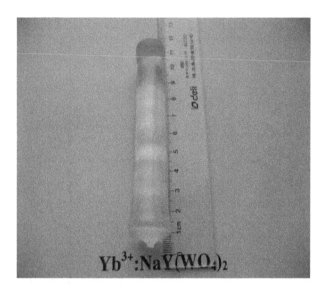

Fig. 2.3 The grown Yb^{3+}:NaY(WO$_4$)$_2$ crystal.

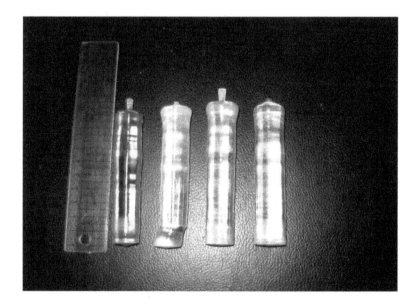

Fig. 2.4 The grown Re^{3+}:NaY(WO$_4$)$_2$ crystals (Re=Yb, Nd,Er/Yb).

Fig. 2.5 The grown Re^{3+}:NaY(WO$_4$)$_2$ crystals (Re=Tm/Ho).

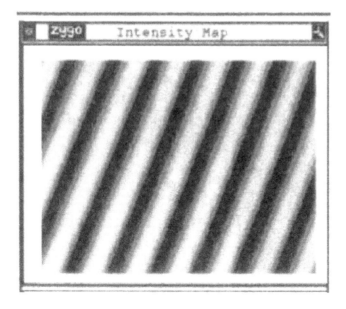

Fig. 2.6 Interference fringe of crystals.

The thermal-expansion coefficient $\left[\alpha_{ij}\right]$ of a crystal is a symmetrical second-rank tensor and it can be described by the representation quadric. The $NaY(WO_4)_2$ crystal belongs to the tetragonal system and 4/m point group. The unique symmetry axis is a fourfold axis along the crystallographic c-axis; the axes of the crystallographic and crystallophysical coordinate systems in $NaY(WO_4)_2$ have the same direction. In this case the value of thermal expansion along a- and b-axis are comparable and the values of α_1 and α_3 can be obtained by measuring the thermal expansion of the a- and c-oriented crystal.

The expansion coefficient in the [001] is about two times larger than that of the [100] direction according to our experimental results, which means that the $NaY(WO_4)_2$ crystal has anisotropic thermal expansion. The reason for the themal expansion coefficient along the c-axis being larger than that along the a- or b-axis can be explained by the structure of the $NaY(WO_4)_2$ crystal. The $NaY(WO_4)_2$ crystal has a scheelite structure according to the XRPD experiment results. According to Fig.3.1, it can be seen that there are five layers and three layers perpendicular to the c- and the a- or b-axis, respectively. According to the XRPD experiment results, the distance of the interlayer of five layers and three layers are c/4 and a/2 (or b/2),which is equal to 2.813×10^{-10} and 2.603×10^{-10} m, respectively. The larger the distance of the interlayer is, the weaker the chemical bonds of the interlayer will be according to the crystal lattice vibration dynamics. It can be seen that the interaction force along the c-axis is weaker than that along the a- or b-axis, and there are more layers in the c-direction than in the a-direction. Thus when the crystal is heated, the thermal expansion of the $Yb^{3+}:NaY(WO_4)_2$ crystal along the c-axis is larger than that along the a- or b-axis.

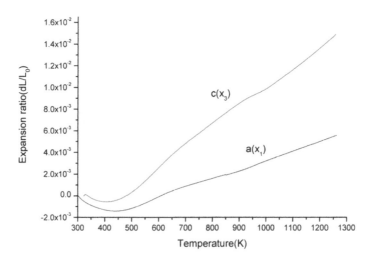

Fig. 3.1 The curve of thermal expansivity of $Yb:NaY(WO_4)_2$.

4. The spectroscopic characteristics

4.1 The spectroscopic characteristic of Nd^{3+}: $NaY(WO_4)_2$ crystal

Fig.4.1 shows the RT absorption spectrum of Nd^{3+}:$NaY(WO_4)_2$ Crystal. Owing to the disordered structure and the high Nd-doping concentration, the strong absorption intensity and broad FWHM of every band are shown, especially for the 806 nm[1]. Its FWHM is about 16 nm and the cross-section is about 2.8×10^{-20} cm² at 806 nm, which is benefit to the pumping of commercial laser diode. Fig.4.2 shows the RT emission spectrum with the pumping perpendicular to (001) planes. There are six emission peaks at follows wavelength: 894, 917,1063,1087,1339 and 1389 nm. The value of emission cross-section at 1063 nm is about 4.6×10^{-20} cm². Fig.4.3 shows the fluorescence decay of $^4F_{3/2}$ level of Nd^{3+} in NYW crystal at RT and the lifetime of $^4F_{3/2}$ level is about 85 μs and relative luminescent quantum efficiency is about 47% . Tab.4.1 presents the integrated absorbance, the line strengths, the experimental and calculated oscillator strengths. Table 4.2 shows the calculated radiative probabilities, radiative branching ratios and radiative time for the emissions from the $^4F_{3/2}$ level of Nd^{3+}:NYW crystal. Table 4.3-4 give the comparison of spectrum parameters in Nd:NYW and other Nd-doped crystals.

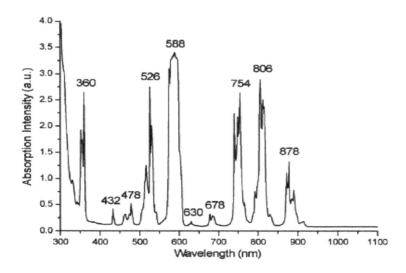

Fig. 4.1 Absorption spectra of Nd:NaY(WO₄)₂ crystal.

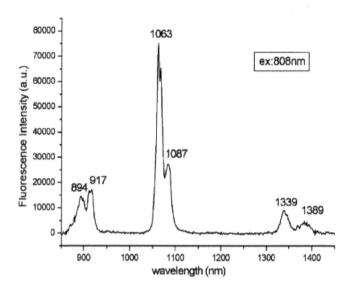

Fig. 4.2 Fluorescence spectra of Nd:NaY(WO$_4$)$_2$ crystal.

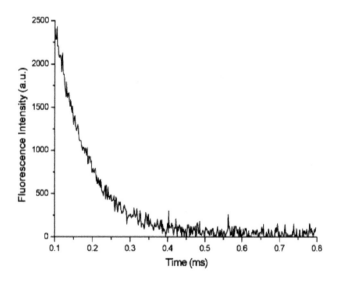

Fig. 4.3 Fluorescence decay of Nd:NYW crystal at RT.

Excited state	Wavelength (nm)	Γ (nm/cm)	S_{mea} (10^{-29}cm^2)	S_{cal} (10^{-29}cm^2)	f_{exp} (10^{-6})	f_{cal} (10^{-6})
$^4D_{1/2}+^4D_{3/2}+^5D_{5/2}$	360	71.2	2.862	3.127	15.874	17.342
$^2P_{1/2}$	432	4.57	0.153	0.221	0.708	1.021
$^2K_{13/2},,^2G_{9/2},,^2P_{3/2},,^2G_{11/2}$	478	18.42	0.558	0.416	2.329	1.739
$^4G_{7/2}+^4G_{9/2}+^2K_{13/2},$	526	130.86	3.600	2.186	13.666	8.296
$^4G_{5/2}+^2G_{7/2}$	588	375.88	9.251	9.345	31.412	31.732
$^2H_{11/2}$	630	1.47	0.034	0.061	0.107	0.193
$^4F_{9/2}$	678	10.2	0.218	0.238	0.641	0.702
$^4F_{7/2}+^4S_{3/2}$	754	159.94	3.070	3.115	8.129	8.249
$^4F_{5/2}+^2H_{9/2}$	806	191.84	3.444	3.681	8.532	9.118
$^4F_{3/2}$	878	78.2	1.289	1.554	2.931	3.533

Table 4.1 The integrated absorbance, the line strengths, the experimental and calculated oscillator strengths of Nd:NYW crystal.

Start levels	Wavelength(nm)	A(S^{-1})	β	τ (μs)
$^4I_{9/2}$	894	2612	0.473	180
$^4I_{11/2}$	1063	2434	0.441	
$^4I_{13/2}$	1339	450	0.082	
$^4I_{15/2}$	1852	23	0.004	

Table 4.2 Calculated radiative probabilities, radiative branching ratios and radiative time for the emissions from the $^4F_{3/2}$ level of Nd^{3+}:NYW crystal.

Crystal	Ω_2 (10^{-20} cm^2)	Ω_4 (10^{-20} cm^2)	Ω_6 (10^{-20} cm^2)	FWHM (nm)	σ_a at 808 nm (10^{-29} cm^2)	Ref.
Nd:NYW	5.8	5.74	4.32	16	2.80	[1]
Nd:NBW	30.9	12	9.3	10	2.6	[2]
Nd:KYW	8.80	3.11	3.16	8	3.13	[3]
Nd:KGW	12.67	10.15	7.48	12	26	[4,5]
Nd:YVO$_4$	5.88	4.08	5.11	8	27	[6]
Nd:GdVO$_4$	12.629	4.828	8.425	4.7	9.396	[7]
Nd:YAG	0.2	2.7	5.0	0.9	7.0	[8,9]
Nd:YAP	0.69	3.69	4.56	3	10.2	[10,11]
Nd:GAB	3.118	2.676	5.343	8.7	4.3	[12]

Table 4.3 Comparison of spectral values in Nd:NYW and other Nd-doped crystals.

Crystal	σ_e at 1064nm (10^{-20} cm^2)	τ_f (μs)	η (%)	References
Nd:NYW	4.56	85	47	[1]
Nd:NBW	16	122	85	[2]
Nd:KYW	4.5	154	78.6	[3]
Nd:KGW	38	110	92.4	[4,5]
Nd:YVO$_4$	100	98	46.8	[6]
Nd:GdVO$_4$	7.6	90	45.5	[7]
Nd:YAG	34	240	91	[8,9]
Nd:YAP	24.4	180.4	88	[10,11]
Nd:GAB	30	55.6	19	[12]

Table 4.4 Comparison of the emission spectroscopic parameters of some Nd-doped laser crystal.

4.2 The spectroscopic characteristic of Yb^{3+}: NaY(WO$_4$)$_2$ crystal

Fig.4.4 shows the RT Polarized absorption spectrum of Yb^{3+}:NaY(WO$_4$)$_2$ Crystal. The largest absorption cross-section is located at 976 nm in the $\sigma-$ and $\pi-$ spectrum, which is the main pump wavelength of the possible Yb^{3+} laser in NaY(WO$_4$)$_2$ crystal using the InGaAs LD, and the value is 1.81×10^{-20} cm^{-2}. This value is smaller than that of Yb^{3+}:KY(WO$_4$)$_2$ (13.3×10^{-20}) crystals at 981 nm [13], but larger than those of Yb^{3+}:YAG crystal (0.8×10^{-20} cm^2) [14] at 942 nm and Yb^{3+}:YCOB crystal (1.0×10^{-20} cm^2) [15] at 976nm. Fig.4.5 shows the RT Polarized emission spectrum of Yb^{3+}:NaY(WO$_4$)$_2$ Crystal. The emission cross-sections of crystal calculated from the fluorescence spectra by the reciprocity method and the Füchtbauer-Ladengurg formula are shown in Fig.4.6 [16~18]. The radiative lifetime τ_r of the ^{2}F$_{3/2}$ manifold is measured to be 0.902 ms. The gain coefficient was calculated for several values of population inversion P (P =0, 0.1, 0.2...) and is shown in Fig.4.7 (a) and Fig.4.7 (b). Positive gain coefficient for P values larger than 0.5, which are encountered in a free-running laser operation, implies a tuning range from 990 to 1070 nm.

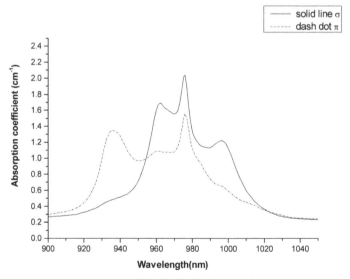

Fig. 4.4 Polarized absorption spectra of Yb:NaY(WO$_4$)$_2$ crystal.

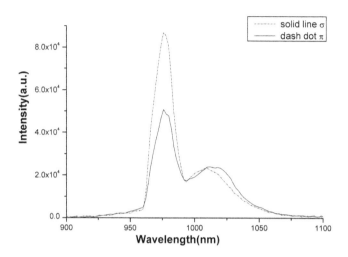

Fig. 4.5 Polarized fluorescence spectra of Yb:NaY(WO₄)₂ crystal.

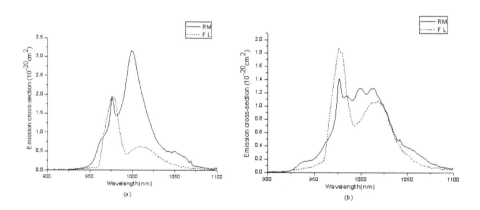

Fig. 4.6 The emission cross-sections of crystal calculated from the fluorescence spectra by the reciprocity method and the Füchtbauer-Ladengurg formula.

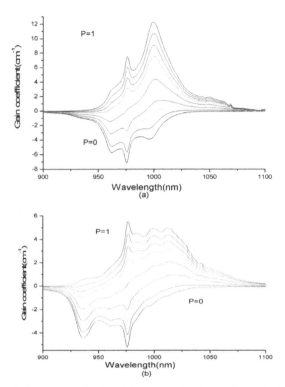

Fig. 4.7 The gain coefficient was calculated for several values of population inversion P
(P =0, 0.1, 0.2...).

4.3 The spectroscopic properties of Tm³⁺, Ho³⁺: NaY(WO₄)₂ crystal

Fig.4.8 shows the Room temperature absorption spectra of Tm^{3+}-, Ho^{3+}-doped and Tm^{3+}/Ho^{3+} co-doped $NaY(WO_4)_2$ crystals (a) in the range 300-850 nm and (b) in the range 1100-2100 nm. The spectrum of Tm^{3+}: $NaY(WO_4)_2$ crystal consists of six resolved bands associated with the transitions from the $^3H_6$ ground state to the $^3F_4$, $^3H_5$, $^3H_4$, $^3F_{2,3}$, $^1G_4$ and $^1D_2$ excited states. It can be seen that the absorption band of the σ polarization is narrower and has a larger peak cross section than the π absorption band. The spectrum of Ho^{3+}: $NaY(WO_4)_2$ crystal consists of ten resolved bands associated with the transitions from the $^5I_8$ ground state to the $^5I_7$, $^5I_6$, $^5F_5$, $^5F_4+^5S_2$, $^5F_3$, $^3K_8+^5F_2$, $^5F_1+^5G_6$, $^5G_5(^3G_5)$,, $^3H_6+^5F_2+^3H_5$ and $^3K_6+^3F_4+^3H_4+^3G_4$ excited states[19]. Some absorption bands of Tm^{3+} and Ho^{3+} ions overlap in the $Tm^{3+}/Ho^{3+}:NaY(WO_4)_2$ crystal. Compared to Ho^{3+} ions concentration in Ho^{3+}: $NaY(WO_4)_2$ and Tm^{3+} ions concentration in $Tm^{3+}/Ho^{3+}:NaY(WO_4)_2$ crystal, the concentration of Ho^{3+} ions in $Tm^{3+}/Ho^{3+}:NaY(WO_4)_2$ crystal is very low; the $^5I_8 \rightarrow ^5I_7$ (Ho^{3+}) transition of $Tm^{3+}/Ho^{3+}:NaY(WO_4)_2$ crystal is extremely weak.

Fig.4.9 shows the absorption cross sections and polarized stimulated emission cross sections associated with the (a) $^3F_4 \rightarrow ^3H_6$ transition for the $Tm^{3+}:NaY(WO_4)_2$ and (b) $^5I_7 \rightarrow ^5I_8$ for $Ho^{3+}:NaY(WO_4)_2$ crystal derived by the reciprocity method. The maximum values of σ_{em}

are 1.399×10^{-20} cm² for σ polarization at 2044 nm and 1.426×10^{-20} cm² for π polarization at 2047 nm. For comparison, the σ_{em} obtained for Tm³⁺ in NLuW are $2.0(\pm0.1) \times 10^{-20}$ cm² at 1798 nm and $1.9(\pm0.1) \times 10^{-20}$ cm² at 1830nm, respectively[20] . The FWHMs of the emission bands for σ and π polarizations are 161 and 130 nm, respectively.

Fig.4.10 presents the gain cross-section calculated for different values of P ($P=0.1\sim0.5$) for (a) the $^3F_4 \rightarrow ^3H_6$ transition of Tm³⁺ in NaY(WO₄)₂ crystal and (b) the $^5I_7 \rightarrow ^5I_8$ transition of Ho³⁺ in NaY(WO₄)₂ crystal. The gain curves at a wavelength longer than 1900 nm are obscure due to the low signal-to-noise ratio of the absorption spectrum. The positive gain cross-section can be obtained at about 2.0 μm when P exceeds 0.2. The positive gains for $P=0.5$ are in a range from 1758 to about 1954 nm for σ polarization and from 1758 to about 1977 nm for π polarization, respectively.

Fig.4.11 presents the room temperature fluorescence spectra of Tm³⁺-, Ho³⁺-doped and Tm³⁺/Ho³⁺ co-doped NaY(WO₄)₂ crystals.

Fig.4.12 gives the decay curves of $^3F_4$ manifold in the samples of bulk and powder in the Tm³⁺ doped NaY(WO₄)₂ crystals. Fig.4.13 also gives the decay curves of Ho: $^5I_7$ level in the (a) samples of bulk and powder in the Ho³⁺:NaY(WO₄)₂ and (b) Tm, Ho:NaY(WO₄)₂ crystals.

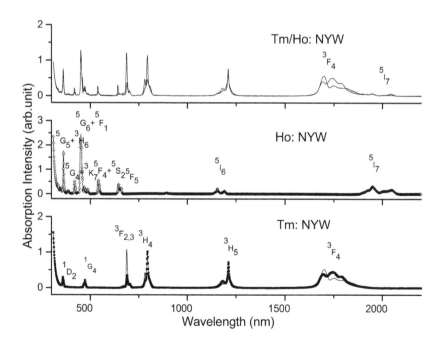

Fig. 4.8 Room temperature absorption spectra of Tm³⁺-, Ho³⁺-doped and Tm³⁺/Ho³⁺ co-doped NaY(WO₄)₂ crystals.

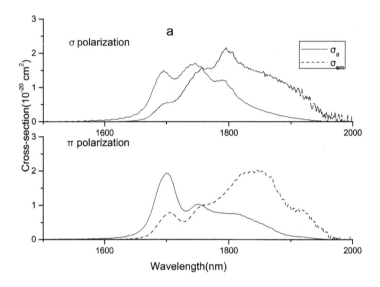

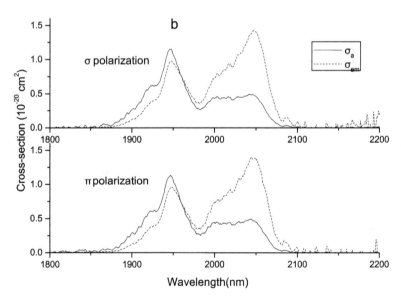

Fig. 4.9 Absorption cross sections and polarized stimulated emission cross sections associated with the (a) $^3F_4 \rightarrow {}^3H_6$ transition for the Tm^{3+}:NaY(WO$_4$)$_2$ and (b) $^5I_7 \rightarrow {}^5I_8$ for Ho^{3+}:NaY(WO$_4$)$_2$ crystal derived by the reciprocity method.

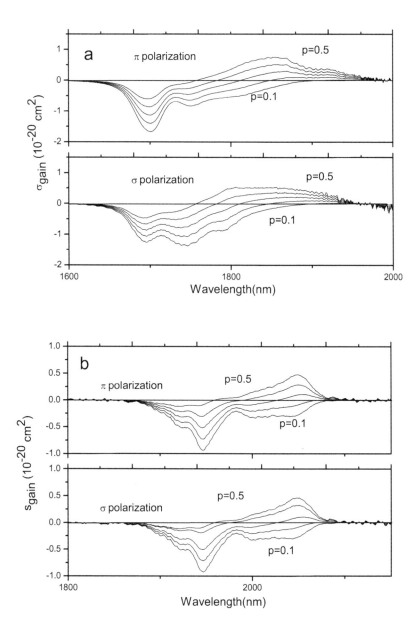

Fig. 4.10 Gain cross-section calculated for different values of P (P=0.1~0.5) for (a) the $^3F_4 \rightarrow{}^3H_6$ transition of Tm³⁺ in NaY(WO₄)₂ crystal ,(b) the $^5I_7 \rightarrow{}^5I_8$ transition of Ho³⁺ in NaY(WO₄)₂ crystal.

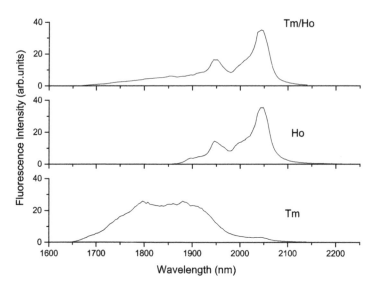

Fig. 4.11 Room temperature fluorescence spectra of Tm^{3+}-, Ho^{3+}-doped and Tm^{3+}/Ho^{3+} co-doped $NaY(WO_4)_2$ crystals.

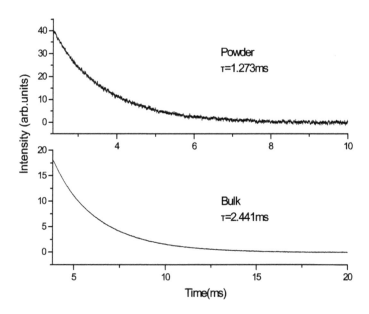

Fig. 4.12 Decay curves of $^3F_4$ manifold in the samples of bulk and powder in the Tm^{3+} doped $NaY(WO_4)_2$ crystals.

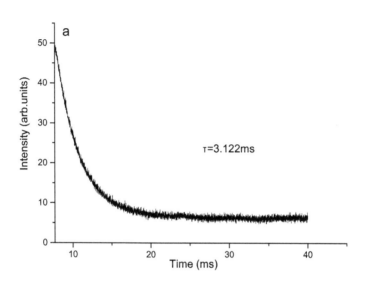

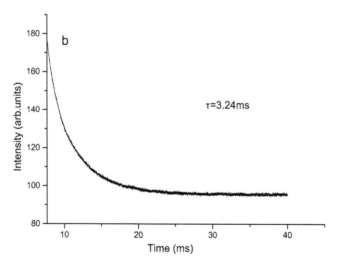

Fig. 4.13 Decay curves of Ho: $^5I_7$ level in the (a) samples of bulk and powder in the Ho^{3+}: NaY(WO₄)₂ and (b) Tm, Ho: NaY(WO₄)₂ crystals.

5. The laser characteristics

5.1 The laser characteristics of Nd³⁺: NaY(WO₄)₂ crystal

The Nd³⁺:NaY(WO₄)₂ crystal was made into laser stick and the laser experiment was performed using a xenon flash lamp as a pump source[1]. Maximum pulse energy of 786 mJ with a repetition rate of 1 Hz has been obtained. A maximum output power of 87 mW at 532 nm has been obtained and the double-frequency conversion efficiency is more than 25% when a LBO optical crystal was used as the frequency-doubling crystal. Table 5.1 shows the data of input and output energy and Fig.5.1 presents the relationship between the Iuput energy and output energy. Table 5.2 presents the Comparison of laser properties of Nd:NYW crystal and Nd:YAG crystal. It can be found that the Nd:NYW crystal has the higher laser efficiency than Nd:YAG crystal. Table.5.3 shows the frequency-doubling laser output power and conversion efficiency and Fig.5.2 presents the relationship between the pump power and output power of SH generation. Fig.5.3 shows the laser facula of the SH generation.

Input V （J）	static state output （mJ）	Wavelength （μm）
440　（9.68）	7	
480　（11.5）	33	
520　（13.5）	67	
560　（15.7）	106	
600　（18.0）	173	
640　（20.5）	215	
680　（23.1）	275	
720　（25.9）	346	1.063
760　（28.9）	376	
800　（32.0）	443	
840　（35.3）	512	
880　（38.7）	572	
920　（42.3）	671	
960　（46.1）	735	
1000　（50）	786	

Table 5.1 Data table of pumping energy and output energy.

Crystal	Size	Input	Output
Nd:NYW Crystal	Φ4.5 mmx52 mm	50 J	786 mJ
Nd:YAG Crystal	Φ8 mmx120 mm	50 J	~800 mJ

Table 5.2 Comparison of laser properties of Nd:NYW and Nd:YAG.

Input	Output/530nm	Threshold	Optical to optical efficiency for SHG	Slope efficiency	Optical to optical efficiency for ground frequency
1500 mW	87 mW	410 mW	5.80%	7.98%	> 25%

Table 5.3 SH generation power and conversion efficiency.

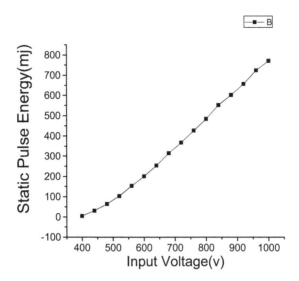

Fig. 5.1 The relationship between the Iuput energy and output energy at 1.062 um.

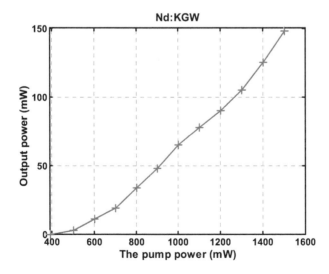

Fig. 5.2 The relationship between the pump power and output power of SH generation.

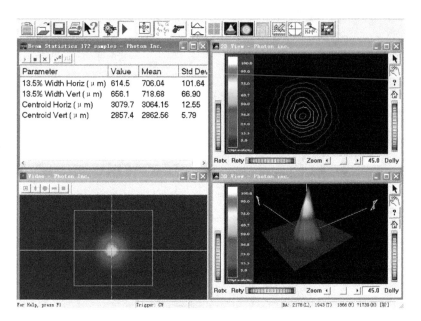

Fig. 5.3 The laser facula of the SH generation.

5.2 The laser characteristics of Tm, Ho: NaY(WO₄)₂ crystal

An infrared laser output at 2.07 μm with Tm, Ho:NaY(WO$_4$)$_2$ crystal end-pumped by 795 nm laser diode at room temperature is reached[2,3]. Fig.5.4 shows the experimental configuration of the LD-end-pumping Tm, Ho: NYW laser. The crystal used with the concentrations of 5 at% Tm^{3+} and 1 at% Ho^{3+}was grown by the Czochralski method. The highest output power was up to 2.7 W corresponding to the crystal temperature being controlled at 283 K. Fig.5.5 presents the output power versus pumping power at different temperatures. The overall optical conversion efficiency was 5.4% and the slop efficiency was 26%. The output characteristics and the laser threshold affected by the pulse duration and temperature have been studied. It can be found that the stability of the output power was correlative with the crystal temperature heavily. In addition, the wider pulse duration of pump could promote the output power efficiently as shown in Fig.5.6, which presents the output power versus pulse duration.

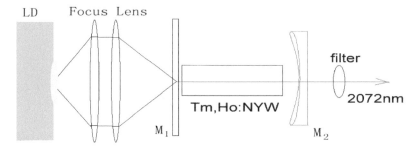

Fig. 5.4 The experimental configuration of the LD-end-pumping Tm, Ho: NYW laser.

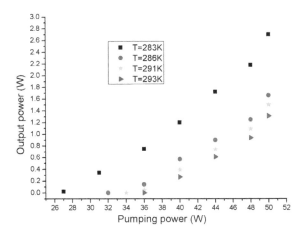

Fig. 5.5 The output power versus pumping power at different temperatures.

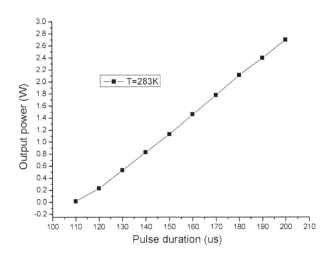

Fig. 5.6 The output power versus pulse duration.

With Ti:sapphire laser pumping at 795 nm, a slope efficiency and a maximum output power as high as 48% and 265 mW, respectively, have been obtained at 2050 nm from a Tm:Ho: NaY(WO₄)₂ crystal by Prof.C.Zaldo[4]. Tuning from 1830 nm to 2080 nm has also been achieved using an intracavity Lyot filter. Fig.5.7 shows Cw laser performance of Tm(5 at%), Ho(0.25 at%): NaY(WO₄)₂ crystal. Fig.5.8 shows Cw and quasi Cw laser performance of Tm(5 at%), Ho(0.5 at%): NaY(WO₄)₂ crystal.

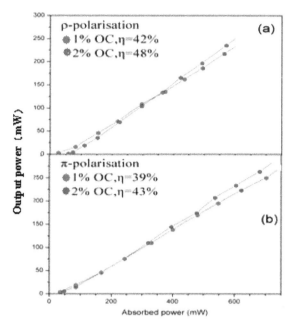

Fig. 5.7 Cw laser performance of Tm(5 at%),Ho(0.25 at%): NaY(WO₄)₂ crystal. (a) σ -pol
(1.55-mm-long crystal).(b) π -pol (3.4-mm-long crystal).

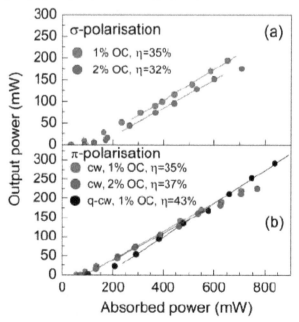

Fig. 5.8 Cw and quasi Cw laser performance of Tm(5 at%),Ho(0.5 at%): NaY(WO₄)₂ crystal.
(a) σ -pol (1.75-mm-long crystal).(b) π -pol (3.5-mm-long crystal).

Prfo.A.A.Lagatsky and C.Zaldo [5] also reported the femtosecond-pulse operation of a Tm:Ho:NaY(WO$_4$)$_2$ laser at around 2060 nm by using an ion-implanted InGaAsSb quantum-well-based semiconductor saturable absorber mirror for passive mode-locking maintenance for the first time. Transform-limited 191fs pulses are produced with an average output power of 82 mW at a 144 MHz pulse repetition frequency. Maximum output power of up to 155 mW is generated with a corresponding pulse duration of 258 fs. Fig.5.9 presents the Input-output characteristics of the mode-locked Tm:Ho:NaY(WO$_4$)$_2$ laser.

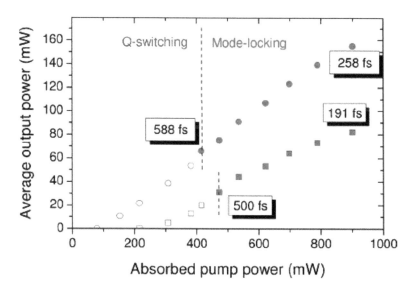

Fig. 5.9 Presents the Input-output characteristics of the mode-locked Tm:Ho:NaY(WO$_4$)$_2$ laser.

Two different operation regimes, shorter-pulse and higher-power, are indicated by squares and circles, respectively. Q-switching and mode-locking regimes are represented by open and closed symbols, respectively.

5.3 The laser characteristics of Tm, Ho, Ce: NaY(WO$_4$)$_2$ crystal

An infrared laser output at 2.07 μm with Tm,Ho,Ce:NaY(WO$_4$)$_2$ single crystal end-pumped by 795 nm laser diode at room temperature[3,6]. The crystal used with the concentrations of 5 at% Tm^{3+}, 1 at% Ho^{3+} and 30 at% Ce^{3+} was grown by the Czochralski method. The highest output power was up to 0.2 W corresponding to the pumping power of 50 W and the threshold was about 40 W at 293 K. Figure 5.10 shows the output power versus the pump power. The introduction of Ce^{3+} brought about a novel phenomenon. End-pumping with the 795 nm LD, it was found the up-conversion was repressed heavily and the green emission disappeared thoroughly in Tm,Ho,Ce:NaY(WO$_4$)$_2$ crystal, which was particularly different from the crystal Tm,Ho:NaY(WO$_4$)$_2$, where the green emission was obvious and weakened the sensitized transition energy.

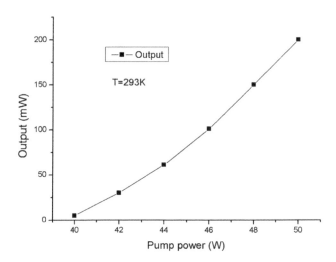

Fig. 5.10 The output power versus the pump power.

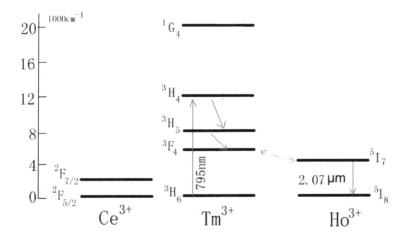

Fig. 5.11 The illustration of Ce^{3+}, Tm^{3+}, Ho^{3+} energy levels.

The original intention of selecting the Ce^{3+} was to compensate the up-conversion loss. The compensation mechanisms of the Ce^{3+} lie in its transition energy. As shown in Fig.5.11, the transition energy of $^2F_{7/2} \rightarrow ^2F_{5/2}$ (Ce^{3+}) is close to that of $^3H_5 \rightarrow ^3F_4$ (Tm^{3+}) and half of the

$^3H_4 \rightarrow {}^3H_5$ (Tm^{3+}). Pumped with 795 nm, the electrons will transit from $^3H_6$ to $^3H_4$, and jump to $^3H_5$, $^3F_4$ depending on the radiationless transition. Because the energy level $^3F_4$ (Tm^{3+}) was close to $^5I_7$ (Ho^{3+}), the electrons will transit from $^3F_4$ (Tm^{3+}) to $^5I_7$ (Ho^{3+}), which is just the sensitized process. At last, the transition $^5I_7 \rightarrow {}^5I_8$ (Ho^{3+}) generates the 2.07 μm laser. In the complex sensitized process, only few of the electrons will transit from the upper pumping energy level $^3H_4$ (Tm^{3+}) into the $^5I_7$ (Ho^{3+}), which is the reason of the lower laser efficiency. By virtue of the Ce^{3+}, the electrons of the $^3H_4$ (Tm^{3+}) can transit fast into the energy level $^3F_4$ (Tm^{3+}). More important, the multiple transition energy can guide the electrons towards $^3F_4$ (Tm^{3+}) instead of irregular radiationless transition. That is to say, in the shorter time, there are more electrons gathering into the energy level $^5I_7$ (Ho^{3+}), which is just the demand of the high laser efficiency. Here, in our experiment, the disappeared green emission is the certification of the function of the Ce^{3+}, which contributes to the improvement of the 2 μm laser.

6. Conclusion

In this review, the growth of rare earth (Tm^{3+},Ho^{3+},Nd^{3+},Yb^{3+}, Er^{3+}/Yb^{3+})-doped NaY(WO$_4$)$_2$ large crystal with the dimensions of Φ25 mm×100 mm is reported. The thermal, optical and spectrum characteristics of these crystals are presented. The laser characteristics of Nd^{3+};Tm^{3+}/Ho^{3+}:NaY(WO$_4$)$_2$ laser crystals are also covered. Maximum pulse energy of 786 mJ with a repetition rate of 1Hz has been obtained from Nd^{3+}-doped NaY(WO$_4$)$_2$ crystal pumped by xenon flash lamp. It can be found that the Nd:NYW crystal has the higher laser efficiency than Nd:YAG crystal. An infrared laser output of 2.7 W at 2.07 μm with Tm,Ho:NaY(WO$_4$)$_2$ crystal end-pumped by 795 nm laser diode at room temperature is also reached. Furthermore, the femtosecond-pulse operation of a Tm:Ho:NaY(WO$_4$)$_2$ laser at around 2060 nm is obtained for the first time. Transform-limited 191fs pulses are produced with an average output power of 82 mW at a 144MHz pulse repetition frequency. Maximum output power of up to 155 mW is generated with a corresponding pulse duration of 258 fs. Also, it is found that the co-doped Ce^{3+} can depress the green up-conversion emission of Tm^{3+} and thus improves the 2 μm laser. All the above performances demonstrate that NaY(WO$_4$)$_2$ crystal can serve as an excellent laser host.

7. Acknowledgements

Some works of this chapter were supported by National Nature Science Foundation of China (No.50902129, 61078076), Major Projects from FJIRSM (SZD09001), the Knowledge Innovation Program of the Chinese Academy of Sciences (Grant No. KJCX2-EW-H03), Science and Technology Plan Major Project of Fujian Province of China (Grant No. 2010I0015).

8. References

1.
[1] Z.X. Cheng, S.J. Zhang, J.R. Han, et al.,, Cryst. Res. Technol. 36 (2001) 135;
[2] A.A. Kaminskii, H.J. Eichler, K. Ueda,, et al., Appl. Opt. 38 (1999) 4533;
[3] Z. L. Zhu, Y. N. Qian, J. H. Liu,et al., J. Cera. Soci. 35(2007) 991-994;
[4] C. Cascales, M. D. Serrano, F. Esteban-Betegón, et al.,, Phys. Rev. B. 74(2006) 174114;

2.

[1] Yan Wang,Chaoyang Tu, Zhenyu You, et al.,Journal of Crystal Growth, 285(1-2)2006:123-129;

[2] Chengli Sun, Fugui Yang, Ting Cao, et al., Journal of Alloys and Compounds, 509(25)(2011): 6987-6993;

[3] Zhenyu You, Master dissertation, The study of Nd^{3+} and Yb^{3+} doped in $NaY(WO_4)_2$ and Cr^{3+}doped in $LiNbO_3$ crystals,Graduated School of Chinese Academy of Sciences,2010.

3.

[1] Zhenyu You, Master dissertation, The study of Nd^{3+} and Yb^{3+} doped in $NaY(WO_4)_2$ and Cr^{3+}doped in $LiNbO_3$ crystals,Graduated School of Chinese Academy of Sciences,2010.

4.

[1] Yan Wang,Chaoyang Tu, Zhenyu You, et al.,Journal of Crystal Growth, 285(1-2)2006:123-129;

[2] A. Mendez-Blas, M. Rico, V. Volkov, et al., Mol. Phys. 101 (2003): 941.

[3] X. Han and G. Wang, J. Crystal Growth 247 (2003):551.

[4] Y. Kalisky, L. Kravchik and C. Labbe, Opt. Commun. 189 (2001): 113.

[5] R. Moncorge, B. Chambon, J.Y. Rivorie,et al., Opt. Mater. 8 (1997):109.

[6] M.H. Randles, J.E. Creamer and R.F. Belt, OSA technical digest ser. Opt. Soc. Am. 10 (1998): 289.

[7] H.D. Jiang, H.J. Zhang, J.Y. Wang, et al., Opt. Commun. 198 (2001):447.

[8] W.F. Krupke, IEEE J. Quantum Electron. QE-7 (1991):153.

[9] W. Koechner, Solid state laser engineering, Optical Science (third ed.), Springer Verlag, New York (1992).

[10] H. Zhang, Z. Luo, A. Zhen, C.M. Wu and G.S. Li, Chin. J. Infrared Res. A 7 (1998): 297.

[11] A.A. Kaminskii, Laser crystals. In: H.F. Ivey, Editor, Springer, Berlin, Heidelberg, New York (1981).

[12] C.Y. Tu, Master dissertation, Study on the blue–green laser crystal Nd3+:GdxY1−xAl3 $(BO_3)_4$, Graduated School of Chinese Academy of Sciences, 2001.

[13] N.V. Kuleshov, A.A. Lagatsky, A.V. Podlipensky, et al., Opt.lett.22(1997)1317.

[14] K.I. Schaffers, L.D. Deloach, S.A. Payne, IEEE J. Quantum electron. 32 (1996) 741.

[15] H.Zhang, X.Meng, P. Wang, et al., Appl. Phys. B 68 (1999) 1147.

[16] K. Ohta, H. Saito, M. Obara, J. Appl. Phys. 73,(1993) 3149.

[17] D.E. McCumber, Phys. Rev. A 134(1964) 299.

[18] S.A. Payne, L.L. Chase, L.K. Smith, et al., IEEE J. Quantum Electron. QE-28 (1992) 2619.

[19] W.T. Carnall, P.R. Fields, K. Rajnak, J Chem Phys, 49 (1968) 4424.

[20] X. M. Han, J. M. Cano-Torres, M. Rico,et al., J. Appl. Phys. 103(2008) 083110.

5.

[1] Zhenyu You, Master dissertation, The study of Nd^{3+} and Yb^{3+} doped in $NaY(WO_4)_2$ and Cr^{3+}doped in $LiNbO_3$ crystals,Graduated School of Chinese Academy of Sciences,2010.

[2] F. G. Yang, C. L. Sun, Z. Y. You, et al., Laser Phys.20(2010)1695;

[3] Fugui Yang, Doctoral dissertation, The study of LD pumped yellow and 2μm laser properties, Graduated School of Chinese Academy of Sciences,2011.

[4] X.Han, F.Fusari, M.D.Serrano, et al. OPTICS EXPRESS 18(6)(2010)5413;

[5] A.A.Lagatsky, X.Han, M.D.Serrano, et al. OPTICS LETTERS 35(18)(2010)3027;

[6] F.G. Yang, F.P. Yan, Z.Y. You, et al., Laser Phys. Lett. 7(2010)867.

Growth and Characterization of Doped CaF$_2$ Crystals

Irina Nicoara and Marius Stef
*West University of Timisoara, Timisoara,
Romania*

1. Introduction

The alkaline-earth fluorides crystallize in the cubic structure and constitute an important class of relatively simple ionic crystals whose optical and lattice-dynamical properties have theoretical and experimental interest. The CaF$_2$ crystals have been used for long time in many optical components due to its exceptional transparency in the UV as well as in the IR spectral domain. CaF$_2$, SrF$_3$ and BaF$_2$ have been among the first solid-state laser hosts and they were lased at the beginning of the 1960s doped with RE^{3+} ions; these rare-earth doped crystals, however, have been abandoned as laser systems during a long time. The reason resides in the charge compensation which is required to maintain the electrical neutrality of crystals. This process gives rise to a rich multisite structure including so-called isolated centers and more or less complex centers [Petit et al., 2008], which leads to broad absorption and emission bands comparable with those of glasses. Rare earth doped CaF$_2$ recently have a new interest firstly because it was found that clustering of ions in these materials could be favorable to produce some infrared laser emission. Yb^{3+} -doped CaF$_2$ has been proved in recent years to be one of the most attractive Yb^{3+} laser materials for different reasons. It is also proved that the proportions of the different luminescent centers vary with the considered RE^{3+} ion and with the nature of the substituted divalent cation. It is expected, indeed that the transition strengths associated with various centers are different; for example, the transition strengths associated with tetragonal centers will be greater than those of the trigonal ones. At high dopant concentrations (about 1 at%) which become interesting for the laser application, the ions generally aggregate and form more or less complex centers that in turn can weaken the emission transitions. Such a detrimental pairing effect can be decreased and some improvement can be obtained by co-doping the crystals with charge compensating buffer ions, such as monovalent ions or non-optically active rare-earth trivalent ions. For example, after Na$^+$ ions were introduced as charge compensators the IR emission intensity of the YbF$_3$-doped CaF$_2$ crystal was enhanced several times [Su et al., 2007].

Taking into account the interesting properties of CaF$_2$ crystals doped with various impurities, in this work we describe our investigations about the optical properties of Pb^{2+}, Yb^{3+} doped CaF$_2$ crystals, as well as the influence of Na$^+$ ions on the formation of various charge compensating defects.

2. Crystal growth

Crystallization process is essentially a phenomenon of periodic arrangement of the constituent ions of a given material obtaining a crystalline state. There are many methods of crystal growth. Controlled solidification in a crucible of a molten material is one of the most used method. First used by Bridgman [Bridgman, 1925], improved by Stockbarger [Stockbarger, 1949] this technique is used successfully (about 40% of production of artificial crystals) because it is a simple technology and does not require complicated control systems. In the design of vertical Bridgman-type technique (VB) it is important to predict the thermal profiles in the growing crystal. The position and shape of the solidification interface and the axial and radial thermal gradients are of particular interest in controlling the growth process. The relatively high axial temperature gradient necessary in order to obtain crystals with large diameters ($\Phi > 20mm$) cannot be obtained by means of a single heating element; two or more independently controlled heaters are necessary. The temperature gradient is directly dependent on the "isolated" zone between the hot and cold zones [Fu&Wilcox, 1980; Mikkelsen, 1980]. In order to obtain crystals with melting temperature higher than 1300^0C, graphite, molybdenum, silicon carbide, etc heaters must be used [Stockbarger, 1949; Gault et al., 1986; Jones et al., 1966]; the high temperature gradient necessary at the crystallization interface is obtained either by supplementary heaters [Stockbarger, 1949; Jones et al., 1966], by thermal screen translation or by using two heaters with a moving temperature gradient and a stationary crucible [Gault et al., 1986]. Taking into account the analysis of the heat transfer in VB technique [Chang&Wilcox, 1974; Fu&Wilcox, 1980; Naumann, 1982; Jasinski et al., 1983], we designed [D.Nicoara, 1975; D. Nicoara et al., 1985a; D. Nicoara et al., 1985b; I. Nicoara et al., 1987; D. Nicoara&I. Nicoara, 1988] several types of graphite heaters whose characteristics would satisfy the growth conditions to obtain crystals with melting point up to 2000^0C. In this paragraph we will focus on the growth particularities in order to obtain pure and various concentrations of PbF_2, YbF_3- doped and NaF-codoped CaF_2 crystals using Bridgman technique [D. Nicoara&I. Nicoara, 1988; Nicoara et al., 2008a, 2008b; Pruna et al., 2009; Paraschiva et al., 2010]

2.1 Shaped graphite heaters

The general view of the crystal growth set-up is shown in figure 1a [D.Nicoara 1975]. Figure 1b illustrates the longitudinal section of three types of heaters and the temperature distribution along them. The model contains an adiabatic zone and a booster heater zone that can be used to increase the thermal gradient near the solidification interface. Built of graphite (see fig 7), the heater produces an almost constant temperature zone (the B-C hot zone) necessary to melt the charge and a high gradient temperature zone (D-E). The C zone (the booster heater) ~ 20 mm long in the case of type I heater is designed as a meander-type resistance with wall thickness of 1.5 mm, by means of which an overheating is obtained in the lower level of the upper zone. The D zone is inserted for the improvement of heat transfer upwards the upper part of the heater, the E zone ~ 10 mm long, which appears like a wall thickening (up to 5 mm), has the role of an „isolated" zone and allows to reach a high temperature gradient. The F zone -the cold zone - is used as the lower cooling chamber. The wall thickness of this zone increases from 1.5 mm to 4 mm in the lower part. The wall thickness of the II type heater in the B zone decreases from 3mm to 2mm; the C zone is ~ 15 mm long with wall thickness of 1.5 mm. The III type heater produces a relatively flat temperature profile in the hot zone B-C. The wall thickness decreases gradually from 3 mm in the upper part of the B zone to 1.5 mm in the C-D zone, then increases to 4mm in the lower part of the F zone. Curve I represents the temperature distribution in I type heater and the curve I_a is the temperature recorded during the growth of a 30 mm diameter CaF₂ crystal. Comparing these two distributions one can see that the furnace parameters were well chosen, the axial temperature distribution remaining approximately the same. In order to avoid thermal loss, the graphite heater is surrounded by a set of concentric screens made of molybdenum, graphite, and stainless steel.

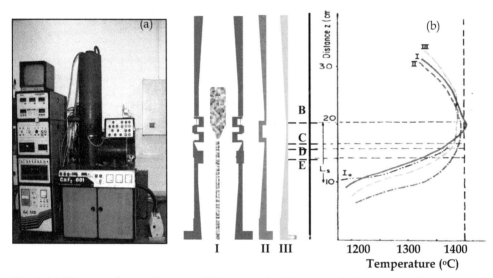

Fig. 1. (a) The crystal growth set-up; (b) Longitudinal sections of three types of heaters and temperature distribution along them, [D. Nicoara, 1975; D. Nicoara et al., 1985a,b].

Fig. 2. (a) Calcium fluoride crystals; (b) multiple type crucible; (c) crucible with seed.

In order to obtain crystals, with diameter up to 10 mm, the C zone can be eliminated, the necessary temperature gradient (7-10⁰ C/cm) can be obtained only by the heater wall thickness adjustments, like III type heater. For crystals with diameter less than 20 mm, type II and III heaters could be employed. Using these types of heaters fluoride single crystals with diameter of up to 50 mm and 150 mm in length were obtained (fig. 2a). With a multiple type crucible [D. Nicoara et al., 1985a; 1985b] several up to 12 mm diameter crystals can be obtained during a single growth process (fig. 2b). Oriented crystal can be grown using a seed (Fig. 2 c) [D. Nicoara et al., 1983].

The thermal stresses can be reduced by gradual cooling rate to room temperature. After the crystal pulling is finished, the crucible is raised in the B-C zone and by using an automatic system the power supply is gradually lowered.

From our experiments it can be concluded that the use of these types of heaters in the VB technique in order to obtain high quality fluoride crystals is advantageous because the same heater can be used for about hundred growth processes, the growth conditions being reproducible and easy to control.

2.2 Growth of PbF₂-doped CaF₂ crystals

PbF$_2$-doped CaF$_2$ crystals were grown in our crystal research laboratory using vertical Bridgman method. Suprapure grade (Merck) calcium fluoride and PbF$_2$ were used as the starting materials. It is known [Stockbarger, 1949; Yonezawa et al., 2002] that Pb^{2+} ions hardly remain in fluoride crystals if the usual VB technique is used for the growth of CaF$_2$ crystals. In order to obtain PbF$_2$-doped CaF$_2$ crystals the following procedure was used. First, pure, oxygen-free CaF$_2$ crystals were grown using the usual growth conditions, namely adding to the starting material an amount of 4 wt% PbF$_2$ as oxygen scavenger [Stockbarger, 1949]. The obtained CaF$_2$ crystals do not contain any undesired amount of lead ions or other impurities, as results from the optical absorption spectrum; then the PbF$_2$-doped crystals were grown from the crushed pure fluoride crystals doped with the desired amount of PbF$_2$. To prevent the evaporation of the PbF$_2$, a thin floating graphite lid was put on the charge in a sealed graphite crucible. Transparent colorless crystals of about 10 mm in diameter over 6–7 cm long were obtained in spectral pure graphite crucible in vacuum (~ 10^{-1} Pa) using a shaped graphite furnace [D. Nicoara&I. Nicoara, 1988]. The pulling rate was 4 mm/h. The crystals were cooled to room temperature using an established procedure. The as-grown single crystals are shown in Fig. 3.

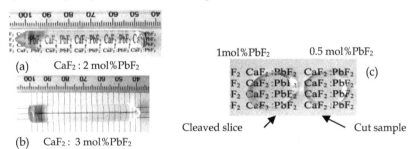

Fig. 3. As-grown x mol%PbF$_2$-doped CaF$_2$ crystals: (a) x = 2; (b) x = 3; (c) cleaved slice of 1 mol%PbF$_2$ doped CaF$_2$ crystal and cut sample 0f 0.5 mol% PbF$_2$ doped CaF$_2$ crystal.

CaF$_2$ crystallizes in cubic structure with a typical fluorite lattice. In order to understand the optical properties and the dopant distribution in CaF$_2$ crystals it is better to see this structure as consisting of a simple cubic lattice of fluorine ions in which every other body center position is occupied by a Ca^{2+} ion. When various ions, such as alkaline metals, rare-earths or heavy metal ions, like Pb^{2+} are introduced into the lattice they usually occupy Ca^{2+} sites. If the introduced impurity ions have other valence than the Ca^{2+} ion, the valence mismatch is compensated in a variety of ways: by vacancy formation, by interstitial fluorine ion, etc. The Pb^{2+} ion has the same valence as Ca^{2+} but with a larger geometric size (0.143 nm) than the Ca^{2+} ion (0.126 nm), and for high dopant concentrations this will lead to distortion of the crystal lattice; this is the reason why the crystals doped with more than 3 mol%PbF_2 reveal structural defects, like blocks with different crystallographic orientations (Figure 4).

Fig. 4. Misoriented high PbF_2 concentration sample; (1) the 2mm thick sample was cut from the bottom of the cystal, (2) after 4mm only two blocks remain and (3) after other 2 mm a single oriented crystal appears.

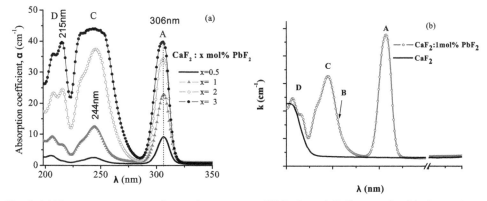

Fig. 5. (a) Room temperature absorption spectra of PbF_2 doped CaF_2 crystals; (b) absorption spectra of pure and 1 mol%PbF_2-doped CaF_2 crystals.

The presence of impurity ions in CaF_2 lattice with the ns^2 ground state configuration (like Pb^{2+} ions) induces absorption bands both in vacuum ultraviolet (VUV) and in UV domain [Jacobs, 1991]. Four characteristic bands are located in UV domain, denoted by A, B, C, and D [Jacobs, 1991]. The absorption spectra of various concentrations CaF_2:PbF_2 samples reveal the four characteristic absorption bands of Pb^{2+} ions (Fig. 5). As Pb^{2+} -ions concentration increases, the shape of the C and D absorption bands modifies due to the overlap of many new bands that appear as a result of the energy levels splitting, only the A band conserves his sharp shape.

2.3 Rare-earth ions-doped CaF₂ crystals

Rare-earth (RE) ions-doped CaF_2 crystals are used as laser active media due to the well-known good optical, mechanical and thermal properties of the CaF_2 host and due to the broad band transitions of the various RE ions. The optical properties of the Yb^{3+} ions in the CaF_2 host are well known and the luminescence properties have been intensively studied due the strong near-IR emission that can be easily pumped with laser diodes. It is known that several of RE ions, such as Sm, Eu, Ho, Tm and Yb, can be stabilized in the divalent state in alkaline earth halide lattices, besides the trivalent state, with which can coexist. The spectroscopic properties of the Yb^{2+} ions have been less investigated, mainly only for its intense and broad yellow-green (540–560 nm) luminescence [Feofilov, 1956; Kaplyanskii&Feofilov, 1962; Kaplyanskii et al., 1976]. When YbF_3 is dissolved in CaF_2, normally the ytterbium ions are in the trivalent state, but it is known that a certain fraction of any of RE^{3+} ions can be reduced to divalent state by various methods.

The change of the valence can be attained by exposing the crystals to ionizing radiation, baking them in a suitable atmosphere [Kirton&McLaughlan, 1967; Kaczmarek et al., 2005] or by electrolytic reduction [Fong, 1964]. The observed emission is weak for the crystals containing Yb^{2+} ions obtained using one of these methods. There are a few reported results [Feofilov, 1956; Kaplyanskii&Feofilov, 1962; Kaplyanskii&Smolyanskii 1976] about the properties of the Yb^{2+} ions, with high concentration in the as-grown crystals. The intense broad yellow-green luminescence has been obtained only for low temperature. Near-UV luminescence at room temperature has not been reported. This was one of the reason why we studied the spectroscopic properties of YbF_3-doped CaF_2 crystals with high divalent ytterbium content in the as-grown crystals. In order to obtain high Yb^{3+} -Yb^{2+} conversion in the as grown crystals, we have developed a special growth procedure.

Rare-earth ions -doped and PbF_2-codoped CaF_2 crystals were grown using vertical Bridgman method [Nicoara et al., 2006a, 2006b, 2008a, 2008b, 2008c; Pruna et al.,2009; Paraschiva, 2010]. As starting materials suprapure grade calcium fluoride, YbF_3 , ErF_3 and PbF_2 have been used. The $Yb:CaF_2$ and $Er:CaF_2$ crystals were obtained by adding YbF_3 and ErF_3 to the melts in molar concentrations varying between 0.07 and 2 mol%. First, pure CaF_2 crystals were grown using the standard growth conditions. In all cases, 0.4wt% PbF_2 was used as oxygen scavenger. Fig. 5a shows the time dependence of the power (in arbitrary units) and temperature in the furnace in order to determine the growth conditions.

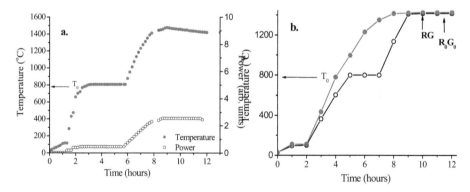

Fig. 6. (a) Time dependence of the temperature in the furnace (the position P in figure 7) against the power (in arbitrary units); (b) various growth conditions.

If the crystal growth process is not correctly established (for example, the T_0 temperature is not reached and then keep the raw material for 3 h at a temperature lower than T_0), the obtained crystals may contain some amount of undesired lead ions. This is why we study the influence of the Pb^{2+} ions on the optical properties of the YbF$_3$-doped CaF$_2$ crystals [Nicoara et al., 2006b; 2008b]. The growth procedure started to run after the whole furnace chamber was vacuumed to 10^{-1} Pa; the temperature was kept at 100^0 C for 3 h to remove water from the raw materials and then 3 h at T_0=800^0 C to eliminate the PbO obtained after the reaction CaO+PbF$_2 \rightarrow$ CaF$_2$+PbO.

Pure CaF$_2$ crystals were grown using the conditions shown in the Fig. 6b, alternative 1. The obtained CaF$_2$ crystals do not contain any undesired lead ions amount, as seen in the optical absorption spectrum (Fig. 5. the inset).

The YbF$_3$ (or ErF$_3$)-doped CaF$_2$ crystals were grown from crushed pure CaF$_2$ crystals obtained using a correct growth process. No oxygen scavenger has been added in order to obtain the doped crystals; in this case the alternative 2 shown in figure 6b was used.

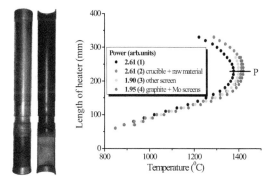

Fig. 7. Axial temperature distribution in the furnace for various powers and growth conditions: (1) without crucible and charge, (2) with crucible and charge, supplementary screen, (3) other positions of the screen, (4) double screen: graphite and molybdenum.

In order to obtain Pb^{2+} codoped crystals, 1 mol% PbF$_2$ was added in the starting mixture in melt. To prevent the evaporation of the PbF$_2$, a thin graphite lid was put on the charge in the sealed crucible. In order to obtain Pb^{2+} codoped YbF$_3$: CaF$_2$ crystals, alternative 2 (fig. 6b) was used. The crystal growth begins by lowering the crucible in the temperature distribution in the furnace shown in Fig. 7. For alternative 1 the growth process starts to run at position R$_0$G$_0$ (the Fig. 6b) and for alternative 2 at RG.

Transparent colorless crystals (CaF$_2$, CaF$_2$: YbF$_3$, CaF$_2$: YbF$_3$ +PbF$_2$; CaF$_2$: YbF$_3$ +NaF; CaF$_2$: PbF$_2$) of about 10mm in diameter over 6–7 cm long were obtained. The crystals were cooled to room temperature using an established procedure. The as-grown single crystals are shown in Fig. 8. The ErF$_3$ doped crystals are transparent pink color.

The optical absorption spectra (190–1090 nm) reveal the characteristic UV absorption bands of the Yb^{2+} ions with more than 10 times higher absorption coefficient than the one of the Yb^{3+} ions (see Fig. 9). We assigned the high Yb^{2+} ions content in the as-grown crystals to the reducing conditions during the growth process due to the presence of the graphite components of the growth set-up and the lack of the oxygen scavenger during the growth of the doped crystals.

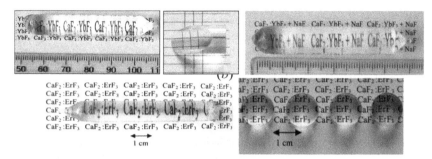

Fig. 8. As-grown crystals: (a) CaF_2: 0.17 mol%YbF_3; (b) CaF_2: 0.7 mol%YbF_3+2.5 mol% NaF; (c) CaF_2:2mol%ErF_3; (d) CaF_2 : x mol%ErF_3, x= 0.69, 0.8, 1.1, 5.

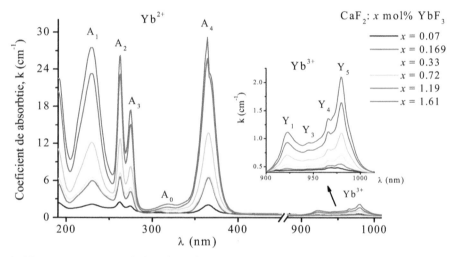

Fig. 9. Absorption spectra of YbF_3 doped CaF_2 crystals.

In order to obtain YbF_3-doped CaF_2 and NaF codoped crystals the following procedure was used. First, pure, oxygen-free CaF_2 crystals were grown using the usual growth procedure, namely adding to the starting material an amount of 4 wt% PbF_2 as an oxygen scavenger. The obtained CaF_2 crystals do not contain any undesired lead ions amount or other impurities, as results from the optical absorption spectrum; these crystals were then crushed and the powder was doped with the desired amount of YbF_3 and NaF codopant. To prevent the evaporation of NaF, a thin floating graphite lid was put on the charge in a sealed graphite crucible.

3. Characterization

Taking into account that these crystals are used as optical materials for various purposes it is important to know how the impurities influence the optical properties of the crystals. Various structural defects appear, from the so-called isolated centers to more or less complex centers. In order to study the structural defects in crystals we used various methods as are described below.

3.1 Dislocations distribution in crystals

The quality of the crystals was studied by examining the dislocations distribution using the chemical etching method. For this purpose the fresh cleavage surface (111) of the crystal were subjected to etching in aqueous solution of 2N-8N HCl.

The effect of temperature and concentration of HCl aqeuos solutions on the etching behavior on the cleavge surface of CaF$_2$ crystals was investigated [Nicoara et al., 1986]. It has been observed that the shape and the evolution of etch pits and the values of dissolution rate depend on the etching conditions. This behavior suggests that the dissolution rate along a certain direction changes with the temperature and the etchant concentrations. Taking into the ionic arrangement in the (111) plan of the CaF$_2$, the experimental observations [Sangval&Arora, 1978] and the theoretical considerations [Benson & Dempsey, 1962] on the repulsive energy between the two fluoride ions, we suppose that the activation energy depends on the crystallographic directions on the (111) surface and on the etching parameters. These explain the shape of the etch pits. Various etch pits shapes were also observed in doped calcium fluoride crystals.

The formation, multiplication and high mobility of the dislocations in ionic crystals lead not only to high densities of individual dislocations, but also cause arrangements of dislocations into well-developed grain sub-boundaries, which are stable and nearly immobile in contrast to individual dislocations (fig 10). Sub-boundaries cannot be easily removed by annealing. Some sub-boundaries appear at the beginning of the growth process.

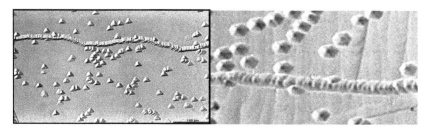

Fig. 10. Individual etch pits and grain sub-boundaries.

An interesting feature is shown in the cross-section of the crystals that are grown in multiple crucibles: the dislocation density is greater in that part of the crystal which is oriented towards the axis support of the crucible [Nicoara et al., 1987]. This fact is explained by the existence of an inadequate radial temperature gradient, the crucible being asymmetrically placed in the thermal field of the furnace, one of its parts near the hot wall. This disadvantage is eliminated by attaching a reflective molybdenum cylindrical screen to the axis, which ensures uniform thermal field for the crucible assembly.

The dislocation density along the sample and for a given cross-section in radial direction was examined. The density of the etch pits and sub-boundaries is quite low and fairly uniform through the sample for crystals with diameter up to 12 mm which were grown in a I type heater with a rate of 6 mm/h and temperature gradient of 7^0 C/ cm. The dislocation density is about 10^3-10^4 dis/cm^2. For higher pulling rate the dislocation density increases [Nicoara et al., 1986; 1987].

The influence of the various dopant and dopant concentration on the dislocations density and etch pits morphologies were studied. The dislocation density and the etch pits shape observed in various crystals are summarized in figure 11and table 1 and 2.

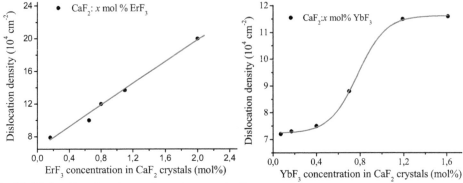

Fig. 11. Dislocation density vs dopant concentration.

CaF$_2$: x mol%ErF$_3$	x=0.17	x=0.69	x= 0.8	x=1.1
Etch pit shape				
Dislocation density	7.9·10^4	10·10^4	11.9 ·10^4	13.7·10^4

Table 1. Dislocation density (dis/cm^2) and etch pit shape in ErF$_3$ doped CaF$_2$ crystals.

Crystal	Etch pit shape	Dislocation density (dis/cm^2)	Etch pit dimensions (µm)
CaF$_2$		5.5·10^4	18; 4.6; 15.7; 3.7; 16.4; 3.7;
CaF$_2$: 1 mol% PbF$_2$		6.0· 10^4	38.9; 33.8; 40.9
CaF$_2$: 2 mol% PbF$_2$		6.6· 10^4	5.1; 15.3; 5.0 12.3; 4.5; 16.2
CaF$_2$: 0.169 mol% YbF$_3$		7.3·10^4	47.6; 37.9; 42;
CaF$_2$: 0.7 mol% YbF$_3$		8.8·10^4	22; 22.7; 22.3;
CaF$_2$:1.2mol%YbF$_3$ + 2.5 mol% NaF		12.3·10^4	12; 3; 11.5; 3; 11.3; 3;

Table 2. Etch pit shape and dislocation density in doped CaF$_2$ crystals.

3.2 Dopant distribution alnog the crystals

The homogeneous distribution of the dopant in laser crystals is important because this affects the efficiency of the laser.

The effective segregation coefficient (k_{eff}) at a given growth rate is defined by

$$k_{eff} = C_S/C_L^0, \tag{1}$$

where C_S is the dopant concentration at the bottom of the as-grown crystal and $C^0{}_L$ is the dopant concentration in the initial melt. The value of C_S can be measured by various methods, or estimated from optical absorption measurements [Kuwano, 1982; Sun et al., 2005].

The effective segregation coefficient determination of the Er^{3+}, Yb^{3+}, Yb^{2+} and Pb^{2+} ions in CaF₂ host by optical absorption method is based on the following two laws.

(a) According to the Beer–Lambert law the absorption coefficient is proportional to the sample concentration (C), $a = aC$ where a is the absorption coefficient for unit ion concentration and unit light path length; a may be recognized as constant for the investigated ions concentration [Kuwano, 1982; Sun et al., 2005]. The various ions concentration can be estimated from the measured optical density,

$$O.D. = \log(I_0 / I) \tag{2}$$

where I_0 is the light intensity incident on the sample, I is the transmitted light intensity. Taking into account the relation

$$I = I_0 \exp(-\alpha \cdot d) \tag{3}$$

where a is the absorption coefficient and d is the sample thickness, the dopant concentration of a slice can be estimated using the relation [Kuwano, 1982]:

$$C = \frac{\alpha}{a} = \frac{O.D. \cdot 2.30258}{a \cdot d} \tag{4}$$

(b) The dopant concentration (along the growth axis) at the distance z from the origin of the crystal can be obtained by using the classical relation [Hurle, 1993]:

$$C_S(z) = C_L^0 k_{eff} \left[1 - g(z) \right]^{k_{eff} - 1} \tag{5}$$

where $g(z)$ is the crystallized fraction of the melt given by $g(z) = V\,t/L = z/L$, V is the crystal growth rate, t is the growth time and L is the crystal length, so Vt is the grown crystal length, z, at the moment t. The more the k_{eff} differs from unity, the larger is the concentration gradient in the crystal.

Taking into account the Beer–Lambert law, the dopant distribution along the crystal length can be estimated using the optical absorption method. In order to determine the effective segregation coefficient, we cut the crystal into i slices with the same thickness (Fig. 12) and calculated the absorption coefficient for a particular absorption peak, from the optical absorption spectrum of every slices ($a(z)$). Using the relations (4) and (5), the following expression is obtained in order to determine the effective segregation coefficient:

$$\lg\alpha(z) = (k_{eff} - 1)\lg(1 - z/L) + \lg(ak_{eff}C_L^0) \tag{6}$$

The effective segregation coefficient can be calculated from the slope $m = k_{eff} - 1$, of the fitting line of log $a(z)$ versus log$(1-z/L)$. This was the method used to calculate the segregation coefficient.

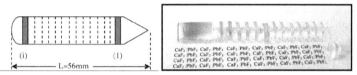

Fig. 12. Crystal cut for segregation coefficient measurement.

Segregation coefficient of Yb³⁺ and Yb²⁺ ions
Room-temperature optical absorption spectra were recorded using a Shimadzu 1650PC spectrophotometer. The absorption coefficient increases as the YbF$_3$ concentration in the CaF$_2$ host increases. Fig. 13 shows the influence of the YbF$_3$ concentration on the absorption coefficient of the peaks at 979 nm and 365 nm.

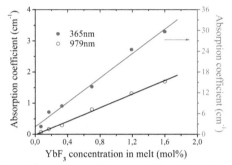

Fig. 13. Dependence of the absorption coefficient of 979 nm and 365 nm on the initial YbF$_3$ concentration in the melt.

The variation of Yb²⁺ and Yb³⁺ ions' concentration distribution along the as-grown crystals has been studied only for two concentrations, namely with YbF$_3$ added to the melt in molar concentrations equal to 0.7 mol% and 1.6 mol%.
The absorption coefficient $\alpha(z)$ of every slice was calculated from the absorption spectra for two characteristic absorption bands: 979 nm for Yb³⁺ ions and 365 nm for Yb²⁺ ions [Nicoara et al., 2008a; Nicoara et al., 2008b]. Fig. 14a shows the Yb³⁺ ions' concentration (characterized by the absorption coefficient, α) distribution along the crystals. For both investigated crystals the absorption coefficient of the slides is almost constant along the crystals; this means that the trivalent Yb ions are distributed homogeneously along the crystals. The slope of the fitting line of the Yb³⁺ ions concentration distribution along the crystals, namely of lg $\alpha(z)$ vs. lg$(1-z/L)$ (Fig. 14b), have been used to calculate the segregation coefficient of Yb³⁺ ions in YbF$_3$: CaF$_2$ crystals. The obtained values of the segregation coefficient are close to the unity for both investigated crystals: k ≈ 1.00 for the CaF$_2$:1.61 mol% YbF$_3$ sample and k ≈ 0.98 for the CaF$_2$: 0.72mol% YbF$_3$ sample; this indicates a rather homogeneous distribution of Yb³⁺ ions in YbF$_3$:CaF$_2$ crystals [Nicoara et al., 2008b]. These values are in good agreement with those reported, for example k ≈ 1.07 for CaF$_2$:1.96 at% YbF$_3$ [Su et al., 2005].

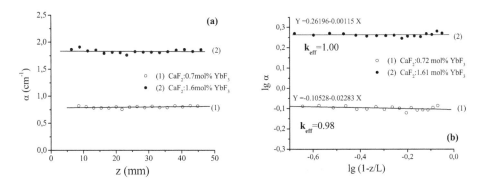

Fig. 14. (a) Yb^{3+} ions distribution along the crystal; (b) fitting lines of lgα vs lg (1-g).

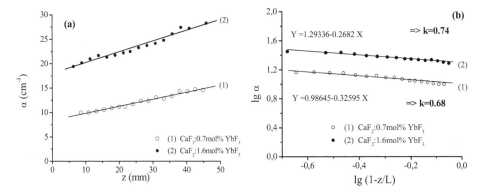

Fig. 15. (a) Yb^{2+} ions distribution along the crystal; (b) fitting line of lg α vs lg (1-g).

Fig. 15a shows the Yb^{2+} ions concentration (characterized by the absorption coefficient, α) distribution along the crystals. The absorption coefficient increases along the crystals. and as the concentration of YbF$_3$ in the initial melt increases the absorbtion coefficient, α, increases too. The fitting lines of $lg\ a\ (z)\ vs.\ lg\ (1-g)$ for the two studied Yb:CaF$_2$ crystals are shown in Fig. 6b. The segregation coefficient is less than unity for both crystals: $k \approx 0.68$ for the CaF$_2$:1.6mol% YbF$_3$ sample and $k \approx 0.74$ for the CaF$_2$:0.72mol% YbF$_3$ sample. The more k differs from unity, the larger the dopant concentration gradient along the crystal is and hence, no homogeneous distribution of Yb^{2+} ions in YbF$_3$: CaF$_2$ crystals can be obtained.

Segregation coefficient of Er^{3+}

The segregation coefficient of the Er$^{3=}$ ion in CaF$_2$ host was determined for two ErF$_3$ concentrations: 0.8mol% ErF$_3$ and 2 mol%ErF$_3$ using the method described above.

The absorption coefficient $a\ (z)_i$ of every slice i was calculated from the absorption spectra for the following absorption peaks: 406nm, 968nm and 979nm. The dopant distribution along the crystals shows some oscillations of the Er^{3+}-ions concentration. These types of oscillations have also been observed for other crystals grown by Bridgman technique; this behavior was not explained yet [Barat, 1995; Mitric et al., 2006]. The strongest oscillatory behavior of the dopant distribution along the crystal has been observed for the CaF_2: 2mol% ErF_3 sample.

The obtained values of effective the segregation coefficient are: k_{eff} = 1.01-0.99 for CaF_2: 0.8mol% ErF_3 sample and 1.03-1.04 for CaF_2: 2mol% ErF_3 crystal [Munteanu et al., 2010]. As the concentration of the ErF_3 in the initial melt increases the effective segregation coefficient, k_{eff}, increases too; the dependence of the effective segregation coefficient on the dopant concentration was also observed in other host [Lifante et al., 1999; Barraldi et al., 2005]. Because the effective segregation coefficient is almost equal with unity, the Er^{3+} ions distribution in CaF_2 crystals is approximately uniform along the crystal and from this point of view, ErF_3- doped CaF_2 crystals could be a good laser material.

Segregation coefficient of Pb^{2+}

Figure 16 shows the Pb^{2+} ions distribution (characterized by the absorption coefficient) along the four investigated crystals, with various PbF_2 concentrations in the initial melt. The dopant distribution along the crystals shows some oscillations of the Pb^{2+} ions concentration. For CaF_2 : 2 mol% PbF_2 sample the dopant distribution along the crystal has the strongest oscillatory behavior. The calculated effective segregation coefficient for the studied samples is: 0.85 for CaF_2 : 0.5 mol%PbF_2 crystal, 0.925 for CaF_2 : 1mol%PbF_2 crystal, 1.002 for CaF_2:2 mol%PbF_2 crystal and 1.15 for CaF_2:3 mol%PbF_2 crystal.

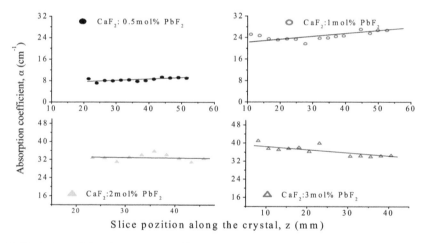

Fig. 16. Variation of the absorption coefficient along the PbF_2 doped CaF_2 crystals

As the concentration of the Pb^{2+} ions in the initial melt increases, the effective segregation coefficient increases, too [Paraschiva et al., 2010]. Dependence of the segregation coefficient of different ions in the CaF_2 host on the dopant concentration is shown in fig. 17.

3.3 Dielectric relaxation

Information on impurity-defect aggregates can be obtained from spectroscopic studies and dielectric relaxation technique. The optical absorption provides information about the nature and site symmetry of the defects. Trivalent RE ions in CaF_2 tend to form pairs of adjacent ions, for charge compensation, even at low RE concentrations. The extra positive charge is usually compensated by interstitial F⁻ ions. Besides the tetragonal (C$_{4v}$) symmetry of the predominant dipolar complex, many other simple or cluster configurations appear. The resultant dipolar complexes can reorient by "jumps" of one of the charges to other lattice sites. In order to use the laser properties of the crystals it is necessary to study the influence of the various type of defects introduced by various impurities, such as the RE activator ions or other ions on the properties of the crystals. Information on impurity-defect aggregates can be obtained from spectroscopic and dielectric relaxation techniques, the last being sensitive to aggregates with a dipole moment which can reorient through migration of the anions. Temperature and frequency dependence of the complex dielectric constant give information about the relaxation processes and permits the determination of the activation energy and the reciprocal frequency factor, τ_0, of the relaxation time, τ, and the number of dipoles that contribute to the relaxation process. [Fontanella& Andeen, 1976; Andeean et al., 1977; Fontanella&Tracy, 1980; Andeen et al., 1979; Andeen et al., 1981].

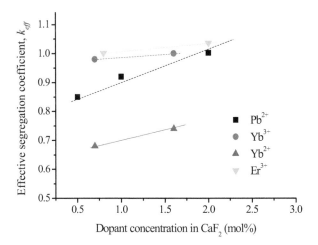

Fig. 17. Dependence of the k$_{eff}$ on the dopant concentration for Pb^{2+}, Yb^{2+}, Yb^{3+} şi Er^{3+} ions.

The impurity-defects in CaF$_2$ crystals doped with various ions-Er^{3+}, Yb^{3+}, Pb^{2+} and Na$^+$ - were studied using the dielectric relaxation method [Nicoara et al., 2006; Nicoara et al., 2006b; Nicoara et al., 2008; Pruna et al., 2009].

Capacitance (C) and dielectric loss (D=tanδ) measurements were performed on the samples using a RLC Meter type ZM2355, NF Corporation, Japan over the temperature range of 150–320 K at seven audio-frequencies. The real part of the complex dielectric constant, ε_1, was calculated from the measured capacitance C. The imaginary part of the complex dielectric constant, ε_2 was then calculated from D=ε_2 /ε_1. Linear heating rates of 2 K/min were employed from liquid nitrogen to room temperature. The dielectric properties have been measured on the 10 mm diameter and 0.6 mm thick disks using an Ag (Leitsilber) contacts. The measurements have been performed on as the (111) cleavage plane as on perpendicular to growth direction cut samples. Using this method we obtained information about the influence of dopant concentration on the formation of various charge compensating defects in doped CaF$_2$ crystals.

After doping some dipolar complex appear which can reorient (relax) by "jumps" of one of the charges to other lattice sites. Such dipoles are usually characterized by a relaxation time, τ given by

$$\tau = \tau_0 \exp(E/kT) \tag{7}$$

In order to determine the activation energy for reorientation, E, and the reciprocal frequency factor τ_0, the complex dielectric constant $\varepsilon(\omega, T) = \varepsilon_1(\omega, T) + i\,\varepsilon_2(\omega, T)$ has to be determined. The real and imaginary parts of the complex dielectric constant are given by the Debye equations. Since ε_2 has a maximum for $\omega\tau = 1$, it follows from Eq. (1) that: $\ln\omega = -(E/kT_{max}) - \ln\tau_0$, where T_{max} is the temperature at which ε_2 has a maximum at a given frequency and E and τ_0 can be determined from the plot of T^{-1}_{max} vs $\ln\omega$.

Dependence of the real and the imaginary part of the complex dielectric constant of CaF_2: 0.17 mol% ErF_3 sample on the temperature and on the frequency is plotted in Fig. 18.

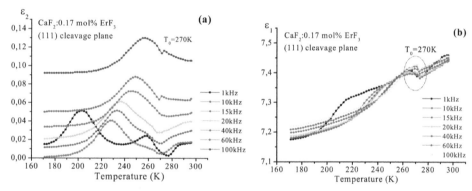

Fig. 18. Temperature and frequency dependence of (a) the imaginary part and (b) real part of the complex dielectric constant. The measurements have been made on (111) cleavage plane.

The peaks occurring at higher temperature as the frequency increases is associated to the R_I, Debye-like relaxation [Fontanella&Andeen, 1976] due to the relaxation of Er^{3+} - F^-_i pair in which the F^- occupies the nearest-neighbor (NN) interstitial position with respect to the Er^{3+} ion. The R_{IV} relaxation [Fontanella&Andeen, 1976] (around 240-260 K, with activation energy > 0.5 eV) increases as the concentration increases and is very clear only for Pb^{2+}-codoped samples [Nicoara et al., 2008]. The values of the relaxation parameters for the observed relaxations are shown in the Table 3.

The value of ε_1 and ε_2 increases as the rare-earth ions concentration increases; around $T_0 = 269K$ an anomaly of ε_1 and ε_2 behavior has been observed which can be assigned with a phase transition of the order-disorder type. The temperature dependence of the dielectric constant and loss tangent for this sample is shown in Fig. 19a. The loss tangent has a maximum at a slightly lower temperature than the ε_1 maximum and a sharp minimum at a slightly higher temperature. The temperature dependence of the reciprocal of ε_1 against $T-T_0$ is shown in Fig. 19b. We can observe that the slopes of the $\varepsilon_1^{-1}(T)$ plots are different on both sides of T_0. These types of anomalies in the dielectric properties have been observed for some perovskite-type compounds [Smolenskii et al., 1958] and are assigned with an order-disorder type phase transition [Strukov&Levanyuk, 1998].

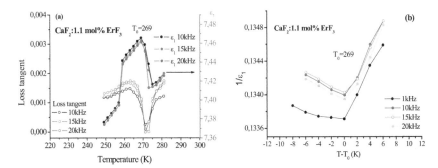

Fig. 19. (a)Temperature dependence of ε_1 and of the loss tangent. (b) Temperature dependence of the reciprocal dielectric constant.

The dielectric spectra of YbF$_3$ doped CaF$_2$ crystals has same behavior as described below, the obtained relaxation parameters are summarized in Table 3.

It has been observed [Doualan et al., 2010;] that the detrimentral pairing effect of the RE ions [Corish et al., 1982; Petit et al., 2007; Petit et al., 2008] can be decreased by co-doping the crystals with charge compensating buffer ions, such as monovalent ions, like Na$^+$ ions [Doualan et al., 2010; Su et al., 2005, 2007] or non-optically active rare-earth trivalent ions. In the case of the double doped (with Yb^{3+} and Na$^+$ ions) CaF$_2$ crystals, the Na$^+$ can work also as charge compensator for Yb^{3+} ions, entering in interstitial (or substitutional) positions near the Yb^{3+} ion and leading to C$_{3v}$ (or C$_{2v}$) symmetry, sites without dipolar properties. In conclusion, codoping with Na$^+$ leads to new (Na$^+$ -V$_F$) and (Yb^{3+}-Na$^+$) centers. These defects can be studied using the dielectric relaxation. Five YbF$_3$ –doped crystals were investigated with the following amount of YbF$_3$ added in the starting mixture in the melt: 0.07, 0.17, 0.72, 1.19 and 1.6mol%YbF$_3$; six NaF co-doped YbF$_3$:CaF$_2$ crystals with different Na:Yb ratios of R=2, 4, 16, 28, 36 have been also grown.

A typical temperature and frequency dependence of the real and imaginary parts of the complex dielectric constant is plotted in Fig. 20 for CaF$_2$: 0.16mol%YbF$_3$ +2.5mol % NaF sample (R=16). The value of ε_1 and ε_2 increases as the YbF$_3$ concentration increases. In the range of the investigated temperatures, for YbF$_3$ doped CaF$_2$ crystals there is only one maximum in the temperature dependence of ε_2, at a given frequency, that corresponds to relaxation of NN (C$_{4v}$) dipoles [Fontanella&Andeen, 1976]. An "anomaly" of ε_1 and ε_2 behavior, like a maximum has been observed around the temperature T_0 = 270-279K for YbF$_3$ doped samples and 275-276K for NaF codoped samples, depending on the sample concentration (Fig. 20 and 21a). The value of the temperature T_0 does not depend on the frequency, so is not a relaxation. This behavior is assigned to an order-disorder type phase transition.

The temperature and frequency dependence of ε_2 for Na$^+$ ions co-doped samples, for ratio R<20 is characterized by three peaks (Fig. 20b, 21a). For all samples the peak occurring at around 195K at 1kHz is associated to the R$_I$ (NN) center relaxation. The second peak is associated to the order-disorder transition and the third peak, that appears only for Na$^+$ ions codoped crystals, is associated to (Na$^+$-V$_F$) dipoles relaxation, formed by a substitutional Na$^+$ ion and a F$^-$ vacancy created to maintain the electrical neutrality of the crystal Johnson et al., 1969; Shelley&Miller, 1970].

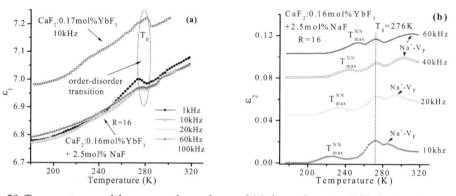

Fig. 20. Temperature and frequency dependence of (a) the real part and (b) the imaginary part of the complex dielectric constant

For heavier NaF- codoped samples, namely for ratio R>20, only one relaxation is observed, associated to the (Na^+-V_F) dipole (see Fig.21b). This confirms the observed suppression of the peaks of optical absorption spectra, corresponding to centers with C_{4v} (NN dipole) symmetry.

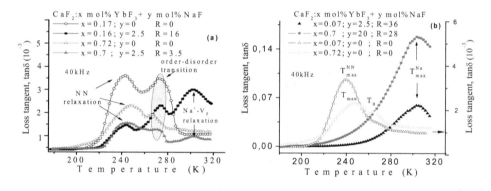

Fig. 21. Influence of Na^+ ions on the loss tangent for (a) low and (b) for high R ratios.

The number of dipoles N_D that contribute to the dielectric relaxation peak can be calculated from the dielectric spectra using the methods described in [Fontanella&Andeen, 1976; Campos&Ferreira, 1974].

Figure 22a shows the variation of the number of NN dipoles (N_{NN}) with the YbF$_3$ concentration for un-codoped and 2.5mol%NaF codoped samples. The number of NN dipoles that contributes to the dielectric loss peak decreases as the Yb^{3+} ions concentration increases. This effect was also observed by Fontanella [Fontanella&Andeen, 1976] for CaF$_2$: ErF$_3$ samples doped with concentration higher than 0.1mol%. This indicates that as the YbF$_3$ concentration increases the predominant dipoles are NNN and/or clusters types (that relax at lower and/or higher temperature then we have investigated), diminishing in this way the number of dipoles that contribute to the dielectric relaxation of NN (C_{4v} symmetry) type dipoles.

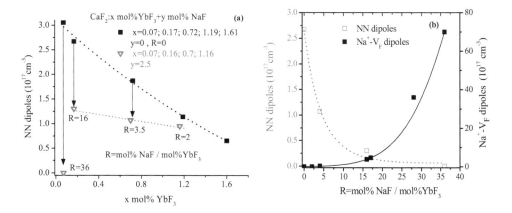

Fig. 22. (a) Influence of YbF$_3$ concentration and ratio R on the NN dipoles concentration for samples doped with 2.5mol%NaF (open symbols); (b) Influence of ratio R on the NN and Na$^+$-V$_F$ dipoles concentration

The Na$^+$- ions co-doped YbF$_3$:CaF$_2$ crystals reveal a decrease of the NN dipoles in comparison with the YbF$_3$- doped samples, indicating that the Na$^+$ ions reduce the formation of NN type dipoles with C$_{4v}$ symmetry; this is confirmed by the absorption spectra. This effect depends on the ratio R= ymol%NaF/ xmol%YbF$_3$, as is shown in Figs.22a,b.; as the ratio R increases, the number of NN (C$_{4v}$) centers decreases and the number of (Na$^+$-V$_F$) dipoles increases (fig.22b); for ratio R=28 and 36 no NN type relaxation was detected (fig.21b), this means a very small concentration of C$_{4v}$ sites.

CaF$_2$:	0.17 mol% ErF$_3$	0.69 mol% ErF$_3$	1.1 mol% ErF$_3$	0.5 mol% PbF$_2$	1 mol% PbF$_2$
dipole	(Er^{3+}- F$^-$)NN	(Er^{3+} - F$^-$) NN	(Er^{3+}- F$^-$)NN	(Pb^{2+}-F$_{vac}$)	(Pb^{2+}-F$_{vac}$)
E(eV)	0.383	0.335	0.345	0.438	0.358
τ_0 (s)	5.1*10^{-14}	26.2*10^{-14}	52* 10^{-14}	1 *10^{-14}	3*10^{-14}
CaF$_2$:	0.07 mol% YbF$_3$	0.17 mol% YbF$_3$	0.72 mol% YbF$_3$	CaF: 0.17 mol%YbF$_3$ +2.5mol% NaF	
dipole	(Yb^{3+}- F$^-$)NN	(Yb^{3+}-F)NN	(Yb^{3+}-F)NN	(Yb^{3+}- F$^-$) NN	(Na$^+$-F$_{vac}$)
E(eV)	0.35	0.42	0.34	0.305	0.515
τ_0 (s)	2.1*10^{-13}	0.055*10^{-13}	4.4*10^{-13}	2.1*10^{-12}	7.45*10^{-15}

Table 3. Relaxation parameters.

3.4 Tailoring the charge compensating defects in YbF₃ doped CaF₂ crystals

When YbF$_3$ is dissolved in CaF$_2$, normally the Yb ions are in a trivalent state. It is also known that a certain fraction of any of RE^{3+} ions can be reduced to divalent state depending on the growth conditions. The Yb^{3+}-ions usually occupy a cation substitutional position, but charge compensation is required to maintain the electrical neutrality of the system; the extra positive charge is compensated by an interstitial fluorine ion (F$_i^-$). For low YbF$_3$ concentration (< 0.1mol%), besides the O$_h$ cubic symmetry with no local charge compensation, the so-called isolated dipolar centers are predominant with tetragonal symmetry (C$_{4v}$) in which the F$_i^-$ ion occupies a nearest-neighbor (NN) interstitial site and trigonal (C$_{3v}$) NNN site [Kirton&McLaughlan, 1967; Corish et al., 1982; Petit et al., 2007; Petit et al., 2008]. At higher Yb^{3+} ion concentrations, the dopant ions aggregate and form more or less complex clusters [Corish et al., 1982; Petit et al., 2007; Petit et al., 2008]. Such a complicated structure leads to broad optical absorption bands. The detrimental pairing effect can be decreased by co-doping the crystals with charge compensating buffer ions, such as monovalent ions [Doualan et al., 2010; Su et al., 2005, 2007] or non-optically active rare-earth trivalent ions. After Na$^+$ ions were introduced as charge compensators the IR emission intensity of the YbF$_3$-doped CaF$_2$ crystal was enhanced several times [Su et al., 2005, 2007].

When a Na$^+$ ion is introduced into the CaF$_2$ lattice, this enters substitutionally and is compensated by a fluorine vacancy (Na$^+$ -V$_F$) [Johnson et al. 1969; Shelley&Miller, 1970; Fontanella et al., 1980]. In the case of the double doped (with Yb^{3+} and Na$^+$ ions) CaF$_2$ crystals, the Na$^+$ can work also as charge compensator for Yb^{3+} ions, entering in interstitial (or substitutional) positions near the Yb^{3+} ion and leading to C$_{3v}$ (or C$_{2v}$) symmetry, sites without dipolar properties. In conclusion, codoping with Na$^+$ leads to new (Na$^+$ -V$_F$) and (Yb^{3+}-Na$^+$) centers.

In order to study the varieties of Yb^{3+} sites in CaF$_2$ host, several YbF$_3$ doped and NaF-codoped CaF$_2$ crystals with different Na:Yb ratios were grown by Bridgman method. Six NaF co-doped YbF$_3$:CaF$_2$ crystals with different Na:Yb ratios of 2, 4, 16, 28, 36 have been also grown. Room temperature absorption spectra and dielectric spectra were measured to study the effect of Na$^+$ ions on the charge compensating defects formation [Pruna et al., 2009].

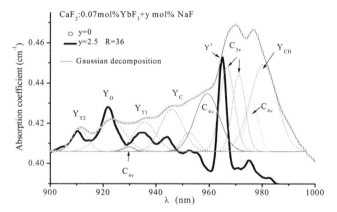

Fig. 23. Absorption spectra of (0.07mol%YbF$_3$+ymol%NaF):CaF$_2$ crystals; the decomposition by Gaussian multi-peak fits is shown for CaF$_2$:0.07mol%YbF$_3$ sample.

Figures 23 and 24 show the absorption spectra of NaF-codoped YbF$_3$:CaF$_2$ crystals in comparison with the YbF$_3$ doped samples; the influence of the Na$^+$ ions on the absorption spectra depends on the ratio R= y mol%NaF/ x mol% YbF$_3$. The absorption spectra show that the codoping with Na$^+$ ions in different R ratios can modulate the spectroscopic properties of Yb^{3+} ions in CaF$_2$ lattice in a large scale. The differences among the spectra corresponding to ratio R<10 and those R>15 are distinct (see Figs. 23 and 24). The absorption spectra of the crystals with R<10 and without NaF are broad and non-structured.

The absorption bands of the samples with R>15 became narrower and clearly resolved into six peaks, (see Figs.23 and 24) corresponding to trigonal T$_2$ (Y$_{T2}$=910nm), cubic O$_h$ (Y$_O$=922nm), trigonal T$_1$ (Y$_{T1}$=936nm), clusters sites (Y$_C$ at 945.6 and 955nm) and trigonal C$_{3v}$ site (Y^1=965nm) with Na$^+$ ion as charge compensator. As it results from the analysis of these spectra, the presence of Na$^+$ ions, at ratio R>20 suppresses the peak Y$_{CH}$=980nm that is attributed to 1-5 transition of hexamer type clusters, and reduces also the intensity of the transitions corresponding to "small clusters", Y$_C$. Another effect of the Na$^+$ ions is to reduce the tetragonal C$_{4v}$ (NN) sites; for samples with R>20 this site is drastically reduced, and this effect is confirmed by the dielectric relaxation measurements described in III.3.

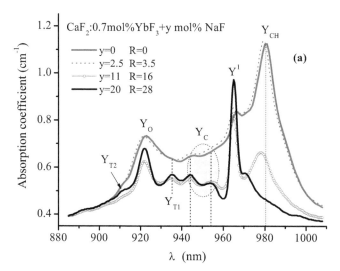

Fig. 24. Optical absorption spectra of 0.7mol%YbF$_3$ doped CaF$_2$ crystals codoped with different amount of NaF.

The peak corresponding to hexamer cluster (Y$_{CH}$=980nm) is completely suppressed as well as the C$_{4v}$ site (see Fig. 23) for the sample with R=36 (0.07mol% YbF$_3$ + 2.5 mol%NaF) : CaF$_2$. The reducing effect depends on the R ratio. Comparing the absorption spectra of Yb doped and Na$^+$ ions codoped crystals it is clear that we can "tailor" the type of the compensating defects by choosing the suitable ratio R of the Na:Yb ions.

The dielectric spectra reveal two peaks that correspond to relaxation of two type of dipoles: NN (Yb^{3+}- F$_i$ centers with C$_{4v}$ simmetry) and (Na$^+$-V$_F$) dipoles (see III.3).The temperature and frequency dependence of ε_2 for Na$^+$ ions co-doped samples, for ratio R<20 is characterized by three peaks. For all samples the peak occurring at around 195 K at 1kHz is associated to the R$_I$ (NN) center relaxation [Fontanella&Andeen, 1976; Andeen et al., 1981].

The second peak is associated to the order-disorder transition [Smolenskii et al., 1959; Nicoara, 2008] and the third peak, that appears only for Na+ ions codoped crystals, is associated to (Na+-V_F) dipoles relaxation, formed by a substitutional Na+ ion and a F⁻ vacancy (V_F) created to maintain the electrical neutrality of the crystal [Johnson et al., 1969; Shelley&Miller, 1970].

For heavier NaF- codoped samples, namely for ratio R>20, only one relaxation is observed, associated to the (Na+-V_F) dipole. This confirms the observed suppression of the peaks of optical absorption spectra, corresponding to centers with C_{4v} (NN dipole) symmetry (see the optical absorption spectra Figs. 23 and 24).

The number of dipoles N_D that contribute to the dielectric relaxation peak can be calculated from the dielectric spectra using the methods described in [Fontanella&Andeen, 1976; Campos&Ferreira, 1974].

The Na+- ions co-doped YbF$_3$:CaF$_2$ crystals reveal a decrease of the NN dipoles in comparison with the YbF$_3$- doped samples, indicating that the Na+ ions reduce the formation of NN type dipoles with C_{4v} symmetry; this is confirmed by the absorption spectra, see figures 23 and 24. This effect depends on the ratio R= ymol%NaF/ xmol%YbF$_3$, as is shown in Fig. 22a,b.; as the ratio R increases, the number of NN (C_{4v}) centers decreases and the number of (Na+-V_F) dipoles increases (fig.22b); for ratio R=28 and 36 no NN type relaxation was detected (fig.21b), this means a very small concentration of C_{4v} sites.

Figure 25 illustrates the variation of the (Na+-V_F) dipoles concentration on the YbF$_3$ concentration for all the studied NaF-codoped crystals, the ratio R for every sample is specified. The insert shows the dependence of the NN dipoles concentration on the NaF concentration for a given YbF$_3$ concentration, namely for 0.7mol% YbF$_3$.

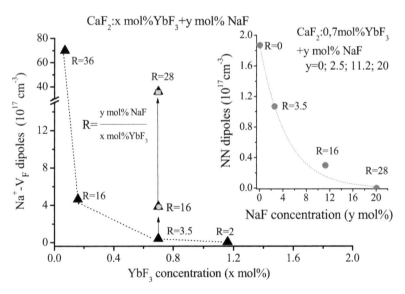

Fig. 25. Influence of YbF$_3$ concentration and of ratio R on the Na+-V_F dipoles concentration. The inset shows the influence of NaF concentration on the NN dipoles concentration for CaF$_2$: 0.7 mol%YbF$_3$ + y mol%NaF samples.

Influence of Na+ ions on the charge compensating defects

From the dielectric spectra results that Na+ ions lead to a decrease of the NN dipoles (see figs.22b and 25), the charge compensating defects with tetragonal C$_{4v}$ symmetry and this behavior is confirmed by the absorption spectra of the samples.

Figure 26 shows the dependence on NaF concentration of the line intensity corresponding to the C$_{4v}$ center and of the NN dipoles concentration for samples doped with 0.7 mol%YbF$_3$. The inset shows the dependence on the NaF concentration of the line intensity corresponding to C$_{3v}$ site with Na+ ion as charge compensator, respectively of the line intensity corresponding to clusters. Taking into account that the peak intensity is proportional with the absorbant centers, we observe that as the Na+ ions concentration increases the charge compensating defects corresponding to C$_{4v}$ site decrease and those corresponding to C$_{3v}$ site increase; the decrease of the cluster type defects is also clear. The decreases of the concentration of the defects corresponding to C$_{4v}$ site is confirmed by the decrease of the NN (C$_{4v}$) dipoles concentration calculated from dielectric spectra.

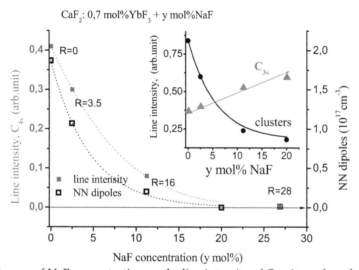

Fig. 26. Influence of NaF concentration on the line intensity of C$_{4v}$ site and on the NN dipoles concentration. The inset shows the influence of NaF concentration on the line intensity corresponding to defects with C$_{3v}$ symmetry with Na+ ion as charge compensator and on the line intensity corresponding to the clusters.

The effect of Na+ ions on the formation of the compensating defects can be explained this way. It is known that for concentration above about 0.05at% the rare-earth ions aggregate to form pairs, to form more or less complex centers, clusters. The most common pairs are the dimmers that consist by two Yb^{3+} ions and two F$_i^-$ ions and the hexametric clusters [Petit et al., 2008]. The Na+ ions substitute one of the partners of the pairs and lead to an increased number of pseudo-isolated centers, like C$_{3v}$ site with Na+ ion as charge compensator, without dipolar properties. This pair-breaking effect is more effective on the hexameric clusters than on the smaller clusters; for R= 28 and 36 the optical absorption peak (at 980nm) associated to hexamer clusters is suppressed, as is also illustrated in figures 23 and 24. As results from our investigation, codoping with Na+ the YbF$_3$ doped CaF$_2$ crystals it is possible to "tailor" the type of the compensating defects by choosing the suitable ratio R of the Na:Yb ions.

The experimental results showed that codoping with Na^+ ions in different Na:Yb ratios we can modulate the spectroscopic properties of Yb^{3+} ions in CaF_2 host. The influence of Na^+ ions on the defects formation is discussed taking into account the optical absorption spectra and the calculated number of NN dipoles whose relaxation are observed. By choosing the suitable ratio of the Na:Yb ions it is possible to "tailor" the type of the compensating defects in YbF_3 doped CaF_2 crystals.

4. References

Andeen, C.; Link, D. & Fontanella, J. (1977). Cluster-associated relaxations in rare-earth-doped calcium fluoride. *Phys Rev. B*, Vol. 16, pp. 3762-3767

Andeen, C.; Matthews, G. E.; Smith, M.K. & Fontanella, J. (1979). Electric dipole relaxation of mixed clusters in double-doped CaF_2. *Phys. Rev. B*, Vol. 19, pp. 5293-5298

Barat, C. (1995), Origines thermiques et convectives des segregations solutales dans des alliages semiconducteurs. PhD. Thesis, University of Rennes

Barraldi, A.; Capelleti, R.; Mazzera, M.; Ponzoni, A.; Amoretti, G.; Magnani, N.; Toncelli, A. & Tonelli, M. (2005). Role of Er^{3+} concentration in high-resolution spectra of BaY_2F_8 single crystals. *Phy. Rev. B*, Vol. 72, pp. 075132

Benson, K.K. & Dempsey, E. (1962). The cohesive and surface energies of some crystals possessing the fluorite structure. *Proc. Roy. Soc. Lond. A*, Vol. 366, pp. 344-358

Bridgman, P.W. (1925). Crystal growth apparatus. *Proc. Am. Acad. Arts Sci.*, Vol. 60, pp. 305

Campos, V.B. & Leal Ferreira, G.F. (1974). Dipolar studies in CaF_2 with Ce^{3+}. *J. Phys. Chem. Solids*, Vol. 35, pp. 905-910

Chang, C.E. & Wilcox, W.R. (1974). Control of interface shape in the vertical bridgman-stockbarger technique. *J. Cryst. Growth*, Vol. 21, pp. 135-140

Corish, J.; Catlow, C.R.A.; Jacobs, P.W.M. & Ong, S.H. (1982). Defect aggregation in anion-excess fluorites. Dopant monomers and dimers. *Phys. Rev. B*, Vol. 25, pp. 6425

Doualan, J.L.; Comy, P.; Brnayad, A.; Minard, V.; Moncorge, R.; Boudeile, J.; Druon, F.; Georges, P. (2010), Yb^{3+} doped $(Ca,Sr,Ba)F_2$ for high power laser applications, In: *Laser Physics*, Pashemin, P., pp.533-537, Springer Publishing, ISSN 1054-660X

Feofilov, P.P. (1956). Absorption and luminescence of divalent rare-earth ions in fluoride crystals. *Opt. Spectrosc.*, Vol. 1, pp. 992-1000

Fong, F.K. (1964). Electrolytic reduction of trivalent rare-earth ions in alkalin-earth halides. *J. Chem. Phys.*, Vol. 41, pp. 2291-2296

Fontanella, J. & Andeen, C. (1976). The dielectric spectrum of erbium doped CaF_2. *J. Phys. C*, Vol. 9, pp. 1055

Fontanella, J. & Treacy, D.J. (1980). The effect of quenching on the defect structure of calcium fluoride doped with erbium. *J. Chem. Phys.*, Vol. 72, pp. 2235-2244

Fontanella, J.; Chadwick, A.V.; Carr, V.M.; Wintersgill, M.C. & Andeen, C.G. (1980). Dielectric relaxation studies of alkali-metal-doped calcium fluoride. *J. Phys. C: Solid State Phys.*, Vol. 13, pp. 3457-3466

Fu, T.W. & Wilcox, W.T. (1980). Influence of insulation on stability of interface shape and position in the vertical Bridgman-Stockbarger technique. *J. Cryst. Growth*, Vol. 48, pp. 416-424

Gault, W.A.; Monberg, E.M. & Clemans, J.E. (1986). A novel application of the vertical gradient freeze method to the growth of high quality III–V crystals. *J. Cryst. Growth*, Vol. 74, pp. 491-506

Hurle, D.T.J. (1993), Crystal pulling from the melt, Springer-Verlag, Berlin, ISBN 3540566767

Jasinski, T.; Rohsenow, W.M. & Witt, A.F. (1983). Heat transfer analysis of the Bridgman-Stockbarger configuration for crystal growth: I. Analytical treatment of the axial temperature profile. *J. Cryst. Growth*, Vol. 61, pp. 339-354

Johnson, H.B.; Tolar, N.J.; Miller, G.R. & Cutler, I.B. (1969). Electrical and mechanical relaxation in CaF$_2$ doped with NaF. *J. Phys. Chem. Solids*, Vol. 30, pp. 31-42

Kaczmarek, S.M.; Tsuboi, T.; Ito, M.; Boulon, G. & Leniec, G. (2005). Optical study of Yb^{3+}/Yb^{2+} conversion in CaF$_2$ crystals. *J. Phys.:Condens. Mater.*, Vol. 17, pp. 3771-3786

Kaplyanskii, A.A. & Feofilov, P.P. (1962). Absorption spectra of divalent rare-earth ions in fluoride crystals. *Opt. Spectrosc.*, Vol. 13, pp. 235-241

Kaplyanskii, A.A. & Smolyanskii, P.L. (1976), Effect of external fields on optical absorption spectra of SrCl$_2$: Yb^{2+}. *Opt. Spectrosc.*, Vol. 40, pp. 528-540

Kirton, J. & McLaughlan, S.D. (1967). Correlation of electron paramagnetic resonance and optical absorption spectra of CaF$_2$:Yb^{3+}. *Phys. Rev.*, Vol. 155, pp. 279-284 .

Kuwano, Y. (1982). Effective distribution coefficient of neodynium in Nd:Gd$_3$Ga$_5$O$_{12}$ crystals grown by the Czochralski method. *J. Crystal Growth*, Vol. 57, pp. 353-361

Lifante, G.; Cantelar, E.; Munoz, J.A.; Nevado, R.; Sanz-Garcia, J.A. & Cusso, F. (1999). Zn-diffused LiNbO$_3$:Er^{3+}/Yb^{3+} as a waveguide laser material. *Opt. Mat.*, Vol. 13, pp. 181-186

Mikkelsen, I.C. Jr. (1980). Three-zone Bridgman–Stockbarger crystal growth furnace. *Rev. Sci. Instrum.*, Vol.51, pp. 1564-1566

Mitric, A.; Duffar, T.; Diaz-Guerra, C.; Corregidor, V.; Alves, L.C.; Garnier, C. & Vian, C. (2006). Growth of Ga$_{(1-x)}$In$_x$Sb alloys by Vertical Bridgman technique under alternating magnetic field. *J. Cryst. Growth*, Vol. 287, pp. 224-229

Naumann, R.I. (1982). An analytical approach to thermal modeling of Bridgman-type crystal growth : I. One-dimensional analysis. *J. Cryst. Growth*, Vol. 58, pp. 554-568

Nicoara, D. (1975). Metal single crystals growth set-up. Romanian Patent nr.62842

Nicoara, D.; Schlett, Z. & Nicoara, I. (1983). Optical crystals growth set-up. Romanian Patent nr. 82663

Nicoara, D. & Nicoara, I. (1984). Crucibles for optical crystals growth. Romanian Patent nr. 85993

Nicoara, D.; Nicoara, I. & Schlett, Z. (1985). Heater for large diameter crystals growth. Romanian Patent nr. 88497

Nicoara, I.; Aczel, O.F.G.; Nicoara, D. & Schlett, Z. (1986). Dissolution kinetics and etch pit morphology of CaF$_2$ single crystals. *Crys. Res. Technol.*, Vol. 21, pp. 647-652

Nicoara, I.; Nicoara, D. & Aczel, O.F.G. (1987). Crystal growth in an improved bridgman variable shape graphite furnace. *Crys. Res. Technol.*, Vol. 22, pp. 1139-1144

Nicoara, D. & Nicoara, I. (1988). An improved Bridgman-Stockbarger crystal-growth system. *Mat. Science and Engineering A*,Vol. 102, L1-L4

Nicoara, I; Munteanu, M.; Pecingina-Garjoaba, N.; Stef, M. & Lighezan, L. (2006a). Dielectric spectrum of rare-earth-doped calcium fluoride crystals. *J.Crystal Growth*, Vol. 287, pp. 234-238

Nicoara, I; Munteanu, M.; Pecingina-Garjoaba, N.; Stef, M. & Lighezan, L. (2006b). Dielectric relaxation in PbF_2–doped and X-ray irradiated CaF_2 crystals. *ECS Transactions*, Vol. 3, pp. 51-58

Nicoara, I.; Pecingina-Garjoaba, N. & Bunoiu, O. (2008a). *J. Crystal Growth*, Vol. 310, pp. 1476-1480

Nicoara, I.; Lighezan, L.; Enculescu, M. & Enculescu, I. (2008b). Optical spectroscopy of Yb^{2+} ions in YbF_3-doped CaF_2 crystals. *J. Crystal Growth*, Vol. 310, pp. 2026-2032

Nicoara, I.; Stef, M. & Pruna, A. (2008c). Growth of YbF_3-doped CaF_2 crystals and characterization of Yb^{3+}/Yb^{2+} conversion. *J. Crystal Growth*, Vol. 310, pp. 1470-1475

Nicoara, I.; Munteanu, M.; Preda, E. & Stef, M. (2008d). Some dielectric and optical properties of ErF_3-doped CaF_2 crystals. *J. Crystal Growth*, Vol. 310, pp. 2020-2025

O'Connor, J.R. & Bostick, H.A. (1962). Radiation effects in CaF_2:Sm. *J. Appl. Phys.*, Vol. 33, pp. 1868-1870

Paraschiva, M.; Nicoara, I.; Stef, M. & Bunoiu, M. (2010). Distribution of Pb^{2+} ions in PbF_2-doped CaF_2 crystals. *Acta Physica Polonica A*, Vol. 117, pp. 466-470

Petit, V.; Camy, P.; Doualan, J.L. & Moncorge, R. (2007). Refined analysis of the luminescent centers in the Yb^{3+}:CaF_2 laser crystal. *J. of Luminesc.*, Vol. 122–123, pp. 5-7

Petit, V.; Camy,P.; Doualan, J.L.; Portier, J. & Moncorge, R. (2008). Spectroscopy of Yb^{3+}:CaF_2: From isolated centers to clusters. *Phys. Rev. B*, Vol. 78, pp. 085131

Pruna, A.; Stef, M. & Nicoara, I. (2009). Dielectric spectra of Li^+ (Na^+) codoped CaF_2:YbF_3 crystals. *Phys. Stat. Sol. A*, Vol. 206, pp. 738-744

Sangwal, K. & Arora, S.K. (1978). Etching of MgO crystals in acids: kinetics and mechanism of dissolution. *J. Mater. Sci.*, Vol. 13, pp. 1977-1985

Shelley, R.D. & Miller, G.R. (1970). Ionic thermocurrent study of the dipole-dipole interaction in CaF_2 doped with NaF. *J. Solis State Chemistry*, Vol. 1, pp. 218-226

Stockbarger, D.C. (1949). Artificial fluorite. *J. Opt. Soc. Am.*,Vol. 39, pp. 731-740

Strukov, B.A. & Levanyuk, A.P. (1998), Ferroelectric Phenomena in crystals, Springer, Berlin, ISBN 3540631321

Su, L.; Xu, J.; Li, H.; Wei, L.; Yang, W.; Zhao, Z.; Si, J.; Dong, Y. & Zhou, G. (2005). Crystal growth and spectroscopic characterization of Yb-doped and Yb, Na-codoped CaF_2 laser crystals by TGT. *J. Crystal Growth*, Vol. 277, pp. 264-268

Su, L. & all. (2005). Sites structure and spectroscopic properties of Yb-doped and Yb, Na-codoped CaF_2 laser crystals. *Chem. Phys. Lett.*, Vol. 406, pp. 254-258

Su, L. & all. (2007). Quaternary doping to improve 1.5 μm quantum efficiency of Er^{3+} in CaF_2 single crystal. *J. of Luminesc.*, Vol. 122-123, pp. 17-20

Sun, D.; Zhang, Q.; Wang, Z.; Su, J.; Go, C.; Wang, A. & Yin, S. (2005). Concentration distribution of Nd^{3+} in Nd:$Gd_3Ga_5O_{12}$ crystals studied by optical absorption method. *Cryst. Res. Technol.*, Vol. 40, pp. 698-702

Yonezawa, T.; Nakayama, J.; Tsukuma, K. & Kawamoto, Y. (2002). Behaviors of trace amounts of metal-oxide impurities in CaF_2 crystal grown by Stockbarger's method. *J. Cryst. Growth*, Vol. 244, pp. 63-69

3

New Class of Apparatus for Crystal Growth from Melt

Aco Janićijević[1] and Branislav Čabrić[2]
[1]Faculty of Technology and Metallurgy, Belgrade,
[2]Faculty of Siences, Kragujevac,
Serbia

1. Introduction

In this chapter, we offer original solutions for crystallization devices by presenting a set of cooling devices that are upgraded models of the existing ones used in a well-known apparatus for crystal growth. Many basic ideas from these articles were used as a starting point for the creation of the new, modern multifunctional devices that may be used both as standard school laboratory tool and as industrial equipment. A number of crystal growth devices previously employed were designed to match contemporary technology level and needs for specific monocrystal growth. This led to the additional engagement on the realization of new working conditions, thereby increasing production costs. In light of this problem, while developing new forms od crystal growth apparatus we also have aimed at making the whole process as economical [1], approachable and efficient as possible.

A brief review of twentieth century devices for the crystal growth from the melt [2, 3, 4] reveals widely accepted remarks on some not so good characteristics of specific apparatus components. Let's mention Tamman's test tube and its tip modification which is essential to the crystal germ formation, realization of suitable apparatus geometry, construction of cooler parts in order to have controlled under-cooling, some specific demands for the adequate temperature gradient. Within this chapter, we have defined certain activities conducted (with the set goal in mind) in order to improve existing and to develop new crystal growth devices. We started with a set of simple steps that allowed for the modeling and construction of school type apparatus [5, 6, 7]. Later on, we came up with original solutions and more complex devices with a number of advantages compared to the known crystal growth devices.

Construction of new devices has as its ultimate goal apparatus standardization. Therefore, in a number of papers we have performed calculations that justify the use of newly designed apparatus. As a matter of a fact, in previous research, a standard and widely used approach in technology of crystal growth was to make a specific prototype of apparatus, and then, through a variety of experimentally gained data, to upgrade and improve the characteristics of crystal growth process, depending on the specific demand set for the purpose [2, 3]. That kind of approach was uneconomical regarding time consumption, and a large number of unsuccessful attempts was something one had to count on. For each specific demand a construction of an apparatus almost identical (with a slight modification only) to the one that failed was necessary. In turn, this led to significant material investments for the research, therefore making the crystal growth research a privilege of financially powerful countries that had the opportunity of gathering the top quality researchers from all over the world. Nevertheless, such huge investments had its justification in the fact that some extraordinary results were achieved. This resulted in production of materials of exceptional purity, as well of some new substances and materials whose crystals were realized for the first time in laboratory conditions.

These new materials found its immediate application in the military industry, where high quality materials are imperative, but also in some industrial branches, making these countries top producers of relevant materials (revolutionary novelties in semiconductor technology, telecommunication and optical devices).

Modern apparatus and its modifications presented here have common characteristics of not being financially [1] demanding (starting with the simple to the complex ones). Secondly, it is desirable to have apparatus that will allow for the large number of repetitions of similar processes (with small modifications only and development of new simple parts of equipment for possible improvements of crystal growth conditions). We went even further by developing models and constructing the devices with suitable geometry that allow for the crystallization of a single substance with different crystallization rates and temperature gradients. In addition, it is possible to achieve crystallization of different materials within the single event by employing materials with similar melting points, while having different crystallization rates and temperature gradients.

Along with previously stated advantages of developed apparatuses, we attempted and applied numerical calculations (whenever possible) to get best possible set of parameters in preparation of a new model for crystallization processes. One such analysis takes into account the dimensions of apparatus parts as well as interrelations among the most relevant crystallization factors that will allow for the optimal quality final product – crystal or monocrystal.

In general, intention of the authors is to intertwine these modern devices (large repeatability and multifunctional aspect of crystallization process being the most important advantages) with relevant numerical calculations and existing software. Computer regulated and monitored crystallization would give us more insight on how different parameter variation (such as temperature variations, heat transfer, crystallization rate etc.) and different apparatus dimensions, influences the crystallization process. In other words, there is a tendency to perform all the possible calculations in order to take necessary steps to modify and improve crystallization, so that we would get a crystal of predefined characteristics in a modern and efficient way by using state-of the-art information technologies within the regime of so-called expert systems.

2. New classes of coolers

In accordance with plans based on the variety of possible choices of data on architecture, construction and reconstruction of crystallization apparatus, we came upon a number of creative ideas that are directed towards the adaptations of apparatus shape within the laboratory conditions, the form of coolers and its more efficient role in crystallization apparatus. Long time experience based on the years of the research led author to the conclusion that the heat conduction is one of the essential factors determining the crystallization rate. When, in the conditions of undercooling, the heat is being released, the undercooling will exist only if the heat is being taken away in a proper manner. The rate of heat conduction is a factor quite responsible for the crystallization rate. Crystal growth rate is constant when there is a balance in heat transfer. The heat transfer is quite a complex problem in the sense of regulating the system that has a continuous and controlled operation in accordance with the predefined phases of the crystallization process. From the very start of the germ formation, it is necessary to get a desired temperature drop that defines the initial state of crystallization, and then, by setting an appropriate temperature gradient one can have optimal conditions for obtaining the crystal of specific characteristics.

The temperature aspect of the crystallization that is so significant for the crystal growth and possibility of programming the process parameters through various shapes and positioning of the coolers (which provide cold fluid flow in crystallization apparatus), demands coolers to have multiple roles: firstly, to enable for more precise crystallization, and secondly, to lead to construction of new coolers made of materials of adequate heat conductivity so to have more convenient conditions for crystal growth from the melt.

Besides, suitably designed coolers have such a shape that they may simultaneously serve as ampoule carriers or test tubes with melt. In this way, the crystallization will be easily controlled. When looking back at the devices previously used, it is easy to see that some parts of the devices were burdened by carriers of pots with melt, as well as due to their heating and operating them in and out of the apparatus. Also, realization of adequate temperature gradient and subcooling through complicated pipe constructions or other forms of the coolers of intricate geometries (positioned within the crystallization apparatus), additionally complicated crystallization apparatus, not to mention the other instruments used in the process. Detailed analysis of these problems gave us very useful data that generated a completely new set of ideas, which ultimately resulted in a new, more complex role that coolers have in process. Their multifunctionallity led to significant simplification in apparatus construction in many of the known methods, which, loosely speaking, were reinvented. In some of author's papers, a demand for cooler improvement was set, and it resulted in design of more efficient and modern generation of crystallization devices.

As a basis for the design of novel or significantly improved and modified standard crystallization devices, we have used a series of originally, for the purpose-constructed coolers presented within the chapter. Cooler models presented in Figs 1- 6, whose forms and functionality gained recognition through presentation in few articles, may be divided in several groups, based on its positioning in the apparatus, cooling fluid flow propagation and its intended method application (Tamman, Stober, Czochralski). The general classification, which arises from the position of the cooler within the apparatus, leads us to two types of coolers: vertical and horizontal.

2.1 Vertical air coolers

Coolers where the cooled air is moving along defined (vertical) tube direction, belong to the group of so called vertical air coolers (Fig. 1). Thanks to the different cross sections of the tube, different speeds of airflow are possible. In that way, various crystallization speeds via heat dissipation are established in test tubes that are attached to the body of the cooler in various manners. There is a whole spectrum of coolers based on the positioning of the test tubes: the ones with fixed test tube position, to the ones with mobile rings on mobile coolers. Large number of test tube positions is available (Fig. 3).

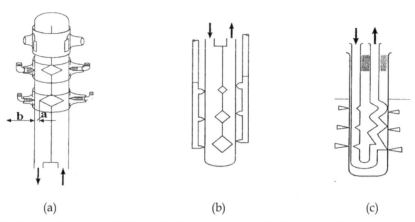

(a)   (b)   (c)

Fig. 1. Vertical coolers: (a) cold "finger", (b) "cold key" and (c) "cold ear-rings".

For the class of coolers presented in Fig. 2, the line of development was the following one: in certain positions, the tubes were constricted and slightly bended, so to achieve the optimal heat dissipation, and to simultaneously allow for an additional number of test tubes to be positioned. This was followed by coolers where the pipes were ring like bended in a couple of independent levels of crucibles, which allows for an increase in crucible operating capacity.

The operating regime of this class of coolers is such that each ring has a direct fluid flow within it and heat dissipation in the environment. The other opportunity are so called spiral coolers where heat generated during the crystallization process from all the rings is being "collected" and dissipated into environment. Detailed analysis of presented models showed some additional possibilities of vertical coolers. These were used for some novel practical solutions. Depending on the geometry of the space the coolers are in, they may be maneuvered (so called movable vertical air coolers) or be fixed while some of the other pipes (with Tamman test tubes) can be maneuvered on order to get a desired temperature gradient or crystallization rate. Whenever the vertical air coolers are employed, whether its orientation is upside down or vice versa, fluid flow is such that it returns in the opposite direction along the same path.

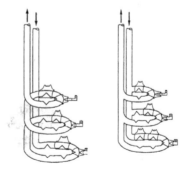

Fig. 2. Air cooler model ("cristallization spiral").

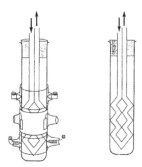

Fig. 3. Air cooler model ("crystallization key").

2.2 Horizontal air coolers

When talking about the horizontal air coolers, there are, basically, two classes with some specific variations:

a. To the first group belong coolers whose fluid flow pipes are horizontal. The cold fluid enters on one side and exits on the other one (single pipe horizontal air cooler, Fig. 4; system may also have two or more horizontal pipes). Couple of horizontal coolers can form an ensemble of instruments in chamber or crucible furnace.

b. Other type of cooler employed in the crystallization purposes, is the one where a horizontal pipe is bended at its end, carrying the fluid in the direction opposite to the initial one, and then the heat is being dissipated into environment (Fig. 4b. and Fig. 5.). If the pipe is bended at 180°, there is a possibility of multiplying initial activities via new conditions and test tube positioning. This allows for a large interval of crystallization rates in direction of the cooler.

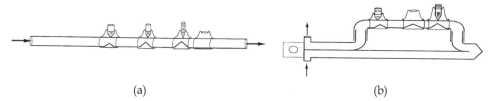

(a)   (b)

Fig. 4. Horizontal coolers: (a) pipe, (b) two-pipe (folding)

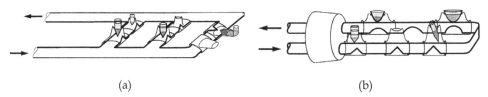

(a)   (b)

Fig. 5. Multifunctional horizontal coolers: a) the standard method, b) for the combined methods.

In such cases, we have come up with an original solution. The flow that convects heat below Tamman's test tubes is now being used for cooling the top layer of the melt that is positioned next to the exit pipes of the cooler. In that way, we have assigned it a new role upon bending the initial pipe. It gives us the opportunity of constructing the apparatus with new combined methods (Tamman's and Stober's). In Fig. 5, we present two solutions from a whole family of coolers whose realization is based on previously presented idea that leads to greater operability and more economical functioning in the crystallization process. Solutions presented give a clearly confirm validity of idea of redesigning some parts of cooler as well apparatus as a whole, and undoubtedly pointout their versatile practical purposes.

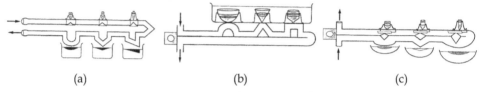

(a)   (b)   (c)

Fig. 6. Horizontal coolers; modification (a) and (c) combined with multivariate methods (b) variation of a method.

In Fig.6, specific horizontal coolers are given. Some parts of pipes are bended in the outer part of the device (unlike the previous ones where constrictions exist on the inner parts only) having endings of different geometrical shapes that allow different flow velocities. We have therefore met the conditions necessary for Stober method crystallization.

In this way, in the course of a single event, we have enabled crystallization based on the two methods, one during the fluid flow in the one direction, and the other for the opposite direction flow. In one case the cooling fluid flows above the crucibles containing crystallization melt. In the other, the flow goes below the melt where, by under-cooling specific capillary endings of test tubes with melt, a new process of germ creation starts all the way to the final crystallization. A geometrical representation of such coolers reminds of "cold horseshoes" and "cold keys". Fig. 11 demonstrates application of the modified cooler, which comprises two horizontal pipes mutually joined to movable pipe, which is an exceptional improvement compared to former examples in a sense of simplified geometry modification and crystallization conditions.

In some of horizontal coolers with one or more pipes containing cold fluid, another innovation is present. The pipe of cooler is introduced into a pipe of greater diameter, which may consist of one or two parts (Fig. 10) with small openings and slots, in which the position of pots and test tubes with melt may be fixed. Such a solution has clear advantages to the previously described ones, since by simply moving the cylindrical pipe (whose function is to move the cooler pipe and to serve as a test tube carrier all at once) a large number of different crystallization conditions and new crystallization geometries is achieved.

3. Original crystallization apparatus

The installation of the innovated systems for cooling, with the aim to monitor heat removal for the regulations of the processes of crystal growth from melted materials, enabled obtaining more devices for crystallization. The new classes of cooling devices, with aforementioned advantages linked to the crystallization processes have an additional quality which is that those cooling devices are very adaptive for installation and operative by application in well known laboratories-crucibles, chamber furnaces and tube furnaces. However, more complex cooling systems with the Tamman's test tubes, as a carriers devices, need to create new forms of crystallization apparatuses. The projecting of the new classes of devices for crystal growth of melts, which will be shown in the following text, is the response to the aforementioned need.

From these methods for crystal growth from the melt, it is estimated that in the school laboratory, the Tamman's method is the most convenient one. If we use the advantages of the horizontal single tube aerial cooling system, an original device, the so called "crystallization bench" can be realized [8]. It consists of a tube furnace and a specially adapted cooling system (Fig 7.).

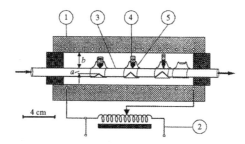

Fig. 7. Crystallization regulation in a tube furnace. (1) electroresistant tube furnace, (2) continuously changeable transformer, (3) air cooler ("cold bench"), (4) Tammann test tubes and (5) rings.

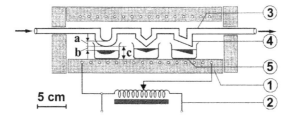

Fig. 8. A chamber furnace for obtaining crystals. (1) laboratory chamber furnace, (2) continuously changeable transformer, (3) air cooler ("cold key"), (4) cold "teeth" (5) crucibles with the floating crystals.

The procedure of choosing the wanted disposition of the test tubes, with the melted material, above the narrowing cooler cross section, is accomplished with moving rings on the cooler tubes. The devices may contain many Tamman's test tubes of various sizes and dispositions. The constructed device enables the simultaneous test of a few various nucleation and crystallization rates. Tamman test tubes of various shapes and dimensions (a family group [2, 3, 4] can be mounted on the test tube rings and thus simultaneously tested).

The variations considering the disposition changes of certain test tubes, as well as simultaneous regulations of some temperature gradients are also possible. The working regime of the devices works as following: at a constant furnace temperature, a weak air flow is turned on through the cooler. There, the crystallization starts on the bottom of the test tube (Fig.9).

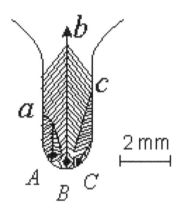

Fig. 9. The beginning of the crystallization at the bottom of the capillary tube.

The bottom of the test tube continues in the capillary, so that in the beginning of the process, only a small amount of the melt is overcooled. Therefore, only certain crystal nucleuses can be formed. The nucleation which grows towards the walls of the capillary, stop growing at a certain time. Only the nucleation which grows towards the axis of the capillary overgrows the other nucleations, and when they exit the capillaries, they expand to the full cross section of the testing tube.

The preparation of crystals of good quality, containing a low concentration of impurities and defects, requires a crystable substance of high purity, test tubes of materials that do not react chemically with the melt, a high degree of temperature stabilization of the furnace, and the absence of shocks [9]. The conditions required to grow crystals of some example substances, wich have low melting temperatures and can be used to obtain single crystals in school laboratory.

The crystallization rate interval [4] in each tube is regulated by the cross section of the air flow (a), i.e. by translation movement of the test tube rings (Fig. 7). The temperature gradient is regulated by distance (b). Different temperature gradients in the tubes can be simultaneously regulated using an inclined cooler, i.e. "inclined cold bench". By varying the internal and external cooler shape and dimensions, a famili of coolers can be modeled for different intervals of temperature gradients and crystallization rates. Different crystallization fronts and rates in crucible columns can also be regulated below the cooler so that crystallization starts on the surface of the melt (Fig. 8). Crystal growth then occurs downward the lower interface on the floating crystal.

By increasing the air flow velocity, the crystallization front spreads to the other end of the testing tube. The interval of the crystallization rates in each of the testing tubes of the devices (Figs 7, 8) is regulated by the air flow, i.e. the cross section a, which increases or decreases by relocating the moving rings, along with the cooler tubes. The temperature gradient is regulated by distance regulators b (Fig. 7).

Besides the standard case of the "Crystallization bench", other geometric solutions are possible in the design of the part of the devices [10]. Different thermal gradients in test tubes can be simultaneously regulated by the inclined cooler (or some other part of the cooler) relative to the axis of the furnace (Fig. 8). The shape of crystallization fronts and the crystallization rates in the crucibles are regulated by the path and the cross section of the air flow (*a*) of the cooler, as well as by the distance regulator from the surface of the melted material (*b*). By such creations and innovations, considering shapes and cooler functioning, the possibility of the Stober method realization in shamber and tube furnaces is accomplished.

The project of the original developed devices, the so called "the moving crystallization bench" (Fig. 10) contains some of the more complex forms of the aerial cooler which is in the shape of a cyllindric tube, which is located in another tube, which can have one or two parts, with a bearing for the Tamman test tubes [11].

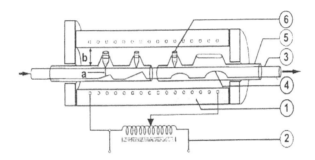

Fig. 10. A tube furnace for obtaining crystals: (1) laboratory tube furnace, (2) continuosly changeable transformer, (3) air coler (telescopic cold bridge), (4) cold "thresholds", (5) cylindrical tube with the mounting holes and grooves (telescopis test sieve) and (6) family group of Tamman test tubes.

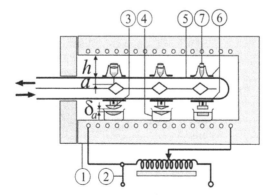

Fig. 11. Apparatus for combining methods: (1) Laboratory chamber furnace, (2) continuosly changeable transformer, (3) movable plugs, (4) columns of crucibles, (5) air cooled toothed tube ("crystallization finger"), (6) movable mounting rings, and (7) Tamman test tubes.

The method of testing crystallization, as well as possible variations of the processes are described. The formula for crystallization rates depending on the parameters of the cooler and the characteristics of the material, as well as respective temperature changes. It creates great possibilities for utilization of various crystallization rates.

Tamman's test tubes of various shapes and sizes can be laid out to the moving cyllindric tube (the so-called "sieve"). One can accomplish the simultaneous test of the crystallization for a great number of different Tamman's test tubes. They are of various temperature gradients, intervals of crystallization rates, and materials. They can be used for obtaining single crystals from the melt by using cheap and practical modular devices-crystallization apparatus with the moving elements.

The development of the models of one group of apparatuses, whose work is based on single tube horizontal cooler, has developed in several phases. Each one of the phases is characterized by innovations in the series of details, and therefore a very high level has been achieved. That level has gained a special, important confirmation by publishing the paper with newly accomplished results in the professional journal [12].

In the published article [13], the original modification of the devices, which is considerably more sophisticated and efficient than the previous class of the device. Is has been created based on the experience and the series of practical conclusions from the previous models. That article initiated the design of a certain number of devices, which are based on the simultaneous unwinding of the Tamman and Stober methods (Fig. 11). The specially adapted cooler, functional for this purpose, has been installed in the laboratory chamber furnace [13]. The cross section of the fluid current and the distance of the cooler from the surface of the vessel where the melt is located ,define the shape of the fronts and the crystallization rates. Some more demanding and economical variations of these devices contain two tube-coolers, for the arm with Tamman's test tubes. The tube which serves as the test tube carrier can be mobile and can contain more than one series of Tamman's test tubes in telescopic test sieves in the previous paper [13], the possibility of simultaneous realizations of Stober and Tamman's methods has been accomplished. The presented solution and the defined modifications, with the aim of improving the conditions of crystallization by these methods can be applied in the tube furnace in the horizontal positions, too. The most sophisticated devices of so called double-tube horizontal models are achieved by flexing one tube by 180 degrees, or two horizontal tubes linked by vertical linking extensions. This is not only focused on accomplishing simultaneous developments of the crystals utilizing the two methods (Tamman's and Stober's), but it is the invention of the quality forms, the positioning of every single test tube, cooler aperture up to the influence on the front crystallization according to certain calculations [14].

The creation of a certain number of functional vertical coolers, which are previously presented, has made the simultaneous realization of the projects with the crystallization devices with vertical coolers possible.

The model of an air cooler, which is vertically positioned in the laboratory crucible furnace (so-called "finger") is presented in the paper [15]. Some bended Tamman's test tubes are positioned on the cooler with the help of rings and sliders of the test tube carrier. The formula of linear crystallization rating in each test tube is derived from using the balance between the latent heat of the solidification and removed heat through the cooler. The possibility of translation of each test tube independently is considered, with the aim of simultaneous probe of the matrices of various crystallization rate intervals.

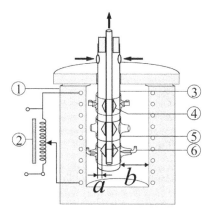

Fig. 12. Multifunction crucibles to obtain crystals: (1) Laboratory crucible furnace, (2) continuosly changeable transformer, (3) air cooler ("cold finger"), (4) movable cold "thresholds", (5) movable mounting rings, and (6) curved Tamman test tubes.

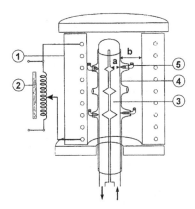

Fig. 13. Apparatus for obtaining crystals: (1) laboratory tube furnace, (2) continuously changeable transformer, (3) air cooler ("cold tree"), (4) movable cylindrical tube with the mouting holes ("test sieve"), (5) family of ("grafted") Tammans test tubes.

The devices in work [16] present the basis for the previous solution of the new apparatus (Fig. 12), but there is a difference in the flexibility of the elements of the devices in the systems as well with the purpose of the geometrical solution of the test tube (with the melt), with many possibilities of the realization of various temperature gradients.

The devices for the crystallizations shown in Fig. 13 presents exactly one level more operative devices [17] than those shown in Fig. 12. These systems usually consist of aerial coolers (with a cold fluid flowing through) and movable cyllindric tubes with placable holes for Tamman's test tubes. The coolers, in this case, are movable, and they give the possibility of definition of certain parameters during the crystallization process. That is how one can very operatively influence the progression of the process and the quality of the obtained crystal.

The mobile test tube carrier, with the melt, can easily enable an adequate position of the test tube tip, depending on the cross section aerial currents through the cooler. It is easier to control the parameters which influence the substantial magnitudes in the crystallization process (crystallization rate, temperature gradient) that way.

An example of an even higher quality of the devices with vertical coolers (Fig. 12), enable us to translate and move the cooler vertically, but also very convenient for quality work due to the various possibilities of the cooler rotation. It enables the regulation of the front crystallization wanted regulation dynamics, which enables the conditions to obtain quality crystals [18]. These devices have emphasized the mobility of elements, and higher potential for the work by choosing the position of the test tube, and the number of the test tube, in the process of crystallization. It is better than that of the devices in Fig. 13, thanks to the fact that it contains a constructive design solution, with a mobile ring and a mobile mechanism of the test tube carrier.

The achieved variety, considering the design, on the crystallization devices, with another cooler class, has brought a new quality in the sense of the possibility ensure a heightened quantity of the melt. It could be in strictly defined and stabilized conditions, which is substantial as an introductory activity, for the crystallization process itself. Such an idea has been realized and has justification in constructive solutions of the devices which are shown in papers [19] and [20].

The original devices for obtaining single crystals from the melt in coherence with the new demands considering the quantity of the melt, as well as obtaining the possibility where more devices can be put in with melted substances with similar melting points. By variation, those crystals are formed in various ways, depending on the conditions. In this case, the idea of applying the Tamman's method with very specific sets of testing tubes, in a laboratory crucible furnace [19] was realized.The regulation and simultaneous crystallization of several substances for a few nucleations of various temperature gradients and crystallization has been made possible.

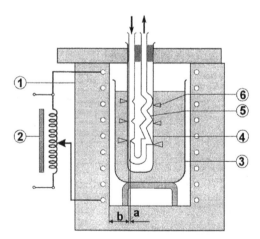

Fig. 14. A crystallization cooler in a crucible furnace. (1) laboratory crucible furnace, (2) continuously changeable transformer, (3) crucible (4) test tube (5) moving air cooler ("cold ear-rings"), and (6) Tamman test tubes.

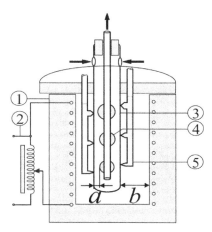

Fig. 15. Crystallization apparatus: (1) laboratory crucible furnace, (2) continuosly changeable transformer, (3) air cooler ("cold key"), (4) movable rings and (5) branched Tamman's test tube ("crystallization test comb").

The combination of several Tamman's test tubes in the form shown in Fig. 15 makes the growth of several crystal from the melt possible, as well as obtaining the conditions for several devices with melted substances who have approximately the same melting points. By variation of the shape and the size of the cooler (inside and out), can model a family of "cold keys" for testing a wider interval of temperature gradients and crystallization rates.

The fusion of the best performance of the devices shown in Figs 12, 13, 15 has benefited the crystallization process in the new devices [14] with a vertical cooler in the crucible furnace, chamber furnace or tube furnace. They also come with a modern form, better functioning and higher economic value, and other important traits which have been greatly improved. Controlled functioning of certain phases during the crystallization process with great reliability for obtaining the crystal's wanted characteristics.

The devices who improve the efficient solution of the form of the apparatus related to the previously described apparatus [21] are shown in Fig. 14 (from the constructional point of view, they are very similar at the first glance). The presence of a larger quantity of the melted substance, but also a bigger number of the test tubes for obtaining adequate crystals with big potential for variation of the conditions, which are very important for the regulation of the crystallization rate, is made possible [22].

3.1 New generation of devices for crystal growth – "Expert systems"
Before mentioned division of coolers on horizontal and vertical ones, and related construction of apparatus could not exist within the given frame. Rich experience in connection with work on crystallization apparatus and vision of development directed towards new possibilities, led to so called "hybrid" solutions for the coolers. In their regime of work, they employ both horizontal and vertical fluid flow.

This, in turn, gives a variety of opportunities for development of original, high quality devices with new possibilities and advantages for crystallization process. A increased efficiency and reduced costs may also be expected.

In [23] a successful realization of combined ("hybrid") device is demonstrated in laboratory chamber furnace. For that purpose, one improved model of crystallization cooler in ladder-like shape on which movable bended Tamman's test tubes are positioned, is presented. By finding the appropriate angle between axes of the test tube and direction of crystallization, defect drainage towards test tube wall may be regulated. To this intermediate group of crystallization devices belongs the apparatus described in [24].

It is quite obvious that fluid current that conducts crystallization heat from a certain level of test tubes, circulates several levels by passing through profiled sections. Therefore, on the remaining semicircular levels, we may, either have the same substance with different cooler cross section on the location of the test tube (in this way we will get the family of crystals of same substance), or, on each level we may have a system of test tubes of different substances from a group of substances having more or less the same crystallization conditions.

In [10], [13] and [25], original modification of devices based on the Stober's method are presented. During the research of methods for crystallization regulation in laboratory chamber furnace for crystal substances with unknown crystallization parameters, we reached the conclusion that combined Tammn-Stober's method can be employed.

Particularly adapted cooler for this purpose was installed in laboratory chamber furnace [26]. The forms of crystallization fronts and crystallization rate in crucibles are regulated via trajectory and the cross section of cooler air flow (d_1), and via the distance of the cooler from the furnace wall (d_2). In more demanding and more economical type of this apparatus, two pipe coolers can contain an array of Tamman's test tubes on one branch [27].

The apparatus presented in [28] is practical realization of combination of more elements from different types of presented devices. It is specific in the sense that it has built in parts of devices that contain several groups of Tamman's test tubes of different shapes, volume and inclination of test tube axes where the formation of crystals is expected. It more complex variant, the apparatus may contain ensembles of coolers and test tubes. In that way, many of the steps can be repeated within single crystallization process [29]. During the research on realization of monocrystals of family of substances with unknown crystallization parameters in laboratory chamber furnace, we have modeled air cooler that enables simultaneous crystallization of several substances at different temperature gradient, shapes of crystallization fronts and crystallization speeds in column of crucibles and test tubes.

By upgrading the existing experiences and improving the characteristics of previously described classes of devices, we have achieved results which give a solid bases for accomplishing the highest goal set during production of new devices for crystal growth from the melt: realization of "smart systems" that control process of crystal growth via computer programming [30].

The first results appeared almost simultaneously in two articles: in Russian journal Instruments and Experimental Techniques [31] (Fig. 16), and another one in American journal American Laboratory [32] (Fig. 17), both published in 2011.

This new class of devices ("superclass") owes its name to its multifunctionality, and ability of its dynamical elements to react almost instantaneously to the tasks regarding regulation and monitoring the crystal growth. It is achieved by establishing permanent connection of devices with computer-controlled programs [33]. At this stage of realization, results of process simulation and apparatus conditions are used. Nevertheless, practical realization of establishing direct connection of computer to apparatus and its movable parts (cooler and test tubes with melt) is a matter of time.

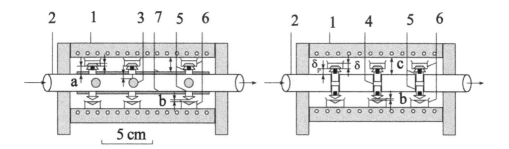

Fig. 16. "Programmed crystallization bench" for crystal growth from melt: (1) laboratory tube furnace, (2) air cooled tube ("crystallization shelf-comb"), (3) radial holes in a horizontal position ("crystallization thresholds"), (4) radial holes in a vertical position ("crystallization sockets"), (5) movable cold plugs, (6) columns of crucibles, and (7) slide bars.

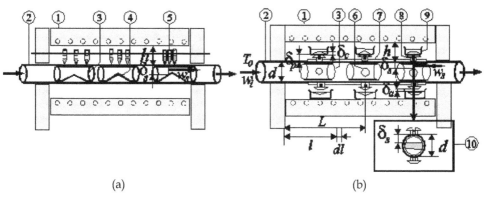

(a)   (b)

Fig. 17. "Smart" coolers for the combined methods of crystal growth from melt: (modular unilateral (a), and bilateral (b) "crystallization comb" in a tube furnace). (1) laboratory tube furnace, (2) air-cooled tube, (3) modular and movable pipes: (4) "Δ" cold thresholds, (5) string of family group of Tamman test tubes, (6) radial holes in a horizontal (or vertical) position ("O" or "I" cold thresholds), (7) slide bars (or movable mounting rings), (8) movable plugs with modular heads, and (9) column of crucibles; (10) cross section of the cooler.

Programmed conditions will be controlled by computer system that directly define the dynamics of movable systems, such as optimal positioning of cooler within the furnace, controlled heat dissipation in the course of crystallization as well as monitoring the position of test tube carriers where crystallization from the melt is taking place.

4. Results of modeling

The rate of melt solidification depends upon extracting the latent heat of solidification. For a time interval t a crystal layer of thickness δ_c is formed (Fig.17). During the formation of an elementary crystal layer of thickness $d\delta_c$ per unit area, the amount of heat released is

$\lambda \rho d \delta_c$ (λ denotes the latent heat of solidification and ρ the crystal density); the latter is being extracted through the cooler for a time interval dt. On this basis the following equation may be written [34].

$$\lambda \rho d \delta_c = \frac{\Delta T(L)}{1 / \alpha_s + \delta_p / k_p + \delta_a / k_a + \delta_c / k_c} dt \tag{1}$$

where $\Delta T(L)$ denotes the difference between the temperature of the melt and that of the air stream; α_s is the coefficient of heat transfer from the cooler wall to the air stream; and k_p, k_a and k_c designates the heat conductivity of the plug, air and crystal respectively (Fig. 17b). Transforming equation (1) we obtain

$$R = \frac{d \delta_c}{dt} = \frac{\Delta T(L)}{\lambda \rho (1 / \alpha_s + \delta_p / k_p + \delta_a / k_a + \delta_c / k_c)} \tag{2}$$

The quotient $d\delta_c/dt$ denotes the rate of crystal layer growth which is usually represented by the symbol R.

The coefficient of heat transfer from the cooler wall to the air stream can be calculated using the following expression [35]:

$$\alpha_s = \left[4,13 + 0,23 \frac{t}{100} - 0,0077 \left(\frac{t}{100} \right)^2 \right] \left(\frac{U}{4A} \right)^{0.25} (w_{s0})^{0.75} \tag{3}$$

where t is average temperature of the air stream in ^0C (up to 1000 ^0C), U and A are circumference and area of the cross section of the airstream, respectively – see (10) in Fig. 17b,

$$w_{s0} = w_s \left(\frac{273}{273 + t} \right),$$

w_s is average velocity of the airstream (0 ^0C, 1.013 bar) in m/s.
On the basis of the continuity of the airstream, and the cross section at the entrance of the tube and on the threshold – see (10) in Fig. 17b, the following expression for the velocity of the air stream on the threshold w_s we have derived:

$$w_s = w_i \left(\frac{\pi}{\varphi - \sin \varphi} \right) \tag{4}$$

where w_i denotes an average velocity of the air stream at the entrance of the tube

$$\varphi = 2 \arccos \left(1 - \frac{2 \delta_s}{d} \right)$$

in rad, δ_s is the width of the air stream, d is diameter of the tube (Fig. 17b).
Based on the fact that the heat removed from the cooler wall is equal to the heat accepted by the air stream, we have derived the following expression (integral equation) for the difference of the temperature between of the melt and that of the air stream along the cooler $\Delta T(L)$:

$$\Delta T(L) = \Delta T_0 - \frac{4}{d}\int_0^L \frac{\alpha_s \Delta T(l)}{\rho_a c_a}dl \qquad (5)$$

where ΔT_0 denotes the difference between the temperature of the melt and that of the air stream at the point $L=0$ (at the entrance of the tube), d is diameter of the tube, α_s is the coefficient of heat transfer from the cooler wall to the air stream - eq. (2), when put

$$\left(\frac{U}{4A}\right) = \left(\frac{1}{d}\right)$$

(i.e. tube without thresholds), $\Delta T(l)$ denotes the difference between the temperature of the melt and that of the air stream at the point l (Fig. 17b), ρ_a designates the air density, c_a is heat capacity of the air.

The crystallization parameters (designed in Fig. 17b) are determined by the numerical analysis of eqs. (2), (3), (4) and (5), in the case of bismuth: $T_{melt} = 271\ ^0C$, $\lambda = 52300\ J/kg$, $\rho = 9800\ kg/m^3$, and $k_c = 7.2\ W/mK$, In all numerical calculation is was taken that: $k_p = 0.756$ W/mK (pyrex i.e. borosilicate glass, softening point $\approx 600\ ^0C$), $k_a = 0.0342\ W/mK$, $\rho_a=0.682$ kg/m³,$c_a = 1.035\ kJ/kgK$, $\Delta T_0 = 251\ ^0C$, $T_0 = 22\ ^0C$ (the temperature at the entrance of the tube), $d = 2cm$, $\delta_p = 5\ mm$, $\delta_c = 0\ mm$ (on the bottom or the surface of the melt).

The dependence of the crystallization rate R on the position of the plug along the cooler L is represented on Fig. 18a, when $w_s = w_i$, eq. (3), i.e. the tube without the thresholds. As can be seen from Fig. 18a, the crystallization rate decreases with increasing L, which is the consequence of the fact that ΔT decreases with the increase of the L, eq. (5). Fig. 18b shows the dependence of the crystallization rate R on the with of the air stream δ_s - see (10) in Fig. 17b. As can be seen from Fig. 18b, the crystallization rate increases with decreasing δ_s, which is the consequence of the fact that w_s, eq. (4), and consequently α_s, eq. (3) and R, eq. (2) increases with decrease δ_s.

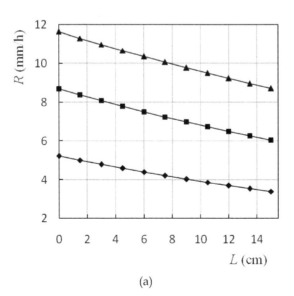

(a)

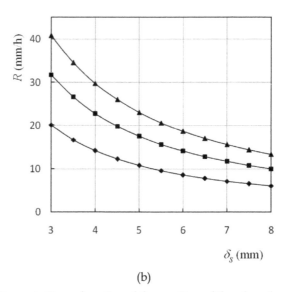

(b)

Fig. 18. Crystallization rate R as a function of the position of the plug along the cooler L, and the width of the air stream δ_s respectively, when: $-\blacklozenge-$ $w_i = 0.3$ m/s, $-\blacksquare-$ $w_i = 0.6$ m/s, $-\blacktriangle-$ $w_i = 0.9$ m/s; $\delta_a = 0$ (crucibles above the plugs) (a) $\delta_s = d$; (b) $L = 9$ cm (Fig. 17).

Fig. 19a represent the possible values of distances of the plug head from the surface of the melt δ_a and the position of the plug along the cooler L, for definite values of crystallization rates. As can be seen, if L is larger, then δ_a must be smaller for the definite crystallization rate, which is the consequence of the fact that ΔT decreases with increasing of the L,- eq. (5) and R increases with decresing δ_a – eqs. (2), (3) and (4).

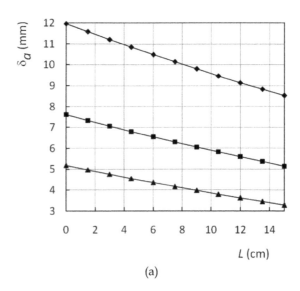

(a)

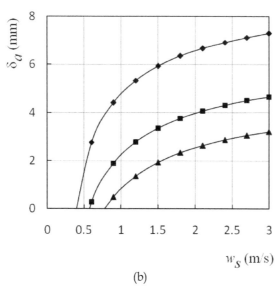

(b)

Fig. 19. Dependence of the distances of the plug head from the surface of the melt δ_a (crucibles below the plugs, Fig. 17b) on the positions of the plug along the cooler L, and the velocities of the air stream w_s respectively, for the crystallization rates –◆– R = 5 mm/h, –■– R = 7 mm/h, –▲– R = 9 mm/h, when: (a) w_s = 2.4 m/s; (b) L = 7 cm.

In Fig. 19b the possible values of distances of the plug head from the surface of the melt δ_a and velocities of the airstream w_s, are presented, for definite values of crystallization rates. As can be seen from Fig. 19b, if w_s is larger then δ_a must be larger for the same crystallization rate, which is the consequence of the fact that a_s increases with increasing of the w_s - eq. (2), and R decreases with incresing δ_a – eq. (1)

5. Conclusion

The subject of our research belongs to the field of crystal growth from the melt, particularly growth conditions depending on the design and construction of crystallization apparatus, which have significant influence on germ formation conditions and controlled crystal growth. A class of new modern devices for crystal growth from the melt, based on the well-known methods and crystal growth techniques, is presented in the paper. Crystal growth from the melt plays an important role in area of electronic technologies, because it includes a major part of most efficient methods for production of semiconductor and electronic monocrystal materials. In this monograph, we have systematically presented results concerning crystal growth from the melt, from both renowned authors and the author of monograph.

Technology of crystal growth depends on apparatus state-of-the-art and on devices with specific characteristics for particular growth method. However, in order to meet specific demands in their subsequent application, some of the apparatus for production of standard materials have additional peculiarities. Consequently, an upgrade in both constructional and functional sense for a variety of apparatus for crystal growth from the melt was necessary. As presented in the paper, this resulted in development of a class of novel devices with notably improved solutions for both some elements of the device and device as a whole.

Basic settings for the new approach in fulfilling desired crystal growth conditions and flexibility of specific devices while varying some parameters, were obtained through realization of whole set of coolers, starting from elementary specifically positioned to the, so called, mobile coolers of different profiles.

Studying of conditionality of crystallization parameters and physical conditions of the process itself, generated an original idea where modern design coolers gain multifunctional role. On the one side, they have become carriers of the ensemble of test tubes with melt, and on the other side they allow for the positioning of melt in desired spots thereby bringing about the needed temperature gradient. Finally, cooler-melt system has a potential of easily being positioned where necessary by moving it in various directions or by rotating it within the space available in furnace. A particular quality in innovations that we came upon, is that idea of multifunctional coolers triggered an idea of incorporating computer system into the crystallization process. Application of computer systems allows one to define crystallization conditions prior to the crystal growth via simulation process. In addition, it is possible to permanently control and monitor quality of crystallization process. Development of some original programs in MATH LAB only confirmed validity of idea. This gives vast opportunities in presented modern approach to growth of crystals and monocrystals.

All the mentioned innovations in both specific parts of crystal growth apparatus and apparatus as a whole, allowed relatively easy reproducibility of crystallization process. This approach enables, for the predefined conditions, simultaneous growth of a family of crystals of single material in same or different conditions on the one side, and simultaneous growth of different material crystals in, more or less, same crystallization conditions on the other side. Along with the described multifunctionality, a new class of crystal growth devices gained in quality and importance in connection with low cost, efficiency, rationalization and modernization of crystallization process. Usage of computer modeling and development of original computer programs are a good basics for achieving the highest goals of this monograph: to incorporate, via application of information technology in the process of crystal growth from the melt (with the use of latest class of devices designed for crystal growth), in a certain way, the expert systems. In this way, efficiency and accuracy is significantly increased due to a possibility of controlling and simultaneously eliminating undesired effects, in a process that is almost fully automatic and that can be influenced essentially in order to get a crystal of desired quality. Results presented here are of great practical interest for theoretical and applied research in solid state physics, as well as in the area of new materials, all this fulfilling high requirements and standards demanded in both laboratory and industry growth of crystals and single crystals.

6. Acknowledgments

The paper (chapter) was supported by Serbian Ministry of Education and Sciences, grant No. 44002.

7. References

[1] H.J. Scheel and T. Fukuda, *Crystal Growth Technology*, John Wiley & Sons, Ltd. ISBN: 0-471-49059-8, (2003).

[2] K.-T. Vilke, *Virashchivanie kristallov* («Nedra». Leningradskoe odelenie, Leningrad, pp. 258-311), (1977).

[3] K.-Th. Wilke and J. Bohm, *Kristallzüchtung* (Verlag Harri Deutsch: Thun, Frankfurt/Main, , pp. 591-647), (1988).

[4] R. A. Laudise, *The Growth of Single Crystals*, (Prentice-Hall, Englewood Cliffs, NJ, pp 159-172), (1970).

[5] B. Čabrić, B. Žižić, and M. Lj. Napijalo, *An apparatus for crystal growth in the undergraduate laboratory*, Eur. J. Phys., 11, 233, (1990).

[6] B. Čabrić, T. Pavlović, and S. Savović, *A simple programming the crystallization rate from the melt*, Cryst. Res. Technol., 29, No. 7, K96, (1994).

[7] B. Čabrić, T. Pavlović, and B. Žižić, *Crystallization in a laboratory chamber furnace*, J. Appl. Cryst., 27, 199 (1994).

[8] B. Čabrić, T. Pavlović, *Krystallization bench*, J. Appl. Cryst., 33, 387-388, (2000).

[9] I.Tarjan and M. Matrai, *Laboratory Manual on Crystal Growth* (Akademiai Kiado, Budapest, pp. 221-238), (1972).

[10] B. Čabrić, A. Janićijević, *A Method For Obtaining Crystals In A Laboratory Furnace*, Program and Contributed Papers of XVI National Symposium on Condensed Matter Physics SFKM 2004, Ed. Institute of Physics Belgrade - Serbia and Montenegro, str. 325-328, (2004).

[11] (46) B. Čabrić, A. Janićijević, *Obtaining crystals in a laboratory furnace*, J. Cryst. Growth, (Holand.), Vol. 267, str. 362-363, (2004).

[12] B. Čabrić, A. Janićijević., *Crystallization shelf*, J. Applied Crystallography, 40, 391, 2007).

[13] B. Čabrić, A. Janićijević, *A laboratory furnace for obtaining crystals*, J. Appl. Cryst. Engleska), Vol.37, str. 675, (2004).

[14] [13] (49) B. Čabrić, A. Janićijević, *Crystalization shelves for laboratory furnaces*, Cryst. Res. Technol., 42, No. 4, 342-343, (2007).

[15] B. Čabrić, T. Pavlović, A. Janićijević, *Regulation of the crystallization in a crucible urnace*, J. Cryst. Growth - Holandija, No. 200, str. 339-340, (1999).

[16] B. Čabrić, T. Pavlović, *Obtaining crystals in crucible furnace*, J. Appl. Cryst., 29, 745, 1996).

[17] B. Čabrić, A. Janićijević, *A method for obtaining crystals in a tube furnace*, Extended Abstracts of XII Conference of the Serbian Crystallographic Society, Serbian Crystallographic Society, Belgrade, p. 20-21, (2004).

[18] B. Čabrić, A. Janićijević, *Crystalization regulation in a crucible furnace*, Program and Contributed Papers of XVI National Symposium on Condensed Matter Physics FKM 2001, Arandjelovac, Ed. Institute of Physics Belgrade - Serbia and Montenegro, str. 44, (2001).

[19] B. Čabric, and A. Janicijevic, *Cooler for obtaining crystals in a crucible furnace*, J. Appl. Cryst., 36, 950, (2003).

[20] A. Janićijević, B. Čabrić, R. Simeunović, *Tamman's method for obtaining crystals in a rucible furnace*, VIII Conference of the Serbian Crystallographic Society, Serbian Crystallographic Society, Kragujevac, p. 92-93, (2000).

[21] B. Čabrić, T. Pavlović, and T. Despotović, *A crystallization cooler*, Czech. J. Phys., 49, No. 7, 1115 (1999).

[22] A. Janićijević, and B. Čabrić, *A crystallization coolers*, Extended Abstracts of X Conference of the Serbian Crystallographic Society, Serbian Crystallographic Society, Belgrade, p. 47.(2002)

[23] B. Čabrić, A. Janićijević, T. Despotović, *Regulation of the crystallization in a tube furnace*, VII Conference of the Serbian Crystallographic Society, Serbian Crystallographic Society, Vrnjačka Banja, p. 62-63, (1998).

[24] A. Janićijević, B. Čabrić, *Curcible furnace for obtaing crystals*, Extended Abstracts of X Conference of the Serbian Crystallographic Society, Serbian Crystallographic Society, Belgrade, p. 42, (2001).

[25] B. Čabrić, A. Janićijević, *A method for obtaining crystals in a chamber furnace*, Extended Abstracts of XII Conference of the Serbian Crystallographic Society, Serbian Crystallographic Society, Belgrade, p. 35-36, (2004).

[26] B. Čabrić, A. Janićijević, *Stober's method for obtaining crystals in Tamman a chamber furnace*, Extended Abstracts of X Conference of the Serbian Crystallographic Society Serbian Crystallographic Society, Belgrade, p. 49, (2002).

[27] B. Čabrić, A. Janićijević, *A laboratory furnace for obtaining single crystals*, Program and Contributed Papers of XVI National Symposium on Condensed Matter Physics FKM 2004, Ed. Institute of Physics Belgrade - Serbia and Montenegro, str. 337, 2004).

[28] B. Čabrić, A. Janićijević, *A chamber furnace for obtaing crystals*, Extended Abstracts of X Conference of the Serbian Crystallographic Society, Serbian Crystallographic Society, Belgrade, p. 45, (2001).

[29] A. Janićijević, N. Danilović, and B. Čabrić, *Crystallization Shelves*, Program and Contributed Papers of XVII Symposium on Condensed Matter Physics SFKM 2007, Ed. Institute of Physics, Belgrade - Serbia, p. 240. (2007).

[30] N. Danilović, A. Janicijević, and B. Cabrić, *Crystallization columns in a chamber furnace*, Kragujevac J. Sci., 32, 41 (2010).

[31] B. Čabrić, N. Danilović, and A. Janićijević, *Tube for obtaining crystals in a laboratory furnaces*, Instrum. Exp. Tech., 54, No. 2, 282 (2011).

[32] B. Čabrić, N. Danilović, and A. Janićijević, *Simultaneous crystallization testing in a laboratory furnace*, Am. Lab., 43, No. 7 (2011).

[33] A. Janićijević, " *Methods, techniques and analysis of physical conditions for the growth of crystals from the melt in the laboratory"*, doctoral dissertations, PMF, Kragujevac, (2008).

[34] B. Čabrić, N. Danilović, and T. Pavlović, *Cooler for obtaining crystals*, Cryst. Res. Technol., 46, No. 3, 292 (2011).

[35] E.-R. Schramek, *Taschenbuch für Heizung + Klimatechnik* (Oldenbourg Industrieverlag GmbH, München, , p. 152. (2007).

4

Defect Engineering During Czochralski Crystal Growth and Silicon Wafer Manufacturing

Lukáš Válek[1,2] and Jan Šik[1]
[1]ON Semiconductor Czech Republic,
[2]Institute of Physical Engineering,
Brno University of Technology,
Czech Republic

1. Introduction

Single crystal silicon has played the fundamental role in electronic industry since the second half of the 20th century and still remains the most widely used material. Electronic devices and integrated circuits are fabricated on single-crystal silicon wafers which are produced from silicon crystals grown primarily by the Czochralski (CZ) technique. Various defects are formed in the growing crystal as well as in the wafers during their processing. This chapter deals with the topic of engineering of crystal defects in the technology of manufacturing silicon single crystals and silicon wafers for the electronic industry. A basic overview of crystal defects found in semiconductor-grade silicon is provided and mechanisms of their formation are introduced. The impact of crystal defects on the manufacturing and performance of electronic devices is outlined and some of the methods of defect analyses are described. Finally, the most important methods for control of defect formation are summarized.

2. Industrial production of silicon for electronics

Single crystals of silicon for today's electronic industry are produced primarily by the Czochralski (CZ) method (Czochralski, 1918; Teal & Little, 1950). Only applications with extreme demands on pure bulk material utilize the float zone (FZ) method (Keck & Golay, 1953). The CZ method is based on crystal pulling from the melt, while the FZ method utilizes recrystallization of polysilicon rod which is locally molten by passage of the RF coil. The processes differ mainly in production cost and speed, which favor the CZ method, and in the purity of produced material, which is higher in case of the FZ method. The lower purity, which was seemingly unfavorable, helped CZ silicon become the dominant material as it makes silicon wafers more resistant against thermal stress and metallic contamination. Furthermore the FZ process could not follow the continual increase in crystal diameter. The diameter of 200 mm is the current size limit for FZ crystals while 450 mm capability was demonstrated for CZ process. As the chapter deals mainly with CZ silicon, the CZ crystal growth will be more closely described.

2.1 Czochralski growth of silicon crystals

Today's Czochralski (CZ) grown silicon single crystals are produced in a mass scale in diameters of up to 300 mm, but the 150 mm and 200 mm processes are still considered as standard. A typical CZ puller is shown in Fig. 1. The puller consists of an upper and lower chamber formed by steel water-cooled shells. The lower chamber contains a graphite hot zone with active central part and thermally insulating outer parts. A silica crucible is placed in the heart of the hot zone which is supported by a graphite susceptor on the pedestal attached to the lower shaft. The seed holder is fastened onto the upper shaft (or affixed to a cable). The heater is a meandering-coil element heated by high electric current. Both chambers are piped to a vacuum system. The puller is typically purged with inert gas (usually argon). In the beginning of the process the quartz crucible is loaded with a charge of polysilicon chunks and the single crystalline seed is fitted into the seed holder. After closing the puller the chambers are evacuated and re-filled by inert gas to the desired process pressure. The process starts with melting the polysilicon charge by applying high power to the heater. Once the charge is molten, the melt flow is stabilized under steady conditions and the seed is lowered towards the melt. After the seed is dipped into the melt, the system is adjusted to achieve a stable interface between the melt and the seed crystal. Pulling the seed upwards crystallizes the melt at the solid-liquid interface and the crystal proceeds to grow.

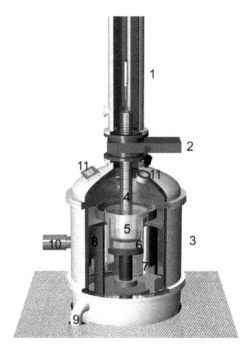

Fig. 1. A typical configuration of a CZ silicon puller. (1) Upper chamber, (2) isolation valve, (3) lower chamber, (4) grown crystal, (5) silica crucible, (6) graphite susceptor, (7) heater, (8) insulation, (9) vacuum pipe, (10) heater pyrometer window, (11) operator and camera windows.

Once the seed touches the melt surface, it is subjected to a huge thermal shock leading to generation of numerous dislocations. In order to achieve dislocation-free growth, "necking" is performed. To achieve this condition, the crystal pulling rate is increased to about 3 to 6 mm per minute, and the crystal diameter is reduced to about 5 to 2 mm, which allows dislocations to partially freeze in the neck and partially move to the crystal surface. Dislocation-free growth is usually achieved after several centimeters of the neck growth. Then the pulling rate is significantly decreased and the diameter is slowly increased. The crystal grows into the form of a cone called the crown. As the diameter increases to the desired crystal diameter the pull rate is gradually increased (the so called "shouldering stage") until the crystal grows with the desired diameter and the proper growth rate. Then, the cylindrical portion of the crystal, the "crystal body", is grown.

The melt and the crystal are in intimate contact at the solid-liquid interface. The melt surface forms a meniscus to the crystal which reflects the light from the hot crucible to the chamber windows. This results in the appearance of a shiny ring on the melt surface around the crystal. As the meniscus height increases with crystal radius, changes in the meniscus height can be sensed and used for crystal diameter control during the growth of the crystal body. Another option for crystal diameter control is a measurement of the meniscus diameter with a CCD camera. The crystal diameter is controlled by the pulling rate and simultaneously the pulling rate is adjusted by the heater power to be within the empirically-determined process window for dislocation-free growth, typically at or below 1 mm per minute. Solidification heat is conducted to the crystal surface and radiated to the chamber. For longer ingots the heat conduction is reduced and therefore the pulling rate has to be reduced. The crystal is rotated to homogenize the distribution of impurities and to suppress inhomogeneities in the temperature field. The crucible is rotated in the opposite sense to the crystal to stabilize the melt flow and control the oxygen concentration in the crystal.

The final stage of the crystal growth is the tail growth where the diameter is slowly decreased and a conical shape is achieved. The diameter of the solidification interface is reduced and dislocation formation is suppressed due to minimization of the thermal shock. Once the crystal has detached from the melt the power to the puller is decreased and the crystal is cooled down while being lifted into the upper chamber. At the end of the process, the crystal is removed from the puller for further processing.

2.2 Manufacturing of silicon wafers

Manufacturing of silicon wafers involves a series of mechanical, physical and chemical processes, all optimized to provide superior properties of the final wafer. A simplified manufacturing flow of the silicon wafer is shown in Fig. 2.

Wafer manufacturing follows the crystal growth process. First, the crystal crown and tail are cut off and the crystal body is divided into several pieces. Then the crystal quality (resistivity, oxygen and carbon content, dislocation-free state) is assessed on test wafers. Each section of the crystal is then surface-ground to the desired diameter, the crystal is oriented, and the flat is ground onto the cylindrical ingot. The flat identifies the orientation of the silicon wafers with respect to specific crystallographic directions; usually it corresponds to the $(1\,1;^-0)$ plane. Silicon wafers are sliced from the crystal sections using wire-saws or the inner-diameter (ID) saws. After edge grinding the wafer is lapped, etched and polished. Finally, the polished silicon wafer may have an epitaxial layer deposited on the prime surface by silicon epitaxy methods. Optionally the wafer backside can be coated with a polysilicon layer and/or a protective layer of silicon oxide. After final cleaning and inspection the silicon wafers are suitable for device or integrated circuit (IC) manufacturing.

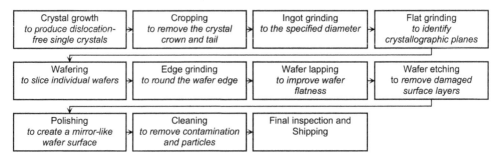

Fig. 2. Schematic manufacturing flow of the polished silicon wafer.

3. Defects in CZ silicon

Single crystalline CZ silicon wafer is a well-defined material of a very high quality. Nevertheless, it still may contain various defects which are formed either during the crystal growth or during processing of the silicon wafer. Defects which arise most frequently will be briefly introduced in the following sections. A detailed overview of defects in silicon can be found e.g. in (O'Mara, 1990).

3.1 Introduction

Soon after the start-up of mass production of dislocation-free silicon in the 1960's it was realized that the highly pure silicon wafers suffered from enhanced formation of slip lines and degraded device yields as compared to formerly-used dislocated silicon. Later it was found that the issues were actually related to impurities and crystal defects.

Swirl-like distributions of agglomerates of silicon self-interstitials (so called A- and B-defects) were described first (Föll & Kolbesen, 1975), and vacancy agglomerates (so called C- and D- defects) were observed soon after (Roksnoer & van den Boom, 1981). Although the initial investigations were focused on FZ silicon, the defects were observed in CZ silicon as well, and obeyed essentially the same rules. The intensive research in the field revealed that formation of crystal defects depended on the crystal growth conditions. Several attempts were made to explain the occurrence of vacancy- and interstitial-type defects in silicon, but a unifying and generally accepted theory was not provided until the work of Voronkov (Voronkov, 1982).

The complete picture of defects in CZ silicon is even more complicated due to influence of impurity inherently tied to CZ silicon, namely oxygen. Oxygen preferentially occupies interstitial sites in the silicon lattice, but below about 1200°C it becomes supersaturated and tends to cluster and precipitate. Moderate oxygen precipitation significantly enhances the mechanical strength of the silicon wafer, whereas too strong precipitation can degrade the mechanical properties. Oxygen precipitates in the active region of an electronic device usually results in degradation of the device performance. On the other hand, oxygen precipitates outside of the active region may have a highly beneficial effect through intrinsic gettering (Rozgonyi, 1976). As there are both negative and positive effects closely related to oxygen precipitation, it is a phenomenon which has to be carefully engineered in the silicon manufacturing technology. Oxygen precipitation depends on the concentration of interstitial oxygen, annealing conditions and oxygen interactions with intrinsic point defects. Hence, a lot of effort was directed to the investigation of defect formation in silicon during past decades.

3.2 Silicon crystal structure

Silicon crystallizes in the diamond cubic lattice (Bullis, 1991), which structure can be constructed as two interpenetrating face-centered cubic (fcc) lattices displaced along the unit cell body diagonal by the quarter of its length. The lattice constant of pure silicon is representing the length of the side of the fcc cube and has the value of 0.5431 nm at room temperature. Each silicon atom from one fcc lattice is covalently bound to its four nearest neighbors which belong to the other fcc lattice. In principle silicon wafer can be manufactured with the silicon lattice being oriented in any arrangement with respect to wafer surface. The common crystallographic orientations of wafer surface, described by Miller indices, are (100), (111) and (110). The crystal planes belonging to the families of these low-index crystal planes, i.e. {100}, {111} and {110} family, are the most important planes in the silicon lattice with the largest differences in various properties among these planes.

The closely spaced planes of the {100} family are the {400} planes with the interplanar distance of 0.1358 nm. The {220} planes with spacing of 0.1920 nm are the closely spaced planes of the {110} system. While spacing of all the {400} and {220} planes is equidistant, the situation is more complicated in case of the {111} system. The {111} planes of the fcc lattice are stacked in an ...ABCABC... manner, where none of the atoms in each layer is aligned with atoms in other layers when viewed perpendicularly to the stack. The {400} planes follow the ...ABCDABCD... scheme and the {220} planes follow the ...ABAB... scheme. Since the diamond lattice consists of two fcc lattices displaced along the direction perpendicular to the {111} system, stacking along <111> follows the ...AaBbCcAaBbCc... scheme. Here atoms in planes of the same letter overlap each other when viewed in <111> direction and capitalization distinguishes planes belonging to the two interpenetrating fcc lattices. The distance between adjacent planes of the same set (A-B, B-C and C-A) is 0.3135 nm; the distance between the closely spaced planes of the system a-B, b-C and c-A is 0.0784 nm and 0.2352 nm for the system A-a, B-b, C-c, which is the length of the covalent bond silicon-silicon. The closely spaced planes are bound strongly and often appear as one double plane. Different spacing of the crystal planes corresponds to different surface atom densities of 6.78 $\times 10^{14}$ cm^{-2} for {100} planes, 7.83 $\times 10^{14}$ cm^{-2} for {111} planes and 9.59 $\times 10^{14}$ cm^{-2} for {110} planes. The {111} plane shows the lowest surface energy and crystallizing silicon therefore tends to be bound by the {111} planes. Also the growth rate of the epitaxial layer is the lowest for the {111} oriented surface.

3.3 Classification and overview

Crystal defects are usually classified according to their shape and dimension. Silicon crystals and wafers may contain:

a. point defects- silicon self-interstitials, vacancies, interstitial impurities such as oxygen, substitutional impurities such as dopants and carbon,
b. line defects- edge and screw dislocations, dislocation loops,
c. planar defects- stacking faults,
d. bulk defects- agglomerates of point defects.

Some of the crystal defects are shown schematically in Fig. 3.

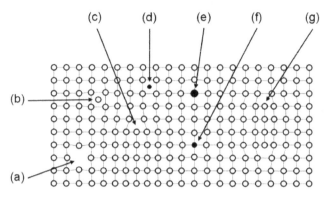

Fig. 3. Schematic 2D representation of crystal defects in silicon. (a) Vacancy, (b) self-interstitial atom, (c) edge dislocation, (d) interstitial impurity atom, (e) substitutional impurity atom of larger atomic radius, (f) substitutional impurity atom of smaller atomic radius, (g) extrinsic stacking fault.

3.4 Intrinsic point defects
Vacancies and silicon self-interstitials (Pichler, 2004) are intrinsic point defects inherent to the material whose occurrence in the lattice arises from thermodynamic equilibrium. Intrinsic point defects are incorporated into the growing crystal at the melt-crystal interface in their equilibrium concentrations. Point defects in the crystal are further created or annihilated by Shottky and Frenkel mechanisms, recombination, interaction with bulk defects, and injection during thermal wafer processing.

A silicon atom removed from its lattice site leaves behind four broken covalent bonds which can be arranged in several configurations. When the dangling bonds reform into molecular orbitals a neutral vacancy is formed. These orbitals are strained and can be relatively easily broken. Interaction with charge carriers results in formation of singly and doubly positively- or negatively-charged vacancies. The silicon self-interstitial is considered as an additional, more or less free atom occupying tetrahedral or hexagonal interstitial sites. Small clusters of vacancies and interstitials may be considered as a point defect. The precise atomic arrangement of vacancies and interstitials in silicon still remains somewhat uncertain, yet their effective thermodynamical properties such as concentration and diffusivity have been experimentally determined.

3.5 Extrinsic point defects
Dopant atoms are the primary extrinsic point defects in CZ silicon. The common dopant species - boron, arsenic, antimony and phosphorus - are introduced deliberately into the silicon melt during the crystal growth. They occupy substitutional sites in the crystal lattice and their concentrations are typically in the range of 10^{15} - 10^{19} cm^{-3}. The effect of dopant atoms on the properties of the crystalline silicon (besides the primary electric effect) originates from different atomic volumes, from interactions with intrinsic point defects, the formation of clusters, and from the effects on the diffusion and redistribution of impurities (Pichler, 2004).

Extremely undesirable point defects are *atoms of metallic elements* in the middle of the periodic table (Graff, 2000). Deep energy levels near the middle of the silicon forbidden band arising from metallic species act as effective generation-recombination centers and pose serious issues for device performance and process yield. The transition metals such as copper, nickel and iron are the most abundant contaminants. At high temperatures the metal atoms are usually distributed in the lattice occupying both interstitial and substitutional sites. As the temperature is reduced the metal atoms can form complexes with dopant atoms, and may eventually precipitate into bulk defects. The state of metals in the lattice depends on their solubility in silicon and on the imposed thermal cycles. The level of metallic contamination in semiconductor silicon for today's processes is typically below 10^{12} cm^{-3}. The steep decrease in solubility with decreasing temperature combined with a high diffusivity at low temperatures allows relaxation-type gettering techniques to remove the metals from the active device regions.

An unavoidable impurity in CZ silicon is *carbon* (O'Mara, 1990) introduced into the crystal growth system mainly from the graphite elements of the hot zone. Typical concentration of carbon in semiconductor silicon is below about 0.1 ppma (5×10^{15} cm^{-3}). Carbon predominantly occupies substitutional sites in the silicon lattice. Due to its four valence electrons substitutional carbon is electrically inactive. Substitutional carbon forms complexes with oxygen. Interactions with silicon self-interstitials may lead to a displacement of the carbon into an interstitial position (Pichler, 2004). Interstitial carbon readily forms complexes with intrinsic point defects, substitutional carbon or dopants, which may become electrically active. Carbon was reported to promote oxygen precipitation. However, due to the low content of carbon in today's silicon, its the role in defect formation is of minor importance.

After dopant elements, *oxygen* is perhaps the most important extrinsic defect in silicon (O'Mara, 1990; Shimura, 1994). Oxygen is incorporated into the CZ silicon during the crystal growth. The silica crucible wall slowly dissolves into the silicon melt and oxygen is transported by diffusion and convection to the melt-crystal interface. Much of the oxygen evaporates from the melt free surface and only about 1% is incorporated into the growing crystal. The solubility of oxygen in silicon is a monotonically increasing function of temperature; its value at the melting temperature is about 2×10^{18} cm^{-3}. Oxygen concentration in the CZ silicon crystal is typically on the order of 10^{17} cm^{-3}. Oxygen atoms occupy interstitial positions in the silicon lattice where they are covalently bound to the two nearest silicon atoms. Interstitial oxygen may form electrically active chains known as "thermal donors". Oxygen also forms complexes with intrinsic point defects (Pichler, 2004) which may play a role during the formation of bulk defects and promote oxygen diffusion.

3.6 Line defects

Today's CZ silicon crystals are grown in a dislocation-free mode (A-defects are not taken into account). Dislocations appear in silicon mainly due to stress generated by high temperature operations during the manufacturing of the wafer and devices on it when the critical resolved shear stress (CRSS) is exceeded. Edge and screw dislocations and dislocation loops are created under the high stress conditions, and have been observed in silicon mainly accompanying oxygen precipitates, self-interstitial agglomerates, dislocation arrays and tangles, and slip.

3.7 Planar defects

Two types of *stacking faults* (SFs) are generally distinguished, namely intrinsic and extrinsic faults. Intrinsic SF is formed by several missing atomic planes while the extrinsic SF is formed by excess atomic planes. SFs in silicon are always of the extrinsic nature (O'Mara, 1990) formed by discs of double planes inserted into the regular AaBbCc order of silicon {111} atomic planes. As the layers in the SF have to be added in pairs, one can drop the double index notation and describe the packing of the sequence with ...ABCABC... In the SF region the order of planes is changed, such as ...ABCACABC.... SFs are considered to be low-energy faults because they involve no change in the covalent bonds of the four nearest neighbors in the lattice (Hirth & Lothe, 1967). On the other hand, faults which disturb the nearest-neighbor covalent bonds are called high-energy faults (these are, for example, dislocations). The SF is bound by a Frank-type partial dislocation. A schematic drawing of such a SF is shown in Fig. 3. Other types of planar defects such as twin planes or grain boundaries are not present in properly-grown single crystal silicon.

(a)   (b)

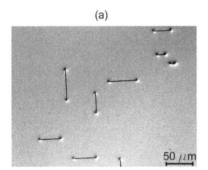 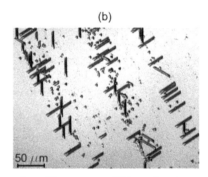

Fig. 4. (a) Bulk stacking faults on a (100) silicon surface. The varying SF length originates from the various depths of nucleation sites below the etched surface. (b) Surface stacking faults and dislocations on a (111) silicon surface. Predominantly equal SFs length confirms nucleation on the surface.

SFs originate from the condensation of silicon self-interstitials (Ravi & Varker, 1974), preferentially on suitable nucleation sites such as oxide precipitates, metal precipitates or mechanically damaged locations in the lattice. Surface and bulk SFs are distinguished according to the location of their nucleation site. Stacking faults can be delineated by selective etching of the sample surface and studied by optical microscopy or other inspection methods. The crystallographic origin determines the appearance of the SFs on the surface of silicon wafers of different orientations (see Fig. 4).

Bulk SFs in the silicon wafer may grow in the regions of strong oxygen precipitation (O'Mara, 1990), which is accompanied by a strong ejection of silicon self-interstitials. Growth of surface SFs can be also enhanced by an increased population of self-interstitials created by surface oxidation. SFs observed on the surface of polished silicon wafer after oxidation are referred to as "oxidation induced stacking faults" (OISFs).

3.8 Bulk defects

Vacancy-type defects. Supersaturated free vacancies present in the silicon crystal at high temperatures can agglomerate into voids (Itsumi, 2002). Voids (Fig. 5) take on an octahedral shape bounded by {111} planes. The octahedral shape may be incomplete, truncated by {100} planes. Under favorable conditions, double- or triple-voids may appear. The inner void surface is covered by an oxide layer typically 2 - 4 nm thick. A typical void dimension is around 100 nm and a typical density in the crystal is about 10^6 cm^{-3}.

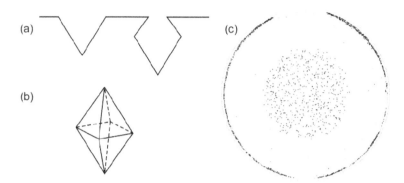

Fig. 5. Schematic representation of (a) COP defects on the wafer surface, and (b) a void defect in wafer bulk. (c) COP distribution observed by laser-type particle counter after dedicated SC1 treatment of the wafer with a vacancy-type core.

A void intersected by the wafer surface creates a pit referred to as the "crystal-originated particle" (COP) (Ryuta et al., 1990). The relation of a COP to a void is demonstrated in Fig. 5. The typical dimension of a COP does not allow direct observation on the polished wafer surface by common industrial equipment; a method such as atomic force microscopy (AFM) has to be used for the analyses. It was found that COPs can be observed after repetitive SC1 cleaning by commercial laser-type particle counters. COPs on the wafer surface can be delineated by other methods such as Secco etching where they appear as the wedge-shaped "flow pattern defects" (FPDs) (Yamagishi et al., 1992). Voids detected by infrared light scattering tomography are denoted as "LSTD" (Vanhellemont et al., 1997). Being vacancy agglomerates, COPs appear in the vacancy-type silicon (Fig. 5c).

Interstitial-type defects. Supersaturated silicon self-interstitials may coalesce into so called A- and B- defects (swirls). The B-defects are considered to be coherent globular clusters; the A-defects are large dislocation loops possibly formed by the collapse of the B-defects (Föll & Kolbesen, 1975). Large dislocation loops appear in size ranging from roughly a micrometer up to a few tens of micrometers with a density typically around 10^8 cm^{-3}.

Oxide precipitates. Due to the rapid decrease of oxygen solubility with temperature and the rather high oxygen concentrations in CZ silicon crystals, oxygen in silicon is usually present in a supersaturated state at most common process temperatures. The resulting precipitation of oxygen interstitials leads to the formation of oxide particles referred to as oxide precipitates or oxygen precipitates (Shimura, 1994). Oxide precipitates are formed of amorphous SiO$_x$, where x ranges from 1 to 2. Oxide precipitates of various morphologies have been observed: rod-like, square platelet, truncated octahedral, and polyhedral to spherical. The morphology of oxide precipitates depends on the formation conditions, mainly temperature and the degree of supersaturation. In principle it is determined by the stress energies associated with the precipitate growth (Borghesi et al., 1995). Growth of oxide precipitates is usually accompanied by the formation of extended defects such as stacking faults and punched-out dislocation loops.

4. Impact of crystal defects on device yield

While the vacancy- and interstitial-type crystal defects have a detrimental effect on yield and the performance of electrical devices, oxide precipitates and associated extended defects can have both a positive and negative effect.

COP defects were shown to degrade gate oxide integrity (GOI) (Yamagishi et al., 1992) causing extrinsic dielectric breakdown of MOS capacitors and failures of the DRAM modules. Voids below the surface can result in excessive leakage currents in power devices. The gate oxide grown across a COP may be locally thinned or stressed on the COP edges where the electric field is locally greater; both these effects may lead to dielectric breakdowns, shifted device thresholds and leakage.

The presence of interstitial-type dislocation loops in the active regions of the devices may cause degradation of the minority carrier lifetime, alter diffusion profiles, and shift p-n junction characteristics (Abe et al., 1983). Due to their relatively large size they are especially harmful for today's processes characteristic by a high degree of device integration. The dislocations act as fast diffusion paths for impurities and dopant species. In particular, when decorated by metals, dislocation cores are sources of shorts and leakage currents.

Oxygen thermal donors can induce resistivity shifts in very lightly-doped silicon which affects numerous resistivity-dependent device parameters. The thermal donors are formed by rather weak binding energies and can be easily dissolved at temperatures above 600°C. As oxygen thermal donors are formed at a peak rate around 450°C the cooling rates during device manufacturing processes should be optimized in order to control the resistivity of the sensitive materials. On the other hand, the oxide precipitates are quite stable defects. Oxide precipitates in the active region of the device result in increased leakage currents in the p-n junctions, reduced refresh times in DRAM memories, reduced breakdown voltages in bipolar devices, degraded minority carrier generation/recombination lifetimes, and other serious failure mechanisms (see Schröder, 1989, and references therein). The root-causes of the failures are the precipitates themselves as well as the associated extended defects. On the other hand, oxide precipitates may have a positive effect through mechanical strengthening through impurity hardening (Sumino, 1980) and intrinsic gettering (Rozgonyi, 1976). However, oxygen precipitation has to be controlled in an optimal range for a given wafer processing flow, and a defect-free near surface region (so called "denuded zone") should be guaranteed.

The effect of crystal defects on manufacturing processes and the properties of silicon wafers and electronic devices is quite varied and the impact is often fundamental. Hence, engineering of defects is inseparable across crystal growth and wafer processing. Due to the continuous evolution of silicon devices it is also a never-ending part of the technology of silicon crystal growth and wafer manufacturing.

5. Formation of grown-in defects in CZ silicon

In order to control formation of crystal defects in silicon wafers one has to understand the mechanisms of their formation from the very early stages of crystal growth. The basic mechanisms of defect formation during silicon crystal growth are described by the Voronkov theory (Voronkov, 1982) which will be briefly introduced in following sections.

5.1 Point defects - A pathway to crystal defects

Voronkov theory describes the formation of crystal defects consisting of several stages:

a.　the incorporation of silicon self-interstitials and vacancies into the crystal,

b.　transport, diffusion and recombination of the point defects,

c.　nucleation and growth of defect clusters.

It is assumed that both silicon self-interstitials and vacancies are incorporated into the growing crystal at the melt-crystal interface. High diffusivities near the melting temperature and the proximity of the crystal surface acting as an effective source/sink allows the intrinsic point defects to exist in their equilibrium concentrations.

It is further assumed that the recombination rate of the point defects is high enough to maintain the product C_IC_V near its equilibrium value $C_I^{eq}C_V^{eq}$. As the equilibrium concentrations of the point defects decrease very rapidly with temperature, the $C_I^{eq}C_V^{eq}$ product quickly drops below the initial concentration of one of the point defects. This defect essentially vanishes and only the second specie survives. The type of the surviving defect is determined by the concentration of vacancies and interstitials prior to recombination, which is governed by the flux of the defects towards the region of effective recombination.

Vacancies and interstitials in the crystal diffuse by Fickian diffusion and thermodiffusion, and are drifted by the growing crystal. Voronkov assumed that the equilibrium concentrations of interstitials $C_I^{eq}(T_m)$, and vacancies $C_V^{eq}(T_m)$, at the melting temperature T_m are comparable, but $C_V^{eq}(T_m)$ is somewhat higher than $C_I^{eq}(T_m)$. Therefore, the drift flux of vacancies should be larger than that of interstitials. The drift (convection) is thus responsible for the supply of vacancies into the crystal, and the flux of vacancies into the crystal is proportional to the crystal growth rate, v. The diffusion flux is proportional to the concentration gradient which arises from the defect recombination processes. This, in turn, is determined by the axial temperature profile above the melt-crystal interface. The diffusion flux is therefore scaled in proportion with the temperature gradient, G. It is assumed that the diffusion coefficient of the self-interstitials is larger relative to that of vacancies. Hence, diffusion is responsible for the supply of interstitials into the crystal and the flux of interstitials into the crystal is proportional to the temperature gradient, G.

The concentration of the point defects above the melt-crystal interface is determined by the competition of the drift flux supplying vacancies and the diffusion flux supplying interstitials. As these fluxes are proportional to the crystal growth rate v, and to the temperature gradient G, respectively, the v/G ratio determines the type and concentration of point defects which survive the recombination stage. Point defect recombination proceeds at a significant rate around 1300°C. A typical temperature profile in the crystals grown in modern CZ crystal pullers results in a recombination length of about 2 - 3 cm from the melt-crystal interface. The v/G ratio in fact determines whether the silicon crystal a few centimeters above the melt-crystal interface (or at about 1300°C) contains vacancies or interstitials.

There is a critical value of the v/G parameter corresponding to the state when the drift and diffusion fluxes (or the vacancy and interstitial fluxes) are roughly balanced. In such a case the concentrations of vacancies and interstitials before the recombination are comparable and recombination leaves essentially defect-free crystal. The critical value of v/G, also referred to as ξ_t, is given by the point defects properties near the melting temperature. The critical ratio ξ_t separates two cases: (a) when the crystal growth process results in $v/G < \xi_t$, defect recombination results in a crystal populated by excess self-interstitials; or (b) when $v/G > \xi_t$, the crystal contains excess vacancies. The two cases are known as the interstitial-type crystal, and the vacancy-type crystal, respectively.

The surviving point defects become further supersaturated during cooling of the growing crystal which results in formation of bulk defects – vacancy-type defects in the vacancy-type crystal, and interstitial-type defects in the interstitial-type crystal. Also other phenomena such as oxygen precipitation may differ substantially between the vacancy-type and interstitial-type crystals. A silicon matrix rich in vacancies can easily absorb silicon interstitials emitted during the growth of oxide precipitates, while emission of interstitials into a matrix already rich in interstitials is energetically unfavorable. Oxygen precipitation in a vacancy-type silicon is generally stronger as compared to oxygen precipitation in an interstitial-type material. Taking into consideration the influence of crystal defects on wafer properties and device performance one realizes the importance of the crystal growth process, which determines the value of v/G.

5.2 Spatial distribution of defects

Principal distribution of defects in silicon crystal is determined by the relation of the v/G value to the critical value ξ_t. The crystal growth rate at a given crystal length can be considered constant across the melt-crystal interface from a macroscopic viewpoint. The temperature gradient, however, strongly varies across the radius. Due to cooling of the crystal surface by radiation and convective heat loss to the atmosphere the axial temperature gradient at the melt-crystal interface increases from the crystal center to the perimeter. The v/G parameter thus decreases from the center to the perimeter (see Fig. 6). The relation of the ξ_t value to the v/G curve determines whether an interstitial-type, a vacancy-type, or a mixed-type crystal is grown, as illustrated in Fig. 6. The transition between the vacancy-type and interstitial-type portion of the crystal is called the vacancy–interstitial boundary (V-I boundary). As the process conditions vary during the crystal growth, the v/G curve changes also with the crystal length.

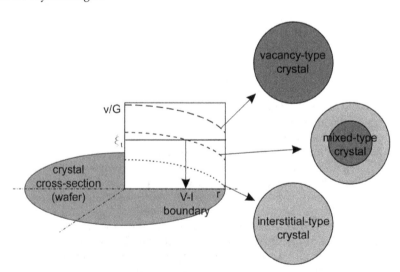

Fig. 6. Left: a typical radial dependence of the v/G ratio and three possible relations of the $v/G(r)$ curve to the critical parameter ξ_t. Right: crystal (wafer) types determined by the relation of $v/G(r)$ to ξ_t. The cross-section image corresponds to a wafer sliced from the crystal.

As noted above, formation of bulk defects in silicon is strongly influenced by the type of the material. The nature of the crystal from the defect point of view (distribution of the vacancy-type and interstitial-type regions) in principle determines the presence and distribution of bulk defects in the silicon wafers. Today's FZ silicon crystals are typically of fully vacancy-type due to high growth rate which is typically 2 - 3 times higher compared to CZ silicon crystals. On the other hand, CZ silicon crystals of all three types shown in Fig. 6 can be found. First CZ silicon crystals were of the interstitial-type, due to a relatively high value of G in the crystals of small diameter. Today's 6" and 8" crystals produced at a rather high pull rates are usually of vacancy-type. The 300 mm crystals may be of the mixed-type and even larger crystals may become fully interstitial-type because of the pull rate which has to be decreased when increasing the crystal diameter up to these levels (von Ammon et al., 1999). Doping elements and their concentration in the crystal can significantly influence formation of crystal defects and shift the process in favor of either vacancy-type or interstitial-type (Borionetti et al., 2002).

5.3 Oxygen precipitation

Oxygen precipitation, i.e., the formation and growth of oxide precipitates from the supersaturated solid solution of oxygen in silicon can be characterized as a two-step process consisting of a nucleation stage and a growth stage. A comprehensive introduction on oxygen precipitation theory can be found in (Shimura, 1994; Borghesi, 1995).

Nucleation is the process of formation of aggregates of a small number of oxygen atoms - the nuclei (or precipitate embryos). Due to relatively low annealing temperatures the supersaturation is high but the diffusivity of oxygen is low. The change in the concentration of interstitial oxygen in the silicon matrix is usually negligible and mainly the precipitate density is established. Precipitate growth (often referred to as the precipitation stage) is caused by diffusion of oxygen atoms and attachment to the existing nuclei at higher temperatures. While the generation of new nuclei during the precipitation stage is negligible the existing nuclei grow substantially while forming oxide particles of sizes reaching up to the micrometer scale. The supersaturation in this stage is lower but the diffusivity of oxygen is much higher than in the nucleation step. During precipitation the reduction in the concentration of interstitial oxygen may be dramatic.

Annealing Step		Precipitated oxygen [%]
750°C/10h	-	1.0 ± 0.5
750°C/40h	-	46 ± 3
750°C/10h	1050°C/20h	91.4 ± 0.5

Table 1. Influence of annealing temperature and time on the amount of precipitated oxygen in a lightly boron doped wafer with an interstitial oxygen content of 9.7×10^{17} cm^{-3}.

The influence of various annealing steps on the reduction of interstitial oxygen is shown in Table 1. Although shorter annealing at low temperature (750°C/10h) reduces the interstitial oxygen concentration only by about 1%, the nuclei have sufficient size for further growth during a precipitation step at 1050°C. It is also shown that even low temperature annealing can reduce interstitial oxygen significantly if the annealing time is extremely long and the initial interstitial oxygen concentration is sufficiently high.

Capturing the nucleation stage has been the most difficult problem over the decades of oxygen precipitation studies. The precipitate nuclei are too small to be observed directly therefore only indirect observations have been used for model validations. Further complications resulted from the uncertainties in the thermo-physical properties of the various defects involved in the nucleation phase. The topic has been treated by numerous authors who assumed both homogeneous and heterogeneous nucleation mechanisms (e.g., Borghesi, 1995, and references therein). Regardless of the particular nucleation mechanism there is a parameter crucial for all theoretical considerations – the critical radius of the nucleus which determines the temperature stability of the nucleus. Nuclei of radius smaller than the critical radius corresponding to the actual temperature tend to dissolve while the larger ones continue to grow.

In order to overcome the lack of analytical methods for identifying the nucleation stage the experimental studies often use various thermal treatments for growing the oxide precipitates to a detectable size and deducing the nucleation processes from the precipitation behavior. The early work resulted in the establishment of procedures for optimization of oxygen precipitation, e.g. for intrinsic gettering. Classical precipitation treatment of silicon wafer consists of a low temperature nucleation step (~700°C) followed by a high temperature precipitation step (~1000°C).

While the nucleation stage of oxygen precipitation remained rather mysterious, the growth of oxide precipitates became experimentally well-described. As shown by many authors oxide precipitates formed during wafer thermal treatments usually show the diffusion-limited growth according to the theory of Ham (Ham, 1958). The precipitate growth controlled by oxygen diffusion results in a square-root dependence of the precipitate size on annealing time at a given temperature. However, due to several assumptions, Ham's theory is not always applicable. The early works usually considered silicon wafer as "clean" input material. However, it has been gradually recognized that some nuclei are formed already during the silicon crystal growth process, and the importance of the thermal history and intrinsic point defects has been recognized.

State-of-the-art investigations of defect formation in silicon thus start from the early beginnings of the crystal growth process, and the interplay of all intrinsic and extrinsic point defects present in the crystal must be considered.

5.4 OISF ring

An interesting feature related to oxygen precipitation during crystal growth is observed on silicon wafers of the mixed-type after oxidation of the surface. Rapid oxidation of the wafer surface (wet oxidation) results in the injection of silicon interstitials below the oxidized surface. These interstitials can condense into stacking faults on grown-in oxide precipitates in the near-surface region. A ring of relatively large oxygen precipitates can be formed near the vacancy-interstitial boundary during the crystal growth process (Voronkov, 2008). This phenomenon results in a ring-like distribution of stacking faults (Hasebe et al., 1989) during subsequent wet oxidations. The observed feature (see Fig. 7) is referred to as the "oxidation-induced stacking fault ring" (OISF ring).

The OISF ring can be easily visualized on an oxidized wafer by stripping the oxide and selectively etching the surface. The etched OISFs are visible by optical microscopy and the OISF ring is detectable even by the naked eye. The OISF ring is located on the edge of the vacancy-type core in mixed-type crystals close to the V-I boundary. Due to the ease of detection it has been widely used for delineating the V-I boundary and investigating the defect distribution in silicon crystals.

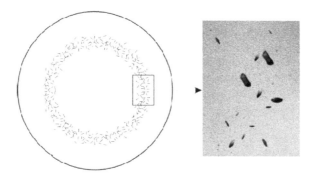

Fig. 7. OISF ring on the surface of a silicon wafer.

6. Methods of studying crystal defects

In order to capture the distinctive nature of crystal defects and the wide range of their size and density, various methods are used for defect analyses. The most common techniques are briefly introduced in following section.

6.1 Preferential etching

Analytical methods based on preferential etching of silicon are frequently used for studying defects in silicon technology. All modern defect etchants for silicon are based on the chemistry of formation and dissolution of silicon dioxide, and therefore all etchants include hydrofluoric acid. The dissolution process is enhanced at and around defects due to weakened bonds and coalescence of impurities around them. The electronic properties may play a role, too. Etching anisotropy with respect to various crystallographic planes (atom density) also has to be taken into account. In general, the etch rate is lowest for the {111} crystal planes. The most common etching solutions are those of Wright, Sirtl, Schimmel, Yang, Secco, MEMC, and Dash. Individual solutions are optimized for various surface orientations, conductivity types or resistivity levels. For more details on use and application see (ASTM standard F 1809–02, 2003).

Preferential etching allows for rather simple detection of dislocations, stacking faults, slip lines, and oxide precipitates as shown in Fig. 8. Individual defects on the etched surface may be observed by optical microscopy or SEM, while defect distributions across a wafer can be observed by the naked eye in most cases.

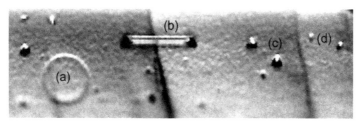

Fig. 8. Defects in silicon revealed by preferential etching of a cleaved surface. (a) A stacking fault in the cleavage plane, (b) a stacking fault tilted with respect to the cleavage plane (note the bounding dislocation), (c) dislocations, (d) oxide precipitates.

6.2 X-ray topography

X-ray topography (Lang, 1978) is based on Bragg diffraction. A monochromatic X-ray beam incident on a surface is reflected by the atomic planes of the crystalline sample. Constructive interference occurs when the Bragg condition is fulfilled. The diffraction pattern is recorded on the detector either in reflection geometry (Bragg geometry) or transmission geometry (Laue geometry).

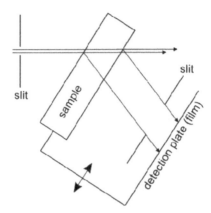

Fig. 9. Schematic layout of the X-ray section and projection topography in Laue geometry.

In the case of section topography, a very narrow beam of dimensions in the order of 10 μm is used. The section topographs therefore investigate only a small volume (section) of the sample. The beam is diffracted at different depths, each one contributing to the image on a different location on the detector. Section topography is therefore used for depth-resolved defect analysis. When the sample and detector are simultaneously moved with respect to the incident beam, the sample is scanned over a larger volume and a projection topograph is formed. The schematic configuration of X-ray topography in Laue geometry is shown in Fig. 9. A homogeneous crystal lattice generally results in a homogeneous distribution of intensity across the topograph. Section topography, however, results in spatial variations in the intensity of diffracted rays even in a perfect crystal (so called Pendelösung fringes) due to the dynamic nature of diffraction and a small investigated volume. Irregularities of the crystal lattice, such as defects and strain, are captured as a distortion of the image of the perfect crystal. X-ray topography can detect irregularities such as phase boundaries, defective areas, cracks, scratches, growth striations, and most of the common crystal defects such as dislocations, oxide precipitates, stacking faults, and interstitial-type defects. An example of X-ray topography results is shown in Fig. 16.

6.3 Fourier transform infrared spectroscopy

Fourier transform infrared spectroscopy (FTIR) is the most common technique for determining the concentration of oxygen in silicon materials. The polychromatic source radiates infrared (IR) light through a Michelson interferometer. The radiation is transmitted through the silicon wafer and directed to a detector. Undoped silicon is transparent to the IR radiation. Impurities cause localized absorption from lattice vibrations. The absorption due to the anti-symmetric vibrational mode at 1107 cm^{-1} has been assigned to interstitial oxygen and it is thus used for determining the oxygen concentration. In order to suppress the influence of the silicon lattice vibrations and to allow the quantitative analysis, a reference spectrum is subtracted from the measured spectrum. The reference spectrum is measured on float-zone silicon prepared with very low oxygen content (less than 1×10^{16} cm^{-3}). After signal processing and taking the wafer thickness into account, the intensity of the oxygen absorption peak is calculated, and thus, the concentration of oxygen may be determined as the product of the peak intensity and a tabulated conversion factor. The selected conversion factor has to be reported together with the concentration value. The oxygen concentrations presented in this work are given in accordance with the ASTM standard (ASTM Standard F1188-02, 2003) commonly referred to as "new ASTM". The FTIR method is not applicable for heavily-doped silicon as this material is not transparent to IR radiation due to the high concentration of free carriers.

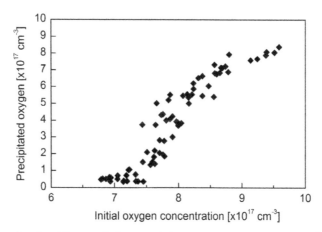

Fig. 10. An example of an "S-curve" characterizing oxygen precipitation in the crystal grown by a particular crystal growth process.

As the contribution of precipitated oxygen atoms to the absorption peak of interstitial oxygen is small (absorption energies are shifted), the FTIR measurements of interstitial oxygen concentration can be used for characterization of oxygen precipitation. The difference in the content of interstitial oxygen prior to, and after, the heat treatment is the measure of oxygen precipitation, relevant for the specific sample material and thermal cycles. For overall characterization of a crystal grown by a particular growth process the dependence of the precipitated oxygen on the initial oxygen concentration is constructed for a standardized thermal cycle. The dependence is called the "S-curve" because of its typical sigmoidal shape (Fig. 10). Another application of the S-curves is the comparison of different thermal cycles (e.g., in IC production) with respect to oxygen precipitation.

6.4 Precipitation test

This test is aimed at assessing the precipitation behavior in the wafers. Oxide precipitates formed in the crystal during growth (grown-in precipitates) are usually too small to be detected directly as noted previously. Oxygen precipitation is therefore evaluated in silicon wafers subjected to a standardized thermal treatment sequence.

Generally, two types of the heat treatment are used: (a) single step annealing at high temperature when the grown-in oxide precipitates are allowed to grow, and (b) two-step annealing which includes also a low-temperature nucleation step. Typical thermal cycles for such heat treatments are the 1050°C/16h annealing, and 750°C/4h + 1050°C/16h annealing. The extent of oxygen precipitation is evaluated as the difference in oxygen concentration before, and after, the thermal treatment as measured by FTIR, or by "cleave-and-etch" analyses. In such cases, the wafer cross-section is prepared by cleaving or grinding, the sample is preferentially etched, and the defect distribution and density is evaluated under the microscope. An example of the cleave-and-etch analysis is shown in Fig. 11.

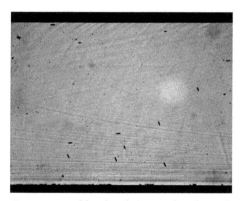

Fig. 11. A wafer cross-section prepared by the cleave-and-etch method. The wafer front surface (polished) is at the top of the figure. The analysis reveals oxide precipitates (dots) and stacking faults (short lines) in the bulk of the wafer.

6.5 OISF test
The OISF test uses the fact that silicon interstitials injected below the wafer surface during oxidation under usual conditions needs a suitable nucleation site for formation of OISFs (Ravi & Varker, 1974). The test wafer is subjected to an oxidation cycle (e.g., 1000°C/1hr in a "wet" atmosphere), then the grown oxide is stripped with HF acid and the surface is preferentially etched. The stacking faults observed on the surface decorate the nucleation centers in the near-surface wafer region. The nucleation sites can arise from the sub-surface damage caused by the wafer manufacturing processes, handling, etc., from wafer contamination, and defects such as dislocations and oxygen precipitates. The OISF test is therefore used as a general method for assessing the quality of the wafer surface. If the manufacturing process is optimized and wafer contamination is avoided, the OISF test can be used for delineation of crystal defects in the wafer as in the case of the OISF ring.

6.6 COP test
Delineation of COPs is based on SC1 cleaning and "particle" inspection methods (Ryuta et al., 1990). As the typical COP dimensions do not allow direct observation on the polished wafer surface by common industrial equipment, the SC1 cleaning chemistry is used to etch the COPs and enlarge them to dimensions greater than the detection limit of the instruments. The size of COP defects after the SC1 treatment is typically 100 - 300 nm and the surface density is of the order of 10 to 100 cm^{-2}. An example of the measurement output is shown in Fig. 5c.

7. Engineering of crystal defects in silicon

Defect engineering in production manufacture of silicon wafers comprises modifications of the crystal growth process and dedicated heat treatment of the wafers. The primary objective is usually the control of oxygen precipitation and the associated defect structures. The specific requirements on the spatial distribution, size and density of oxide precipitates may differ for various devices, but the basic features are essentially the same. It is desirable to achieve a defect-free region near the surface of the wafer so that the devices fabricated there do not suffer from the electrical effects resulting from the presence of the defects. On the other hand, a high density of oxide precipitates is usually required in the bulk of the wafer in order to provide effective intrinsic gettering. The near-surface region free of oxide precipitates is called a "denuded zone".

7.1 Denuding and precipitation of oxygen

The classical procedure for forming the desired denuded zone and achieving intrinsic gettering utilizes a three-step thermal treatment (Nasagawa et al., 1980) demonstrated in Fig. 12a. The first step is the denudation at high-temperature which results in dissolution of the grown-in oxide precipitates and subsequent out-diffusion of oxygen from the wafer surface. Oxygen diffusivity above 1100°C is sufficiently high to result in a depleted zone of several tens of micrometers depth, while the oxygen concentration in the bulk far from the wafer surface retains its initial value. The shape of the transition from the solubility limit value at the surface to the bulk value is determined by the annealing time (Fig. 12b, $C_0 = 8 \times 10^{17}$ cm^{-3}).

The second step causes the nucleation of new oxide precipitates at temperatures around 700°C. During the third step, called the precipitation stage, nucleated oxide precipitates grow in the regions of sufficiently high oxygen concentration, while no oxide precipitates grow in the depleted region near the surface and the denuded zone is formed. The experimentally observed limit of oxygen concentration necessary for appreciable precipitation is around 6.6 - 7 $\times 10^{17}$ cm^{-3} (see Fig. 12b) and varies depending on the type and concentration of dopant species.

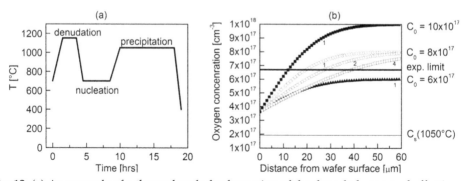

Fig. 12. (a) An example of a thermal cycle for formation of the denuded zone and effective intrinsic gettering. (b) The concentration of oxygen in the wafer with initial oxygen concentration C_0 denuded at 1150°C. Denuding duration in hours is noted in the plot below the curves. The thin horizontal line in (b) represents the equilibrium concentration of oxygen at the temperature of the precipitation step (1050°C); the thick solid line represents the experimental limit of oxygen precipitation. Out-diffusion profiles in panel (b) were calculated after (Andrews, 1983).

The depth profile of the precipitated oxygen in the silicon wafer can be calculated on the basis of the interstitial oxygen depth profile (Borghesi et al., 1995). As shown in Fig. 12b, the oxygen concentration after out-diffusion provides certain supersaturation even in the denuded zone, but the amount of precipitated oxygen depends more strongly on the density of nuclei than on the level of oxygen supersaturation during precipitation. This latter point is critical in making material stable to large thermal budget wafer processing. The nucleation rate during the second step is very strongly reduced for low oxygen concentrations in denuded zone, e.g., the rate was reduced by more than ten times for oxygen concentration reduced by 2×10^{17} cm^{-3} (Borghesi et al., 1995). As a result the region near the surface is precipitate-free and the bulk region is precipitate-rich. The transition between the bulk and the surface is not sharp; it follows the initial oxygen profile (see the dependence in Fig. 14 below).

7.2 Wafer annealing

It was found that high temperature annealing can help control and manage also other types of crystal defects, not simply oxide precipitates. The COP defects on the wafer surface can be annihilated during high temperature annealing in hydrogen (Nadahara et al., 1997), or in argon ambient (Adachi et al., 1998). As shown in Fig. 13, proper treatment of the polished wafer can significantly reduce or even eliminate COP defects on the surface of polished silicon wafers. However, attention has to be paid to oxygen precipitation in the bulk of the wafer, since during the high temperature annealing applied for annihilation of COPs the grown-in oxide precipitates tend to dissolve. Subsequent oxygen precipitation in the annealed wafer may become suppressed and the wafer can loose its gettering capability.

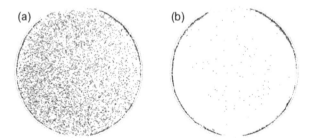

Fig. 13. COP distributions measured by a particle detector (lower dimensional detection limit is 0.13 μm) on vacancy-type wafers (a) before and (b) after hydrogen annealing (1180°C/2hr).

7.3 Optimization of oxygen concentration in the crystal

The denuded zone and intrinsic gettering are closely related to oxygen precipitation, which in turn strongly depends on the oxygen concentration in the wafer. Fig. 12 demonstrates the thickness of the denuded zone, and Fig. 14 shows the density of oxide precipitates, both as the function of oxygen concentration. As the oxygen concentration in silicon is determined during the crystal growth process, control of the oxygen concentration in the growing crystal is one of the primary tasks for crystal growers.

As was previously discussed, oxygen is incorporated into the crystal at the melt-crystal interface, while its concentration in the crystal relates to concentration in the melt. The concentration of oxygen in the melt below the crystal is determined by the source (crucible materials), by the effectiveness of the transport in the melt, and by the strength of the sink, i.e., by the dissolution rate of the silica crucible, by the melt flow, and by the evaporation from the melt surface. Critical process parameters driving these phenomena are the hot zone design, the crucible rotation rate, and gas atmosphere parameters. All of the parameters are usually optimized to reach a desired oxygen concentration in the whole crystal with limited variation and good reproducibility. Typical concentrations of oxygen in CZ silicon crystals range from about 11×10^{17} cm^{-3} down to about 6×10^{17} cm^{-3}; special designs of the hot zone (usually utilizing shields for gas flow control) allow decreasing oxygen concentration slightly below 5×10^{17} cm^{-3}.

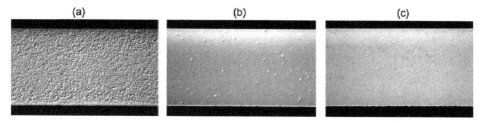

Fig. 14. Oxide precipitates delineated by preferential etching after 750°C/4hr + 1050°C/16hr annealing in lightly phosphorus-doped silicon wafers with oxygen content of (a) 8×10^{17} cm^{-3}, (b) 7×10^{17} cm^{-3}, (c) 6×10^{17} cm^{-3}.

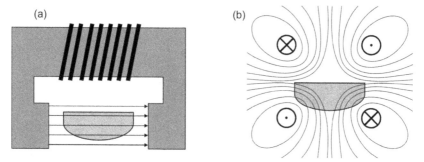

Fig. 15. Schematic drawing of (a) transverse and (b) cusp magnetic fields for control of the melt convection during crystal growth.

There are applications which require extremely low oxygen concentrations or very uniform distributions of oxygen in the crystal. These requirements are beyond the capability of most standard Czochralski crystal growth processes. Magnetic Czochralski process (MCZ) has to be used to achieve reduced concentrations of oxygen in the silicon crystals. The principal aspect of MCZ lies in the influence of the Lorentz force on the convection in the melt, which arises from the interaction of magnetic field with the fluid flow of the highly electrically conductive silicon melt. The effect of various kinds of magnetic fields has been studied extensively (Hurle, 1993). Based on many investigations, only transverse (horizontal) and cusp magnetic fields are used in industrial applications. The two configurations are schematically shown in Fig. 15. The horizontal magnetic field is used for growth of large diameter crystals, while for smaller diameter crystals the cusp field is applied. The two most important effects of the magnetic field generally consist in damping of micro-scale growth rate instabilities at the melt-crystal interface through damping of melt temperature fluctuations, and in the possibility of tuning the incorporation of impurities into the growing crystal. Application of the magnetic field adds additional degrees of freedom to the crystal growth process which widens the capability of the process, but simultaneously brings complexity related to the higher technological level of the process.

7.4 Nitrogen doping

Extremely low oxygen concentration allows for the formation of very deep denuded zones, but also results in negligible oxygen precipitation in the bulk. Oxygen precipitation is generally low also in heavily-doped n-type wafers due to enhanced evaporation of dopant atoms in the form of oxides. For applications requiring effective intrinsic gettering this drawback can be solved by nitrogen doping. Introducing nitrogen in silicon crystal results in enhanced precipitation of oxygen during subsequent wafer annealing due to stable nuclei formed during the crystal growth, an increased concentration of free vacancies available for oxygen precipitation and consumption of the interstitials emitted during the precipitate growth (von Ammon et al., 2001). Effective intrinsic gettering then can be achieved even at low oxygen concentrations or under conditions of unfavorable thermal treatments, such as argon or hydrogen annealing (Ikari et al.). The effect of nitrogen doping on oxygen precipitation is demonstrated in Fig. 16.

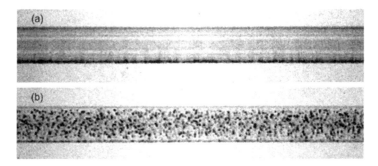

Fig. 16. An X-ray section topograph showing the effect of nitrogen doping on oxygen precipitation in heavily antimony-doped wafers with low oxygen concentration. (a) Standard wafer. (b) A wafer co-doped with nitrogen at the level of 5×10^{14} cm^{-3}. Wafer (a) is "defect-free". The horizontal Pendelösung fringes are clearly visible (see Section 6.2). Strain related to the polysilicon layer on the wafer backside (bottom) results in dark features in the topograph. A section topograph of wafer (b) shows oxide precipitates (dark spots) and residual damage on the wafer backside (bottom). The wafer thickness is 625 μm.

7.5 Vacancy-controlled denuded zone

An elegant solution to overcome the dependence of denuded zone parameters on the oxygen concentration was developed by Falster et al. (Falster et al., 1998). The concept of so-called "Magic Denuded Zone" (MDZ) utilizes the importance of vacancies for oxygen precipitation. The concentration profile of vacancies in the wafer is modified via rapid thermal annealing (RTA). Annealing at temperatures of about 1200°C forms a population of vacancies and interstitials in equal concentrations through the generation of Frenkel pairs. Point defects reach their equilibrium concentrations in a short period (a few tens of seconds) throughout the thickness of the thin wafer by diffusion to and from the surface. During cooling of the wafer to low temperatures, the equilibrium concentrations of point defects decreases, providing a driving force for enhanced diffusion and recombination. This favors diffusion to the surface, where equilibrium is maintained. However, when the cooling rate is fast enough, the diffusion is effective only in the near-surface region. In the bulk of the wafer, only point defect recombination takes place leaving behind a population of excess vacancies at a concentration given by $C_v^{eq}(T_a)$ – $C_i^{eq}(T_a)$, where T_a is the annealing temperature. The MDZ wafer finally contains an essentially vacancy-free surface and a transition region below the surface with steadily increasing vacancy concentrations up to the bulk level.

When an MDZ wafer passes through a high temperature anneal oxide precipitates nucleate and grow only in the region with sufficient vacancy concentrations which results in the denuded zone at the wafer surface. The depth of the Magic Denuded Zone can reach even deeper into the bulk than the classical denuded zone, but the solution has another major advantage– the result is practically independent of oxygen concentration in the wafer. Although the MDZ comprises an extra treatment of the wafer it suppresses the need for complicated oxygen control in CZ silicon crystals.

7.6 Optimization of v/G

Sometimes annealing of the wafer cannot meet the requirements of the wafer processing thermal budget. In such cases, formation of defects in the wafer can be adjusted by controlling the distribution of point defects in the growing crystal. As described above, vacancy-type or interstitial-type regions are formed in the crystal depending on the v/G parameter. The different regions within the crystal contain different types of defects and also substantially differ in oxygen precipitation characteristics. Tuning of the crystal growth rate v, and/or the temperature gradient G, can shift the V-I boundary and so influence defect formation.

There are often issues with excessive oxygen precipitation in heavily-boron doped wafers which results in electronic devices with excessive leakage currents. Analyses showed that the strong precipitation is constrained within the vacancy-type region in the portion of the crystal with the highest oxygen concentration (Valek et al., 2007, 2008). If the oxygen profile of the particular crystal growth process cannot be altered, then the crystal growth rate may be optimized to eliminate the vacancy-type region from the crystal (Fig. 17). The low growth rate shifts the v/G curve below the critical value ξ_t (see Fig. 6), the material becomes interstitial-type in full cross-section and oxygen precipitation is suppressed.

7.7 Computer simulations of crystal growth

Distribution of crystal defects in silicon wafers is significantly predetermined by the formation of defects during the crystal growth. Incorporation of point defects into the crystal and formation of crystal defects are driven by the time evolution of the temperature field, which governs the solubility and the diffusivity of the various species. However, measurement of the temperature field (necessary for its control) in the growing crystal is practically impossible. The analysis of grown-in defects after the crystal growth is very difficult due to their small size which is often below the detection limit of common techniques. Therefore, detailed numerical models have been developed to overcome these obstacles by computer simulations which help to describe and predict defect formation from the crystal growth up to the point of wafer processing and device manufacturing (Brown et al., 2001; Dupret & Van den Bogaert, 1994; Kalaev et al., 2003).

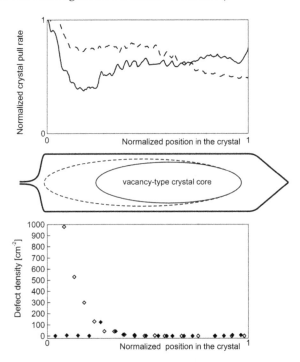

Fig. 17. An example of defect engineering through v/G control. The axial profile of the crystal pull rate (top), the distribution of point defects (middle), and the corresponding defect density (bottom) of the original (dashed lines, open symbols) and optimized (solid lines, solid symbols) crystal growth process.

Simulations of v/G are the simplest method for prediction of defect formation in the crystal. Comparing the simulated v/G values with the critical v/G value allows the identification of the vacancy-type and interstitial-type regions of the crystal separated by the vacancy-interstitial boundary. For example, a complicated distribution of OISFs on the oxidized surface of heavily boron-doped wafers was identified (Valek et al., 2007, 2008). The OISFs formed a pattern of two concentric rings (radii R_1 and R_2 in Fig. 18) on wafers from the beginning of the crystal. With increasing distance from the seed end of the crystal the OISF density decreased and the rings gradually disappeared.

In order to identify origin of the rings the temperature gradient, G, at the melt-crystal interface was modeled at several positions in the crystal using FEMAG code (Dupret & Van den Bogaert, 1994) and the corresponding radial v/G curves were calculated. Using the ξ_t values the V-I boundary was constructed. In Fig. 18 the V-I boundary was found to go through the whole crystal. It was concluded that also the OISF pattern should be formed throughout the whole crystal. This conclusion was supported by a special OISF test with pre-annealing of the wafers. This annealing was performed to enlarge the grown-in oxide precipitates to a supercritical size for the OISF test. Finally, the origin of the "strange" OISF pattern was explained in terms of oxygen concentration and thermal history, see (Valek et al., 2007, 2008), for more details. This work demonstrated the value of computer simulations of crystal growth as a necessary tool for defect engineering.

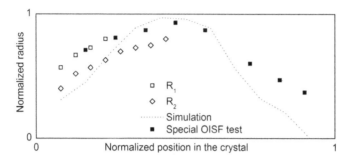

Fig. 18. Plot of the radius of the rings forming the OISF pattern, the simulated V-I boundary, and the radius of the OISF pattern delineated by a special OISF test (with pre-annealing).

8. Conclusion

Crystal defects are formed in silicon during the growth of the single crystals and processing of the wafers. Depending on their nature, density and size, these defects may substantially influence silicon material properties and consequently strongly impact process of device manufacturing. We provided the reader with a brief outlook of crystal defects in silicon and introduced the mechanisms of their formation. Knowledge of these mechanisms is the fundamental requirement for the art of defect engineering.

Crystal defects can be controlled from the early beginning by influencing their formation during the crystal growth. We discussed several aspects of controlling oxygen concentration in Czochralski silicon crystals, which is the basic method for control of oxygen precipitation in silicon wafers. Defect formation can be adjusted by optimization of the crystal growth process with respect to the v/G parameter. Tuning the crystal pull rate v and the hot zone design, which determines the temperature gradient G, one can produce either silicon containing vacancy-type defects and showing enhanced oxygen precipitation or silicon containing interstitial-type defects and showing suppressed oxygen precipitation. Sometimes it can be advantageous to modify the processes of defect formation by doping of the silicon crystal with an extra element like carbon or nitrogen to promote oxygen precipitation. Since the material properties obtained by application of the above mentioned methods may not be invariable or sufficient, it may be necessary to treat also the silicon wafer.

Another reason for treatment of the wafer can be an effort to modify properties of wafer surface. For example oxygen precipitation and also appearance of the vacancy-type defects, the COPs, can be controlled by high temperature annealing of polished silicon wafers in optimized ambient. In this manner a wafer which contains a defect-free near-surface denuded zone and an optimized defect density in the bulk of the wafer providing effective gettering of metal impurities can be obtained.

We showed that various methods of defect engineering can be utilized during the growth of silicon crystals and manufacturing of wafers to produce silicon wafers of desired quality. Properties of the wafers have to be optimized for specific requirements of individual types of electronic devices and integrated circuits. Following the development of electronic industry, engineering of crystal defects remains an inseparable and never-ending part of the silicon manufacturing technology.

9. Acknowledgment

This work was partially supported by grant FR-TI3/031 awarded by the Ministry of Industry and Trade of the Czech Republic. Critical reading of the manuscript by Dr. John M. Parsey, Jr., is highly acknowledged.

10. References

Abe, T.; Harada, H. & Chikawa, J. (1983). Swirl Defects in Float-Zoned Silicon Crystals, *Physica B+C*, Vol. 116, No. 1-3, (February 1983), pp. 139-147, ISSN 0378-4363

Adachi, N.; Hisatomi, T.; Sano, M.; & Tsuya, H. (1998). Reduction of Grown-in Defects By High Temperature Annealing, In: *Semiconductor Silicon 1998*, Huff, H.R.; Gösele, U.; Tsuya, H. (Eds.), pp. 698-706, The Electrochemical Society, ISBN 978-1-56677-193-1698, Penington, NJ, USA

Andrews, J. (1983). Oxygen Out-Diffusion Model for Denuded Zone Formation in Czochralski-Grown Silicon with High Intersitital Oxygen Content, In: Defects in Silicon, Murray Bullis, W. & Kimerling, L. C. (Eds.), pp. 133-141, The Electrochemical Society, Pennington, NJ

ASTM Standard F1188-02 (2003). *Standard Test Method for Interstitial Atomic Oxygen Content of Silicon by Infrared Absorption with Short Baseline*, ASTM International, West Conshohocken, PA, USA

ASTM standard F 1809–02 (2003), *Standard Guide for Selection and Use of Etching Solutions to Delineate Structural Defects in Silicon*, ASTM International, West Conshohocken, PA, USA

Borghesi, A.; Pivac, B.; Sassella, A. & Stella, A. (1995). Oxygen Precipitation in Silicon, *Journal of Applied Physics*, Vol. 77, No. 9, (May 1995), pp. 4169-4244. ISSN 0021-8979

Borionetti, G.; Gambaro, D.; Porrini, M. & Voronkov, V.V. (2002). In: *Semiconductor Silicon 2002*, Huff, H.R.; Fabry, L. & Kishino S. (Ed.), pp. 505-516, The Electrochemical Society, ISBN 1-56677-374-1, Penington, NJ, USA

Bullis, W.M. (1990). Silicon Material Properties, In: *Handbook of Silicon Technology*, O'Mara, W.C.; Herring R.P.; & Hunt, L.P., (Eds.), 347-450, Noyes Publications, ISBN 0-8155-1237-6, Park Ridge, NJ, USA

Brown, R.A.; Wang, Z. & Mori, T. (2001). Engineering Analysis of Microdefect Formation During Silicon Crystal Growth, *Journal of Crystal Growth*, Vol. 225, No. 2-4, (May 2001), pp. 97-109, ISSN 0022-0248

Czochralski, J. (1918). Ein neues Verfahren zur Messung des Kristallisationsgeschwindigkeit der Metalle, *Zeitschrift für Physikalische Chemie*, Vol. 92, (1918), pp. 219-221

Dupret, F. & Van den Bogaert, N. (1994). Modelling Bridgman and Czochralski growth, In: *Handbook of Crystal Growth*, Vol. 2B, Chapter 15, Hurle, D.T.J. (Ed.), 875-1010, North Holland / Elsevier, ISBN 0-444-81554-6, Amsterodam, Holand

Falster, R.; Gambaro, D.; Olmo, M.; Cornara, M. & Korb, H. (1998). The Engineering of Silicon Wafer Material Properties Through Vacancy Concentration Profile Control and the Achievement of Ideal Oxygen Precipitation Behavior, *Material Research Society Syposium Proceedings*, Vol. 510, pp. 27-36, ISSSN 0272-9172

Föll, H. & Kolbesen, B.O. (1975). Formation and Nature of Swirl Defects in Silicon. *APPLIED PHYSICS A: MATERIAL SCIENCE AND PROCESSING*, Vol. 8, No. 4, (December 1975), pp. 319- 331, ISSN 0947-8396

Graff, K. (2000). *Metal Impurities in Silicon-device Fabrication* (Second, Revised Edition), Springer-Verlag Berlin Heidelberg New York, ISBN 3-540-64213-7, Berlin, Germany

Ham, F.S. (1958). Theory of Diffusion-Limited Precipitation, *Journal of Physics and Chemistry of Solids*, Vol. 6, No. 4, (September 1958), pp. 335-351, ISSN 0022-3697

Hasebe, M. Takeoka, Y. Shinoyama S. & Naito S. (1989), Formation Process of Stacking Faults with Ringlike Distribution in CZ-Si Wafers, *Japanese Journal of Applied Physics*, Vol. 28, No. 11, (January 1980), pp. L1999-L2002, ISSN 0021-4922

Hirth, J.P. & Lothe, J. (1967). *Theory of dislocations*, McGraw-Hill, ISBN 0-521-86436-4, New-York, US

Hurle, D.T.J. (1993). *Crystal Pulling from the Melt*, Springer-Verlag Berlin-Heidelberg New York, ISBN 3-540-56676-7, Berlin, Germany

Ikari, A.; Nakai, K.; Tachikawa, Y.; Deai, H.; Hideki, Y.; Ohta, Y.; Masahashi, N.; Hayashi, S.; Hoshino, T. & Ohashi, W. (1999). Defect Control in Nitrogen Doped Czochralski Silicon Crystals, *Solid State Phenomena*, Vol. 69-70 (1999), pp. 161-166, ISSN ISSN 1662-9779

Itsumi, M. (2002). Octahedral Void Defects in Czochralski Silicon, *Journal of Crystal Growth*, Vol. 237-239, No. 3 (April 2002), pp. 1773-1778, ISSN 0022-0248

Kalaev, V.V.; Lukanin, D.P.; Zabelin, V.A.; Makarov, Yu.N.; Virbulis, J.; Dornberger, E. & von Ammon, W. (2003), *Materials Science in Semiconductor Processing*, Vol. 5, No. 4-5, (August-October 2003), pp. 369–373, ISSN 1369-8001

Keck, P.H. & Golay, M.J.E. (1953), Crystallization of Silicon from a Floating Liquid Zone, *Physical Review*, Vol. 89, (March 1953), pp. 1297

Lang, A. R. (1978), Techniques and interpretation in X-ray topography. In: *Diffraction and Imaging Techniques in Materials Science*, Amelinckx, S.; Gevers, R. & Van Landuyt, J. (Eds.), 2nd ed. rev. (1978), pp 623–714, North-Holland, ISBN 9780444851307, Amsterdam, Holand

Nadahara, S.; Kubota, H. & Samata, S. (1997). Hydrogen Annealed Silicon Wafer, *Solid State Phenomena*, Vol. 57-58 (1997), pp. 19-26, ISSN 1662-9779

Nagasawa, K.; Matsushita, Y. & Kishino, S. (1980). A New Intrinsic Gettering Technique Using Microdefects in Czochralski Silicon Crystal: A New Double Preannealing Technique, *Applied Physics Letters*, Vol. 37, No. 7, (October 1980), pp. 622-624, ISSN 0003-6951

O'Mara, W.C. (1990). Oxygen, Carbon and Nitrogen in Silicon, In: *Handbook of Silicon Technology*, O'Mara, W.C.; Herring R.P.; & Hunt, L.P., (Eds.), pp. 451-549, Noyes Publications, ISBN 0-8155-1237-6, Park Ridge, NJ, USA

Pichler, P. (2004). *Intrinsic Point Defects, Impurities, and Their diffusion in Silicon*, Springer-Verlag Wien New York, ISBN 3-211-20687-6, Wien, Austria

Ravi, K.V. & Varker, C.J. (1974). Oxidation-Induced Stacking Faults in Silicon. I. Nucleation phenomenon, *Journal of Applied Physics*, Vol. 45, No. 1, (January 1974), pp. 263-271, ISSN 0021-8979

Roksnoer, P.J. & van den Boom, M.M.B. (1981). Microdefects in a Non-Striated Distribution in Floating-Zone Silicon Crystals. *Journal of Crystal Growth*. Vol. 53, No. 3, (June 1981), pp. 563- 573, ISSN 0022-0248.

Rozgonyi, G.A.; Deysher, R.P. & Pearce, C.W. (1976). The Identification, Annihilation, and Suppression of Nucleation Sites Responsible for Silicon Epitaxial Stacking Faults, *Journal of The Electrochemical Society*, Vol. 123, No. 12, (December 1976), pp. 1910-1915, ISSN 0013-4651.

Ryuta, J.; Morita, E.; Tanaka, T. & Shimanuki, Y. (1990). Crystal-Originated Singularities on Si Wafer Surface after SC1 Cleaning, *Japanese Journal of Applied Physics*, Vol. 29, No. 10, (October 1990), pp. L1947-L1949, ISSN 0021-4922

Schröder, D.K. (1989). Lifetime in Silicon, In: *Gettering and Defect Engineering in the Semiconductor Technology*, Kittler, M. (Ed.), pp. 383-394, Sci-Tech, ISBN 3-908044-04-9, Vaduz, FL

Shimura, F. (1994). *Oxygen in Silicon*, Academic Press, ISBN 0-127-52142-9, London, UK

Sumino, K.; Harada, H. & Yonenaga, I. (1980). The Origin of the Difference in the Mechanical Strengths of Czochralski-Grown Silicon and Float-Zone-Grown Silicon, *Japanese Journal of Applied Physics*, Vol. 19, No. 1, (January 1980), pp. L49-L52, ISSN 0021-4922

Teal, G.K. & Little, J.B. (1950). Growth of Germanium Single Crystals, *Physical Review*, Vol. 78, (1950), pp 647

Válek, L.; Lysáček, D. & Šik, J. (2007). OISF Pattern and Grown-in Precipitates in Heavily Boron Doped Silicon, *Journal of The Electrochemical Society*, Vol. 154, No. 10 (August 2007), pp H904-H909, ISSN 0013-4651

Válek, L.; Šik, J. & Lysáček, D. (2008). Enhanced Oxygen Precipitation during the Czochralski Crystal Growth, *Solid State Phenomena*, Vol. 131-133, (2008), pp. 167-174, ISSN 1662-9779

Vanhellemont, J.; Senkader, S.; Kissinger, G.; Higgs, V.; Trauwaert, M.-A.; Graef, D.; Lambert, U. & Wagner, P. (1997). Measurement, Modelling and Simulation of Defects in As-grown Czochralski Silicon, *Journal of Crystal Growth*, Vol. 180, No. 3-4 (October 1997), pp. 353-362, ISSN 0022-0248

von Ammon, W.; Dornberger, E. & Hansson, P.O. (1999). Bulk properties of very large diameter silicon single crystals, *Journal of Crystal Growth*, Vol. 198/199, No. 1, (March 1999), pp. 390-398, ISSN 0022-0248

von Ammon, W.; Holzl, R.; Virbulis, J.; Dornberger, E.; Schmolke, R.; Graf, D. (2001). The Impact of Nitrogen on the Defect Aggregation in Silicon, *Journal of Crystal Growth*, Vol. 226, No. 1, (June 2001), pp. 19-30, ISSN 0022-0248

Voronkov, V.V. (1982). The Mechanism of Swirl Defects Formation in Silicon, *Journal of Crystal Growth*. Vol. 59, No. 3, (October 1982), pp. 625- 643, ISSN 0022-0248.

Voronkov, V.V. (2008). Grown-in defects in silicon produced by agglomeration of vacancies and self-interstitials, *Journal of Crystal Growth*, Vol. 310, No. 7-9, (April 2008), pp. 1307-1314, ISSN 0022-0248

Yamagishi, H.; Fusegawa, I.; Fujimaki, N. & Katayama, M. (1992). Recognition of D Defects in Silicon Single Crystals by Preferential Etching and Effect on Gate Oxide Integrity, *Semiconductor Science and Technology*, Vol. 7, No. 1A, (January 1992), pp A135-A140, ISSN 0268-1242

5

Growth and Characterization of Ytterbium Doped Silicate Crystals for Ultra-Fast Laser Applications

Lihe Zheng, Liangbi Su and Jun Xu
Shanghai Institute of Ceramics, Chinese Academy of Sciences,
P. R. China

1. Introduction

Diode-pumped solid-state lasers (DPSSL) have predominated over waveguide lasers and fiber lasers when considering the efficiency and operability since the first realization of laser-diode pumped Yb-doped laser at room temperature (Lacovara et al., 1991). As a rule of thumb, DPSSL are preferable for devices operating with high peak power, whereas low-threshold and high-gain operation is much easier to be achieved with waveguide lasers and amplifiers. Besides the application in the fields of double-frequency, remote sensing and biomedical, ultra-fast DPSSL with diversified wavelength and stable system is widely exploited in the fields of mechanics, micro-electrics and ultra-fast photo-communication. DPSSL are composed of laser resonator which is mostly formed with discrete laser mirrors placed around gain medium with an air space in between. Bulk crystals or glasses doped either with rare earth ions or transition-metal ions are adopted as gain medium. With the development of DPSSL industries, the demand for laser crystals with the advantageous physicochemical properties such as efficient energy absorption, high optical uniformity and favorable thermal behavior has dramatically increased over the past few decades (Keller, 2003). With the rapid development of InGaAs laser diodes emitting from 900nm to 980nm, Yb^{3+} doped laser crystals are expected to alternate the traditional Nd^{3+} doped for generating efficient broad tunable and ultra-fast DPSSL in near-IR spectral range (Krupke, 2000).

Yb^{3+} ion with simple quasi-three energy level scheme of $^2F_{7/2}$ and $^2F_{5/2}$ is provided with high quantum efficiency, long lifetime of metastable $^2F_{5/2}$ level and large crystal-field splitting which is beneficial for reducing thermal load and enhancing Yb^{3+} doping level bringing about the realization of compact device without luminescence quenching caused by cross relaxation and excited-state absorption (Giesen & Speiser, 2007; Pelenc et al., 1995). However, the strong re-absorption at emission wavelengths leads to high pump threshold since the thermally populated terminal level of Yb^{3+} lasers at the ground state manifold is contemporary the laser terminal level. To reduce the re-absorption losses at laser emission wavelengths, strong splitting of ground sublevels of $^2F_{7/2}$ in Yb^{3+} ion is required to form a quasi-four-level system as that of Nd^{3+}. Thus, laser crystal hosts with low symmetry structure and strong crystal field splitting are the central issues in exploiting new Yb doped gain media (Du et al., 2006). Crystal hosts such as aluminum, tungsten, oxides, fluorides and vanadates were explored for diode-pumped ultra-fast lasers (Uemura & Torizuka, 2005; Liu et al., 2001; Griebner et al., 2004; Su et al., 2005; Kisel et al., 2005).

The chapter is devoted to the systematical investigation on Yb doped oxyorthosilicate crystals such as gadollinium silicate (Gd$_2$SiO$_5$, GSO), yttrium silicate (Y$_2$SiO$_5$, YSO), Lutetium silicate (Lu$_2$SiO$_5$, LSO) and scandium silicate (Sc$_2$SiO$_5$, SSO) obtained by the Czochralski Crystal Growth System with Automatic Diameter Control, which encompassing distinctive low symmetry monoclinic structure, excellent physicochemical properties and favorable spectroscopic features for DPSSL. Besides, the chapter summarizes the structure properties of the obtained silicate crystals. Afterwards the chapter discusses the optical properties of silicate crystals available for ultra-fast lasers, together with the calculation of spectroscopic parameters such as pump saturation intensities I_{sat}, minimum pump intensities I_{min} and gain spectra of laser medium. Finally, the laser performance of the studied silicates is briefly outlined.

2. Experimental details of silicate crystals

The basic properties of rare earth oxyorthosilicate crystals were presented in Table 1*. Silicate Crystals were obtained by Czochralski method. Structure and spectra method were outlined as well as the laser experiments in the following elaboration.

2.1 Basic properties of rare earth oxyorthosilicate crystals

The monoclinic orthosilicate crystals RE$_2$SiO$_5$ could be stably formed according to the binary phase diagram of RE$_2$O$_3$-SiO$_2$, where RE stands for Lu^{3+}, Gd^{3+}, Y^{3+} and Sc^{3+}. RE^{3+} ions occupy two different low symmetry crystallographic sites which Yb^{3+} ion could substitute with selectively. The Rare Earth silicates with larger Rare Earth ion radius from La^{3+} to Tb^{3+} manifest the space group of P2$_1$/c with typical compounds of La$_2$SiO$_5$ and GSO, while those with smaller Rare Earth ion radius from Dy^{3+} to Lu^{3+} as well as Y^{3+} and Sc^{3+} hold the space group of C2/c with typical compound of YSO and LSO.

Researchers show solitude for the crystal growth, structure, opto-electrical properties on Rare Earth doped RE$_2$SiO$_5$ crystals (Eijk, 2001; Kuleshov et al., 1997; Melcher & Schweitzer, 1992; Ivanov et al., 2001). Table 1 reveals the comparison of the basic physicochemical properties of monoclinic silicate crystals (Gaume, 2002; Smolin & Tkachev, 1969; Camargo et al., 2002). As seen from Table 1, LSO, YSO and SSO crystals are with Monoclinic C2/c structure occupying two different low symmetric and distorted crystallographic sites which would provide strong crystal field for Yb^{3+} ions with quasi-four-level laser operation scheme. SSO crystal retains the highest thermal conductivity among silicate crystals together with the striking negative calorescence coefficient which is crucial for laser operation regarding laser power resistance (Petit, 2005). The crystallographic sites, coordination and mean distance of RE-O in RE$_2$SiO$_5$ crystals are extended for ongoing discussion on energy level splitting as shown in Table 2 (Gaume et al., 2003; Ellens et al., 1997).

2.1.1 Lu$_2$SiO$_5$

LSO is a positive biaxial crystal with n$_Y$ axis along the b direction and two other axis a and axis c lying in plane (010) (Wang, 2004). The structure of monoclinic LSO crystal with isolated ionic SiO$_4$ tetrahedral units and non-Si-bonded O atoms in distorted OLu$_4$ tetrahedron were determined by neutron diffraction (Gustafsson et al., 2001). The OLu$_4$ tetrahedron form edge-sharing infinite chains and double O_2Lu_6 tetrahedron along the c axis. The edge-sharing chains are connected to O_2Lu_6 double tetrahedron by isolated SiO_4 units.

Crystal Molecular Formula	LSO	GSO	YSO	SSO	YAG
Space group/simple	C2/c	P2$_1$/c	C2/c	C2/c	Ia3d
Site symmetry	C_1	C_s+C_{3v}	C_1	C_1	D2
Lattice constant (Å)	a=10.2550Å, b=6.6465Å, c=12.3626Å, β=102.422°	a=9.132Å, b=7.063Å, c=6.749Å, β=107.56°	a=10.41Å, b=6.72Å, c=12.49Å, β=102.65°	a=9.961Å, b=6.429Å, c=12.03Å, β=103.8°	a=12.01 Å
Density / g·cm^{-3}	7.4	6.7	4.44	3.52	4.56
Melting point / °C	2100	1950	2000	1920	1950
Mohs hardness	5.8	5.5	5.6	-	8.5
Thermal conductivity undoped@30°C/W·m^{-1}·K^{-1}	4.9	4.5	4.4	7.5	10.7
dn/dT(*10^{-6} K^{-1})	-	-	7.2	-6.3	7.9
Refractive indices	1.82	1.85	1.782(a) 1.785(b) 1.806(c)	n_x=1.82, n_y=1.84, n_z=1.86	1.83

Table 1. Physicochemical property of monoclinic silicate crystals.

Host	Label	Coordination number	Re^{3+} (Å)	Re–O$_{mean}$ (Å)
LSO	Lu(I)	7	1.001	2.277
	Lu(II)	6		2.229
GSO	Gd(I)	7	1.05	2.49
	Gd(II)	9		2.39
YSO	Y(I)	7	0.910	2.3080
	Y(II)	6		2.2520
SSO	Sc(I)	6	0.885	2.2289
	Sc(II)	6		2.2149

Table 2. Crystallographic sites, coordination and mean distance in RE-O in RE$_2$SiO$_5$ crystal.

2.1.2 Gd$_2$SiO$_5$

As seen from Table 1, Gd^{3+} in monoclinic GSO crystal is coordinated with oxygen atoms of 7 and 9, respectively (Fornasiero et al., 1998). The Gd(I) coordinated with 7 oxygen is linked with three isolated oxygen ions and three [SiO^{4-}] ions. The Gd(II) coordinated with 9 oxygen is bonded with one isolated oxygen and six tetrahedral [SiO^{4-}] ions. As shown in Table 2, the average distance of 2.39 Å in Gd(I)–O is shorter than that of 2.49 Å in Gd(II)-O (Felsche, 1973) The symmetry and intensity of crystal field would be affected by the coordination field after doping with active Rare Earth ions into the host (Suzuki et al., 1992; Cooke et al., 2000).

2.1.3 Y$_2$SiO$_5$

The distorted crystallographic sites of Y^{3+} in YSO crystal are with coordination number of 6 and 7 which is similar to those in LSO. Crystallographic site Y(I) forms polyhedron with 5 oxygen from that in tetrahedron (SiO$_4$)$^{4-}$ and 2 isolated oxygen. Crystallographic site Y(II) comprises reticular formation of pseudo octahedron with 4 oxygen from that in tetrahedron (SiO$_4$)$^{4-}$ and 2 isolated oxygen. The reticular formation of OY$_4$ is composed of edge-sharing infinite chains along c axis together with the network structure of OSi$_4$ (IEM Databases and Datasets).

2.1.4 Sc$_2$SiO$_5$

The structure of SSO crystal is analogous to those of YSO and LSO, with monoclinic structure and space group of C2/c. Although the large sized Yb:SSO crystal is difficult to obtain due to crystal diameter controlling, the impressive thermo-mechanical properties in SSO crystal make it an excellent performer in high power laser applications (Ivanov, 2001; Gaume et al., 2003; Campos et al., 2004). As indicated in Table 1, SSO crystal possesses a much stronger thermal conductivity of 7.5W•m^{-1}•K^{-1} than that in YSO (4.4W•m^{-1}•K^{-1}), LSO (4.5 W•m^{-1}•K^{-1}) and GSO (4.9 W•m^{-1}•K^{-1}). Furthermore, SSO crystal preserves the characterization of minus calorescence coefficient dn/dT of -6.3×10^{-6} K^{-1} which is quite different from that of YSO (7.2×10^{-6} K^{-1}) and favorable for high power laser operation.

2.2 Crystal growth

Czochralski method (Cz) is one of the major crystal growth methods of obtaining bulk single crystals with high optical quality and fast growth rates from melt for commercial and technological applications. Cz is named after Polish scientist Jan Czochralski, who discovered the method in 1916 while investigating the crystallization rates of metals.

Yb doped silicate crystals such as Yb:LSO, Yb:GSO, Yb:YSO and Yb:SSO studied in this chapter were grown by the Cz method in inductively heated crucibles under inert atmosphere of 5N nitrogen. The starting materials were Rare-Earth Oxide (Lu$_2$O$_3$, Gd$_2$O$_3$, Y$_2$O$_3$ or Sc$_2$O$_3$), Yb$_2$O$_3$ and SiO$_2$ powders with purity higher than 99.995%. The powders were mixed and pressed into tablets followed by sintering at 1400°C for 24 h before loading into the iridium crucible. The chemical reaction is shown in Equation (1). The doping level of Yb ions in the melt was 5at.% with respect to that of Rare-Earth ions in the crystal host. Accordingly the compound formulae of the crystals could be written as (RE$_{0.95}$Yb$_{0.05}$)$_2$SiO$_5$.

$$(1-x)RE_2O_3 + SiO_2 + xYb_2O_3 \rightarrow RE_{2(1-x)}Yb_{2x} SiO_5 \qquad (1)$$

A precisely and oriented rod-mounted seed crystal with diameter of 4.5mm and length of 40mm was introduced for growth. The seeds for Yb:LSO, Yb:YSO, Yb:SSO were oriented along b-axis while that for Yb:GSO crystal was along [100]. The seed crystal was pulled upwards around 0.8-3mm·h^{-1} and rotated simultaneously at 10-30rpm. To keep convex solid–melt interface was important in growing silicate crystals to eliminate engendered dislocations. Temperature gradients and velocity fields were accurately controlled to gain stable settings (Zheng et al., 2007).

Automatic diameter control (ADC) with computer control system and weight sensor or so-called load cell is applied to detect the weight during crystal growth process which is wholesome for increasing crystal yield as well as reducing thermal stress. Strain gauge weight sensor with sufficient resolution was employed comparing with the total weight of obtained crystals. The input signal of weight sensor via an A/D converter is collected t o calculate the diameter of the generative crystal. A proportional–integral–derivative controller (PID controller) was adopted to better control the loop feedback during temperature regulation.

The difficulty in controlling diameter is originated from encapsulated materials and the high temperature atmosphere which would initiate diameter fluctuation bringing about dislocations besides irregular crystal shape and polycrystallization. The diameter deviation is regulated by controlling the temperature gradients where the double layered zircon cover was designed to optimize the temperature gradients. The heater temperature was increased to reduce the diameter when the measured diameter is larger than expected and vice versa. Crystal boules were finally obtained as shown in Fig. 1.

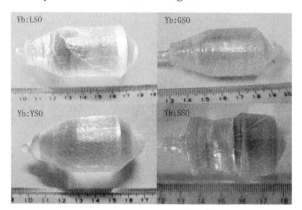

Fig. 1. Bulk crystals of 5at.%Yb doped silicate crystals obtained by Czochralski method.

2.3 Segregation coefficient characterization in silicate crystals

The segregation coefficient of Yb ions in silicate crystal hosts was measured by the inductively coupled plasma atomic emission spectrometer (ICP-AES). Crystal samples adjacent to the seed crystal position were cut and ground into fine powder in an agate mortar. The results of ICP-AES analysis are shown in Table 3.

Sample	seed crystal	Yb^{3+} c_0 (at.%)	Atom content (wt.%)					Yb^{3+} c_{top} (at.%)	k_m
			Yb	Lu	Gd	Y	Sc		
Yb:LSO	LSO	5.0	3.31	70.76	0	0	0	4.515	0.903
Yb:GSO	GSO	5.0	3.06	0	76.22	0	0	3.52	0.704
Yb:YSO	YSO	5.0	5.22	0	0	50.06	0	5.085	1.017
Yb:SSO	SSO	5.0	7.19	0	0	0	36.9	4.82	0.964

Table 3. Segregation coefficients of Yb ions in silicate crystals.

As acquired from Table 3, the segregation coefficient of Yb is 0.903 in Yb:LSO, 0.704 in Yb:GSO, 1.02 in Yb:YSO and 0.964 in Yb:SSO. The solubility of Yb ion in the LSO, YSO and SSO host lattice is higher than that in GSO indicating that Yb ions are liable to incorporate into the crystals with structure of C2/c comparing with that of P2₁/c. Meanwhile, the congenial radius of Yb^{3+} (0.868Å) with Lu^{3+}(1.001Å), Y^{3+} (0.910Å) and Sc^{3+} (0.885Å) makes ideal adulteration squaring up that of the Gd^{3+} (1.05Å).

2.4 X-Ray diffraction measurement

The crystal structure of silicate crystals were characterized by powder X-ray diffraction (XRD; Model D/Max 2550V, Rigaku Co., Tokyo, Japan) using Cu Kα radiation (λ = 0.15418 nm). The XRD patterns were inspected using PCPDF software package. Fig. 2 presents the XRD pattern for Yb doped silicate crystals which demonstrated that Yb:LSO, Yb:YSO and Yb:SSO crystals would maintain the primitive monoclinic structure with space group of C2/c, while Yb:GSO maintain that of $P2_1/c$.

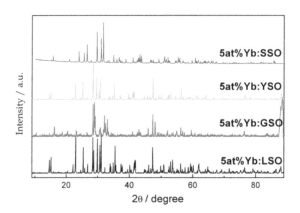

Fig. 2. Comparison of the XRD pattern for Yb doped silicate crystals.

2.5 Absorption and emission spectra

The unpolarized absorption spectra of silicate crystals were measured by JASCO Model V-570 UV/VIS/NIR spectrophotometer at a resolution of 1nm in the range between 860-1100nm with Xe light as pump source.

Fluorescence spectra were measured at a resolution of 1nm from 950nm to 1150nm by TRIAX 550-type spectrophotometer (Jobin-Yvon Company) with 940nm laser pumping source. The fluorescence lifetime with sample thickness of 1mm was pumped with a Xenon lamp and detected with an S-photomultiplier tube, while the data of emission decay curve was collected by a computer-controlled transient digitizer simultaneously.

2.6 Laser experiment

Kerr-lens mode-locked is well-developed for ultrafast pulses in efficient and compact lasers which is initiated by the self-focusing effect yielded from the nonlinear refractive-index variation in laser crystal. Diode-pumped mode-locked Yb:LSO laser with a W-typed cavity was developed based on Kerr-lens effect. The experimental setup for mode-locked Yb:LSO lasers was represented in Fig. 3. The maximum output power of the fiber-coupled diode-pumped semiconductor laser reached 30W around the emitting wavelength of 978 nm. The radius and the numerical aperture of the fiber were with 200 μm and 0.22, respectively. Yb:LSO host was cut into the dimension of 3×3×2 mm³ and end-coated with antireflection at lasing wavelength of 1030-1080 nm and pump wavelength of 978 nm. Yb:LSO wrapped with indium foil was placed with a small angle to suppress the Fabry-perot etalon effect and mounted in a water-cooled copper block with temperature maintained at 14 °C.

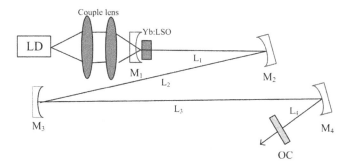

Fig. 3. Schematic of the Kerr-lens Yb: LSO laser.

Laser experiments on Yb:GSO crystal was performed with aperture of 5×6 mm^2 and length of 5 mm. The crystal was end pumped by a 30 W fiber-coupled laser diode with emission centered at 974 nm. The diameter and numerical aperture of the fiber core were 400 µm and 0.22, respectively. The laser cavity consisted of two mirrors M1 and output coupler OC, where dichroic mirror M1 was with anti-reflection coating at 974 nm and high-reflection coating in a broadband of 1020-1120 nm besides OC with various transmission.

The schematic of the Yb:SSO laser is shown in Fig. 4. The b-cut Yb:SSO crystal is with dimensions of 2.2 × 3 × 3 mm^3. The crystal was coated at both the pump and laser wavelengths with high transmission coatings to minimize the Fresnel reflection losses. The crystal was titled to about 6° with respect to the principle axis of the cavity to suppress etalon effects and improve the stability of the mode-locking operation. A single emitter diode with central wavelength at (975 ± 3) nm was used as the pump source.

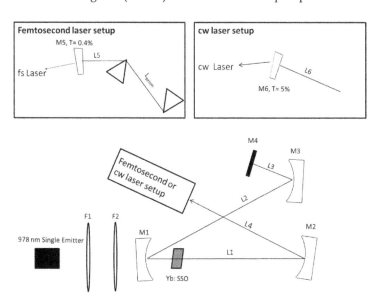

Fig. 4. Schematic of the experimental setup for Yb:SSO laser.

3. Spectra analysis of silicate crystals

Absorption and fluorescence spectra in Yb doped silicate crystals were presented. Based on the above spectra, key parameters were calculated for predicting the laser performance of the studied hosts.

3.1 Absorption and fluorescence spectra

Fig. 5 showed the absorption spectra of Yb doped silicate crystals with main absorption peak located around 920nm and 978 nm which is suitable for InGaAs diode pump.

The absorption bands were corresponding to the typical transitions from ground state $^2F_{7/2}$ to the sublevels of $^2F_{5/2}$. The absorption peak around 978nm belongs to the zero-line transition between the lowest sublevels of $^2F_{7/2}$ and $^2F_{5/2}$ manifolds. Broad absorption bands are beneficial for enhancing the efficiency of diode-pump operation since the laser diodes are typically emitting at a wide spectral range of 5 nm and presenting a thermal shift at the peak wavelength. As manifested in Fig. 5, the broad fluorescence behavior was associated with the typical transitions from the lowest level of $^2F_{5/2}$ to the sublevels of $^2F_{7/2}$ manifold. The emission band above 1080nm was due to transition from the lowest levels of $^2F_{5/2}$ manifold to the highest levels of $^2F_{7/2}$ manifold.

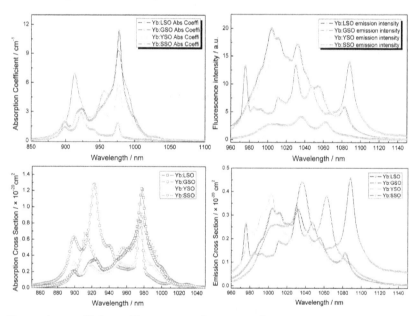

Fig. 5. Absorption coefficients, Fluorescence Intensity, Absorption cross section and emission cross section of Yb doped silicate crystals.

$$\sigma_{abs} = \frac{\alpha}{N} \tag{2}$$

$$\sigma_{em}(\lambda) = \frac{\lambda^5 I(\lambda)}{8\pi n^2 c \tau_{rad} \int \lambda I(\lambda) d\lambda} \tag{3}$$

Parameter	5 at. % Yb:LSO	5 at. % Yb:GSO
Main absorption peaks λ_{abs} (nm)	899, 924, 977	897, 922, 940, 976
Absorption bandwidth (nm)	13, 25, 15	17, 26, 24, 10
Absorption cross-section $\sigma_{abs}(10^{-20}\ cm^2)$	0.21, 0.36, 1.3	0.33, 0.60, 0.39, 0.51
Fluorescence peak (nm)	1004, 1012, 1032,1055, 1083	976, 1011, 1030, 1047, 1088
Emission cross-section σ_{em} $(10^{-20}\ cm^2)$	0.34, 0.32, 0.33, 0.21, 0.14	0.17, 0.45, 0.31, 0.46
Emission bandwidth (nm)	73	72
Fluorescence lifetime (ms)	1.68 @ 1083nm	1.56 @ 1088nm
Parameter	5 at. % Yb:YSO	5 at. % Yb:SSO
Main absorption peaks λ_{abs} (nm)	899, 917, 950, 977	914, 956, 976
Absorption bandwidth (nm)	15, 24, 31, 13	18, 19, 24
Absorption cross-section $\sigma_{abs}(10^{-20}\ cm^2)$	0.31, 0.28, 0.32, 0.64	0.67, 0.49, 0.91
Fluorescence peak (nm)	980, 1003, 1040, 1056, 1081	1006, 1036, 1062, 1087
Emission cross-section σ_{em} $(10^{-20}\ cm^2)$	0.24, 0.39, 0.23, 0.18, 0.12	0.26, 0.44, 0.38, 0.1
Emission bandwidth (nm)	48	57
Fluorescence lifetime (ms)	1.74 @ 1081nm	1.64 @ 1035nm

Table 4. Spectroscopic parameters of Yb doped silicate crystals.

Fig. 5 also revealed the calculated σ_{abs} of Yb^{3+} according to Equation (2) where α means the absorption coefficient and N presents the concentration of Yb^{3+} ions. σ_{em} of Yb^{3+} in silicate hosts explicated in Fig. 5 were calculated by Füchtbauer-Ladenburg formula referring to Equation (3) where τ_{rad} stands for the radiative lifetime of the excited manifold $^2F_{5/2}$ of Yb^{3+}, $I(\lambda)$ for the emission intensity at wavelength λ, n for refractive index and c for light velocity (Caird et al., 1991; Deloach et al., 1993). σ_{em} is fit for describing available gain at given inversion density. The lifetime of excited manifold $^2F_{5/2}$ was measured to be 1.68ms, 1.56ms, 1.74ms and 1.64ms in Yb:LSO, Yb:GSO, Yb:YSO and Yb:SSO crystals respectively. The lifetime is much longer than that of 0.8ms in Yb:YAG. Table 4 outlined the spectroscopic parameters of Yb^{3+} in silicate hosts based on absorption and emission spectra.

The large overlapping between absorption and emission spectra was observed in Fig. 5 due to the transitions among the multiplets of $^2F_{7/2}$ and $^2F_{5/2}$ manifolds of Yb^{3+}. The emission bandwidth FWHM of Yb:LSO crystal is appropriately 73nm which is much larger than that of Yb:YSO and Yb:GSO crystal. Although Yb:LSO possessed the largest σ_{em} around 1004nm, the reabsorption losses was consequently strong which would detrimentally affect laser action. σ_{em} of the terminal laser level at 1083nm was with 1.4×10^{-21} cm^2 containing the smallest thermal populating as well as the least re-absorption losses. In the case of 1032nm, the emission cross section is large enough to obtain low threshold and high efficient laser operation.

The absorption spectrum in Yb:GSO crystal was mainly composed of four strong bands centered at 897 nm, 922 nm, 940 nm and 976 nm. σ_{abs} at 922 nm and 976 nm were estimated as 6.0×10^{-21} and 5.1×10^{-21}, cm^2 with the absorption bandwidths of 26 nm and 10 nm, respectively. The emission spectrum mainly included four bands centered at 1011 nm, 1030 nm, 1047 nm and 1088 nm with fluorescence lifetime of 1.56 ms at 1088 nm which would be treated as radiative lifetime since the re-absorption loss could be negligible at 1088 nm. The σ_{em} at 1088 nm were 4.2×10^{-21} which was with the strongest cross section as well as the lowest pumping threshold. Referring to the spectra in Yb:YSO crystal, the strongest absorption peak was located at 978nm with furthest and weakest emission band centered at 1080 nm. The strongest σ_{em} centered at 1003 nm was severely affected by strong re-absorption losses. Yb^{3+} ions in Yb:SSO crystal show an inhomogeneous broad absorption band with peak absorption coefficient of 9.43cm^{-1} at 976nm which is higher than that in Yb:GSO crystal. The furthest σ_{em} centered at 1087 nm was with 1.0×10^{-21}, cm^2 in Yb:SSO which was at the same level of that in Yb:LSO and Yb:YSO. In conclusion, Yb:SSO with wide spectra bandwidth and long lifetime is favorable for high-efficiency ultra-fast diode-pump lasers.

3.2 Gain cross-section σ_g and minimum pump intensity I_{min}

The pump threshold and energy extraction would affect the application of Yb doped crystals on laser devices. Gain cross section σ_g was calculated according to Equation (4) to predict the laser performance (Deloach et al., 1993).

$$\sigma_g = \beta \, \sigma_{em} - (1-\beta) \, \sigma_{abs} \qquad (4)$$

$$I_{sat} = h\upsilon / (\sigma_{abs} \cdot \tau_{em}) \qquad (5)$$

The inversion coefficient β is defined as the population ratio on the $^2F_{5/2}$ level over the total Yb^{3+} ions population density. σ_g is equal to the emission cross-section σ_{em} when β is equal to 1. To effectively assemble the diode-pumped Yb-laser systems, a substantial fraction of Yb ions on the ground state should be driven to the excited state in order to achieve adequate gain and to overwhelm the absorption losses. As indicated in Equation (5), Pump saturation intensity I_{sat} would be reduced along with stronger σ_{abs} and longer emission life time τ_{em}. The excitation fraction of Yb ions is given by I_{abs}/I_{sat} where I_{abs} stands for the absorbed pump power. Long τ_{em} would promote accumulation of population inversion at fixed power as a complement of the peak-power-limited InGaAs diode sources.

The minimum excitation fraction β_{min} of Yb^{3+} ions calculated by Equation (6) must be balanced for laser output (Moulton, 1983). The parameter of minimum pump intensity I_{min}, obtained by Equation (7), represents the minimum pumping intensity required to invert population in a quasi-three-level system. The fraction I_{abs}/I_{sat} of excited Yb ions equalling to the required fractional excitation of β_{min} permits the equation of I_{min} and I_{abs}. Hence, I_{min} may be interpreted as the absorbed pump intensity required to surpass threshold in a lossless oscillator or to balance at selected extraction wavelength λ_{ext} in amplifier configuration.

$$\beta_{min} = \frac{\sigma_{abs}(\lambda_{ext})}{\sigma_{ext}(\lambda_{ext}) + \sigma_{abs}(\lambda_{ext})} \quad (6)$$

$$I_{min} = \beta_{min} \cdot I_{sat} \quad (7)$$

Key spectroscopic parameters are interpreted for designing high power rare-earth lasers. Strong σ_{abs} permits low Yb^{3+} ion doping level which in turn leads to a lower resonance absorption loss and a higher small-signal gain at given pump flux. I_{min} was accounted as a figure-of-merit to evaluate the potential Yb doped gain media for laser application by describing the suitable inversion density. Altogerther, I_{sat} as well as the minimum pump intensities I_{min} would be deduced from the absorption cross sections σ_{abs}.

3.2.1 Spectra parameters on Yb: LSO crystal

Fig. 6 showed the gain cross-section with various inversion ratio β. The emission spectra extending from 1000nm to 1092nm at β values of 0.5 was beneficial to laser output. The minimum pump intensities I_{min} at the selected extraction wavelengths, as well as the saturable pump density I_{sat} and the β_{min} under different laser output were also presented.

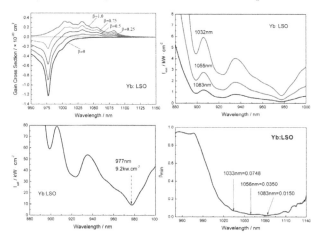

Fig. 6. σ_g, I_{min}, I_{sat} and β_{min} in Yb:LSO crystal.

3.2.2 Spectra parameters on Yb: GSO crystal

Fig. 7 presented the gain cross section σ_g of Yb:GSO crystal at β values of 0, 0.25, 0.5, 0.75 and 1. The emission spectra extending from 999nm to 1150nm at β values of 0.5 was beneficial to ultra-short pulse laser output. The minimum pump intensities I_{min} of 1.9kW/cm², 0.71kW/cm² and 0.094kW/cm² were achieved for selected extraction wavelengths at 1030nm, 1048nm and 1088nm as shown in Fig. 7.

The pump threshold power at the furthest wavelength of 1088nm was much lower than that of 0.13kW/cm² at 1083nm in Yb:LSO crystal which predicted low threshold and high efficient laser output in Yb:GSO crystal. Fig. 7 indicated the saturable pump density I_{sat}. The β_{min} at various laser output wavelength in Yb:GSO crystal was shown in Fig. 7 and the comparatively lowest was located at 1088nm with the β_{min} value of 0.0071 which was half less than that of 0.015 at 1082nm in Yb:LSO crystal.

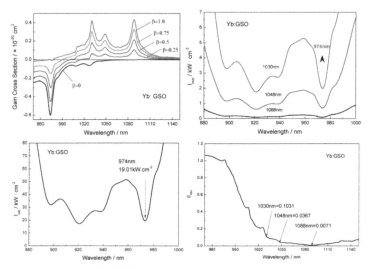

Fig. 7. σ_g, I_{min}, I_{sat} and β_{min} in Yb:GSO crystal.

3.2.3 Spectra parameters on Yb: YSO crystal

Fig. 8 presented the gain cross section of Yb:YSO crystal at β values of 0-1. The emission spectra stretched from 990nm to 1100nm at β values of 0.5. The minimum pump intensities I_{min} of 1.38kW/cm², 0.52kW/cm² and 0.2kW/cm² were achieved for selected extraction wavelengths at 1035nm, 1056nm and 1081nm as shown in Fig. 8. Fig. 8 indicated the saturable pump density I_{sat} in Yb:YSO crystal as well as the β_{min} at various laser output wavelength. The lowest β_{min} was located at 1081nm with the value of 0.015 which was almost the same as that in Yb:LSO crystal.

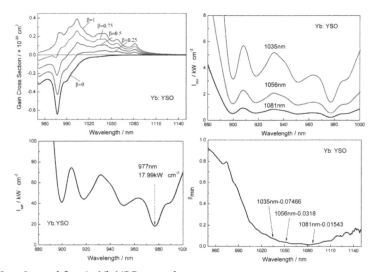

Fig. 8. σ_g, I_{min}, I_{sat} and β_{min} in Yb:YSO crystal.

3.2.4 Spectra parameters on Yb: SSO crystal

Fig. 9 presented the gain cross section of Yb:SSO crystal at β values of 0-1. The emission spectra elongated from 990nm to 1120nm at β values of 0.5. The minimum pump intensities I_{min} of 0.261kW/cm², 0.077kW/cm² and 0.2033kW/cm² were achieved for selected extraction wavelengths at 10356nm, 1062nm and 1087nm as shown in Fig. 9.

Fig. 9 indicated the saturable pump density I_{sat} in Yb:SSO crystal. The $β_{min}$ at various laser output wavelength in Yb:SSO crystal was shown in Fig. 9. The lowest $β_{min}$ was located at 1087nm with the value of 0.007148 which was almost half of that in Yb:LSO crystal. As concluded in Table 5, the laser parameters of Yb doped silicate crystals were listed and compared with that of Yb:YAG crystal (Haumesser et al., 2002).

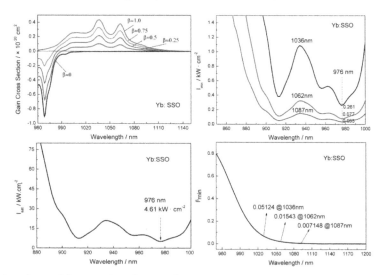

Fig. 9. $σ_g$, I_{min}, I_{sat} and $β_{min}$ in Yb:SSO crystal.

Host	$λ_{pump}$ (nm)	$β_{min}$	I_{sat} (kW/cm²)	I_{min} (kW/cm²)	$λ_{ext}$ (nm)	$σ_{ext}$ (10⁻²⁰cm²)
Yb:LSO	977	0.015	9.2	0.138	1083	0.14
Yb:GSO	974	0.0071	19.01	0.093	1088	0.46
Yb:YSO	977	0.0015	17.99	0.18	1081	0.12
Yb:SSO	976	0.0071	4.61	0.033	1087	0.1
Yb:YAG	942	0.055	28.8	1.53	1030	2.2

Table 5. Calculated Laser parameters of Yb doped Crystals.

3.3 Energy level of Yb in silicate crystal hosts

The energy level schemes for Yb doped silicate crystals were determined by optical spectroscopic analysis and semi-empirical crystal-field calculations using the simple overlap model. The stark levels distributed in the two $^2F_{7/2}$ and $^2F_{5/2}$ manifolds are labeled from one to four in the ground state and five to seven in the excited state. The maximum splitting manifolds of Yb in silicate hosts was calculated according to the spectra shown in Fig. 5.

Fig. 10 explicated the overall energy splitting of $^2F_{7/2}$ manifold of Yb^{3+} in silicate crystal hosts with 1012cm^{-1} in Yb:LSO crystal, 1076cm^{-1} in Yb:GSO crystal, 984 cm^{-1} in Yb:YSO crystal and 1027cm^{-1} in Yb:SSO crystal indicating much stronger crystal-field interaction in silicate hosts than that in Yb:YAG with energy splitting of 785cm^{-1} (Yan et al., 2006).

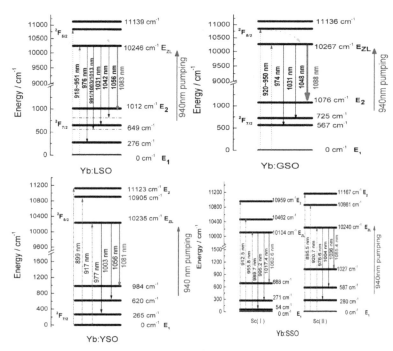

Fig. 10. Energy level of Yb^{3+} in Yb doped silicates.

As indicated in Fig. 10, the manifold splitting for Yb^{3+} in Yb:GSO crystal reached 1076 cm^{-1} which was much larger than that in Yb:LSO, Yb:YSO and Yb:SSO with C2/c structure thanks to the anisotropic and compact structure in GSO crystal. The strong crystal field in Yb:GSO crystal was simultaneously the largest ground-state splitting among Yb doped silicate crystals as obtained by absorption spectra starting from around 897 nm and the emission spectra extending till 1090 nm in Yb:GSO host.

To further elaborate the energy level in Yb doped silicate crystals, we found that the absorption band centered at 976nm in Yb:GSO belongs to the zero-line transition between the lowest level of $^2F_{7/2}$ and $^2F_{5/2}$ manifolds. The emission band at the longest wavelength around 1088 nm corresponds to the transition from the lowest levels of $^2F_{5/2}$ manifold to the highest levels of $^2F_{7/2}$ manifold. The strong electron–phonon coupling among Yb^{3+} ions together with the considerable interaction between Yb^{3+} ions and lattice vibrations give arise to strong vibration sidebands. The supplementary stark level splitting was incited by the resonance between stark levels and phonons which can be easily mistaken for the assignment of electronic transitions. Large crystal-field splitting of the fundamental manifold $^2F_{7/2}$ is critical for limiting the thermal population of the terminal laser level. In other words, less thermal populating of the terminal laser level of $^2F_{7/2}$ ground state would reduce re-absorption losses and decrease laser threshold as well. The strongest emission cross section was around 1000 nm in either Yb:YSO or Yb:LSO, while that in Yb:GSO was located around 1088 nm suggesting weakest re-absorption losses and least thermal populating of the terminal laser level. Accordingly, Yb:GSO around 1088 nm would produce ultra-fast laser with lowest pumping threshold.

The energy-level diagram of Yb^{3+} ions in crystallographic site Sc(I) and crystallographic site Sc(II) located in Yb:SSO crystal was approximately denoted in Fig. 10. The overall splitting of the $^2F_{7/2}$ manifold of Sc(II) reached about 1027cm^{-1} which was approaching that in Yb:GSO but much larger than 984 cm^{-1} in Yb:YSO and 1012 cm^{-1} in Yb:LSO. Yb ions encountered a stronger crystal-field interaction in SSO host indicating that Yb:SSO crystal would be the suitable solid-state laser gain for high efficient and high power application.

4. Laser peformance of Yb doped silicate crystals

High efficient laser output of 198fs/2.61W, 260fs/2.6W and 343fs/400mW were obtained in Yb:YSO, Yb:LSO and Yb:GSO respectively by Chinese and French researchers (Thibault et al., 2006; Li et al., 2007). Based on laser setup in Fig. 3, diode-pumped Kerr-lens mode-locked Yb:LSO laser was achieved without additional components. The mode-locked laser pulses were obtained in five-mirror cavity with average output power of 2.98 W and repetition rate of 103 MHz under incident pump power of 14.44W. Short pulse of 8.2 ps was realized at wavelength centered at 1059 nm (Wang et al., 2010). With an OC of T=2.5%, the cw Yb:GSO lasers with the slope efficiency of 75% at 1094 nm were achieved. Self-pulsed Yb:GSO lasers were achieved with low pumping threshold from 1091nm to 1105 nm (Li et al., 2006). Efficient diode-pumped laser performance of Yb:SSO was demonstrated with slope efficiency of 45% and output power of 3.55W (Zheng et al., 2008). The passive mode-locking and cw lasing performance of Yb:SSO was carried out in an x-fold cavity end-pumped by a 978 nm single emitter. The laser produced a maximum cw output power of 2.73 W with a slope efficiency of 70%. Preliminary tests regarding the laser operation of the thin-disk Yb:SSO laser were presented with 9.4 W of output power with an optical efficiency of 25.3 % (Wentsch et al., 2011). Passive mode-locking of Yb: SSO was initiated using a semiconductor saturable absorber mirror (SESAM) while dispersion compensation was introduced using a pair of SF10 prisms. The laser mode-locked at 1041 nm, 1060 nm and 1077 nm with near Fourier transformed limited pulse width of 145 fs, 144 fs and 125 fs, and average output power of 40 mW, 52 mW and 102 mW, respectively (Tan et al., 2010).

5. Conclusion

Yb doped oxyorthosilicate single crystals as technologically important laser host family owing to quasi-four level scheme were successfully obtained with high transparency and high quality by Czmethod. The efficient diode pumped ultra-fast laser was achieved with distinguished ground-state splitting up to 1000 cm^{-1}. Key spectroscopic parameters of Gain cross-section σ_g and minimum pump intensity I_{min} for designing high power lasers are specifically interpreted. In future, bulk silicate crystals with favorable thermal properties and multi-crystallographic sites for introducing distorted and broad spectra behavior are among the hot spots in the development of ultra-fast lasers.

6. Acknowledgment

The authors thank the supports from National High Technology Research and Development Program of China(2009AA03Z435), National Natural Science Foundation of China (60938001, 60908030, 61178056, 61177037), Innovation Project of Shanghai Institute of Ceramics (Y04ZC5150G), Hundred Talents Project of Chinese Academy of Sciences and National Science Fund for Distinguished Young Scholars. Special thanks to Dr. Heping Zeng, Dr. Xiaoyan Liang, Dr. Jie Liu, Dr. Weide Tan ,Dr. Guangjun Zhao, Dr. Chengfeng, Yan, Dr. Kejian Yang, Dr. Nengyin Sheng, Ms. Jiao Wang.

7. References

Caird J., Ramponi A. & Staver P. (1991). Quantum efficiency and excited-state relaxation dynamics in neodymium-doped phosphate laser glasses. *Journal of the Optical Society of America B*, Vol.8, No.7, pp. 1391-1403, ISSN 0740-3224

Camargo A., Davolos M. & Nunes L. (2002). Spectroscopic characteristics of Er^{3+} in the two crystallographic sites of Gd_2SiO_5, *Journal of Physics: Condensed Matter*, Vol.14, No.12, pp. 3353-3363, ISSN 0953-8984 (print)

Campos S., Denoyer A., Jandl S., Viana B., Vivien D., Loiseau P. & Ferrand B. (2004). Spectroscopic studies of Yb^{3+}-doped rare earth orthosilicate crystals. *Journal of Physics: Condensed Matter*, Vol.16, No.25 pp. 4579-4590, ISSN 0953-8984 (Print)

Cooke D., Bennett B., Mcclellan K., Roper J., Whittaker M. (2000). Oscillator strengths, Huang-Rhys parameters, and vibrational quantum energies of cerium-doped gadolinium oxyorthosilicate. Journal of Applied Physics, Vol.87, No.11, pp. 7793-7797, ISSN 0021-8979 (Print)

Deloach L., Payne S., Chase L., Smith L., Kway W. & Krupke W. (1993). Evaluation of absorption and emission properties of Yb^{3+} doped crystals for laser applications. *IEEE Journal of Quantum Electronics*, Vol.29, No.4, pp. 1179 - 1191, ISSN 0018-9197

Du J., Liang X., Xu Y., Li R., Xu Z., Yan C., Zhao G., Su L. & Xu J. (2006). Tunable and efficient diode-pumped Yb^{3+}:GYSO laser. *Optics Express*, Vol.14, No.8, pp. 3333-3338, ISSN 1094-4087

Eijk C. (2001). Inorganic-scintillator development. *Nuclear Instruments and Methods in Physics Research Section A*, Vol.460, No.1, pp. 1-14, ISSN 0168-9002

Ellens A., Andres H., Meijerink A. & Blasse G. (1997). Spectral-line-broadening study of the trivalent lanthanide-ion series.I. Line broadening as a probe of the electron-phonon coupling strength. *Physical Review B*, Vol.55, No.1, pp. 173-179, ISSN 1098-0121

Felsche J., (1973). The crystal chemistry of the rare-earth silicates, *Structure and Bonding*, pp. 99-197, ISSN 0081-5993 (Print), Springer-Verlag, New York, 1973

Fornasiero L., Petermann K., Heumann E. & Huber G. (1998). Spectroscopic properties and laser emission of Er^{3+} in scandium silicates near 1.5 μm, *Optical Materials*, Vol.10, No.1, pp. 9-17, ISSN 0925-3467

Gaume M. (2002). Relations structures - propriétés dans les lasers solides de puissance à l'ytterbium. Elaboration et caractérisation de nouveaux matériaux et de cristaux composites soudés par diffusion, These De Doctorat de l'universite pierre et marie curie-paris VI

Gaume R., Viana B., Derouet J. & Vivien D. (2003). Spectroscopic properties of Yb-doped scandium based compounds Yb:CaSc$_2$O$_4$, Yb:SrSc$_2$O$_4$ and Yb:Sc$_2$SiO$_5$. *Optical Materials*, Vol.22, No.2, pp. 107-115, ISSN 0925-3467

Giesen A. & Speiser J. (2007). Fifteen Years of Work on Thin-Disk Lasers: Results and Scaling Laws. *IEEE Journal of Selected Topics in Quantum Electron*, Vol.13, No.3, pp.598-609, ISSN 1077-260X

Griebner U., Petrov V., Petermann K. & Peters V. (2004). Passively mode-locked Yb:Lu$_2$O$_3$ laser. *Optics Express*, Vol.12, No.14, pp. 3125-3130, ISSN 1094-4087

Gustafsson T., Klintenberg M., Derenzo S., Weber M., Thomas J. (2001). Lu$_2$SiO$_5$ by single-crystal X-ray and neutron diffraction. *Acta Crystallographica Section C*, Vol.57, No.6, pp. 668-669, ISSN 1600-5759 (Online)

Haumesser P., Gaumé R., Viana B. & Vivien D. (2002). Determination of laser parameters of ytterbium-doped oxide crystalline materials. *Journal of the Optical Society of America B*, Vol.19, No.10, pp. 2365-2375. ISSN 0740-3224

http://database.iem.ac.ru/

Ivanov V., Petrov V., Pustovarov V., Shulgin B., Vorobjov V., Zinevich E., Zinin E., (2001). Electronic excitations and energy transfer in A$_2$SiO$_5$–Ce (A=Y, Lu, Gd) and Sc$_2$SiO$_5$ single crystals, *Nuclear Instruments and Methods in Physics Research Section A*, Vol.470, No.1-2, pp. 358-362, ISSN 0168-9002

Keller U. (2003). Recent developments in compact ultrafast lasers. *Nature*, Vol.424, No.6950, pp. 831-838, ISSN 0028-0836

Kisel V., Troshin A., Shcherbitsky V., Kuleshov N., Matrosov V., Matrosova T., Kupchenko M., Brunner F., Paschotta R., Morier-Genoud F. & Keller U. (2005). Femtosecond pulse generation with a diodepumped Yb^{3+}:YVO$_4$ laser. *Optics Letters*, Vol.30, No.10, pp. 1150-1152, ISSN 0146-9592

Krupke W. (2000). Ytterbium solid-state lasers-the first decade. *IEEE Journal of Selected Topics in Quantum Electron*, Vol.6, No.6, pp. 1287-1296, ISSN 1077-260X

Kuleshov N., Shcherbitsky V., Lagatsky A., Mikhailov V., Minkov B., Danger T., Sandrock T. & Huber G. (1997). Spectroscopy, excited-state absorption and stimulated emission in Pr^{3+}-doped Gd$_2$SiO$_5$ and Y$_2$SiO$_5$ crystals. Journal of Luminescence, Vol.71, No.1, pp. 27-35, ISSN 0022-2313

Lacovara, P., Choi, H., Wang, C., Aggarwal, R. & Fan T. (1991). Room-temperature diode-pumped Yb:YAG laser. *Optics Letters*, Vol.16, No.14, 1089-1091, ISSN 0146-9592 (print)

Li W., Hao Q., Zhai H., Zeng H., Lu W., Zhao G., Zheng L., Su L. & Xu J. (2007). Diode-pumped Yb:GSO femtosecond laser. *Optics Express*, Vol.15, No.5, pp. 2354-2359, ISSN 1094-4087

Li W., Pan H., Ding L., Zeng H., Lu W., Zhao G., Yan C., Su L. & Xu J. (2006). Efficient diode-pumped Yb:Gd$_2$SiO$_5$ laser. *Applied Physics Letters*, Vol.88, No.22, pp. 221117(1-3), ISSN 0003-6951 (Print)

Liu H., Nees J. & Mourou G. (2001). Diode-pumped Kerr-lens mode-locked Yb:KY(WO$_4$)$_2$ laser. *Optics Letters*, Vol.26, No.21, pp. 1723-1725, ISSN 0146-9592

Melcher C. & Schweitzer J. (1992). Cerium-doped lutetium oxyorthsilicate: A fast, efficient new scintillator, *IEEE Transactions on Nuclear Science*, Vol.39, No.4, pp. 502-505, ISSN 0018-9499

Moulton P. (1983). Paramagnetic ion lasers, in: *Handbook of Laser Science and Technology*, M.J. Weber, (Ed.), CRC, ISBN: 978-0-8493-3508-2 , Boca Raton, FL

Pelenc D., Chambaz B., Chartier I., Ferrand B., Wyon C., Shepherd D., Hanna D., Large A. & Tropper A. (1995). High slope efficiency and low threshold in a diode-pumped epitaxially grown Yb:YAG waveguide laser. *Optics Communication*, Vol.115, No.5-6, pp. 491-497, ISSN 0030-4018

Petit J., Goldner P. & Viana B. (2005). Laser emission with low quantum defect in Yb:CaGdAlO$_4$. *Optics Letters*, Vol.30, No.11, pp. 1345-1347, ISSN 0146-9592

Smolin Y. & Tkachev S. (1969). Determination of the structure of gadolinium oxyorthosilicate (Gd$_2$SiO$_5$). *Kristallografiya*, Vol.14, pp.22, ISSN 0023-4761

Su L., Xu J., Li H., Yang W., Zhao Z., Si J., Dong Y. & Zhou G. (2005). Codoping Na$^+$ to modulate the spectroscopy and photo-luminescence properties of Yb^{3+} in CaF$_2$ laser crystal. *Optics Letters*, Vol.30, No.9, pp. 1003-1005, ISSN 0146-9592

Suzuki H., Tombrello T., Melcher C. & Schweitzer J. (1992). UV and gamma-ray excited luminescence of cerium-doped rare-earth oxyorthosilicate. *Nuclear Instruments and Methods in Physics Research Section A*, Vol.320, No.1-2, pp. 263-272, ISSN 0168-9002

Tan W., Tang D., Xu X., Zhang J., Xu C., Xu F., Zheng L., Su L. & Xu J. (2010). Passive femtosecond mode-locking and cw laser performance of Yb^{3+}:Sc$_2$SiO$_5$. *Optics Express*, Vol.18, No.16, pp. 16739-16744, ISSN 1094-4087

Thibault F., Pelenc D., Druon F., Zaouter Y., Jacquemet M. & Georges P. (2006). Efficient diode-pumped Yb^{3+}:Y$_2$SiO$_5$ and Yb^{3+}:Lu$_2$SiO$_5$ high-power femtosecond laser operation. *Optics Letters*, Vol.31, No.10, pp. 1555-1557, ISSN 0146-9592

Uemura S. & Torizuka K. (2005). Center-wavelength-shifted passively mode-locked diode-pumped ytterbium (Yb):Yttrium aluminum garnet (YAG) laser. *Jpn. J. Appl. Phys.*, Vol.44, No.12-15, pp. L361-L363, ISSN: 0021-8979 (print)

Wang W., Liu J., Chen W., Lu C., Zheng L., Su L., Xu J. & Wang Y. (2010). Diode-Pumped Passively Mode-Locked Yb:LSO/SESAM Laser. *Laser Physics*, Vol.20, No.4, pp. 740-744, ISSN 1054-660X (Print)

Wang X., (2004). *Optical Crystallography*, Nanjing University Press, ISSN 7-305-04088-6

Wentsch K., Weichelt B., Zheng L., Xu J., Abdou-Ahmed M. & Graf T. (2011) Contnious-wave Yb doped Sc$_2$SiO$_5$ thin-disk laser. *Optics Letters*, Vol. Accepted 11/08/2011, Doc. ID 153527, ISSN 0146-9592

Yan C., Zhao G., Su L., Xu X., Zhang L. & Xu J. (2006). Growth and spectroscopic characteristics of Yb:GSO single crystal. *Journal of Physics: Condensed Matter*, Vol.18, No.4, pp. 1325-1333, ISSN 0953-8984 (Print)

Zheng L., Xu J., Zhao G., Su L., Wu F., Liang X. (2008). Bulk crystal growth and efficient diode-pumped laser performance of Yb^{3+}:Sc$_2$SiO$_5$. *Applied Physics B*, Vol.91, No.3-4, pp. 443-445, ISSN 0946-2171 (Print)

Zheng L., Zhao G., Yan C., Yao G., Xu X., Su L. & Xu J. (2007). Growth and spectroscopic characteristics of Yb:LPS single crystal. *Journal of Crystal Growth*, Vol.304, No.2, pp. 441-447, ISSN 0022-0248

6

Pattern Selection in Crystal Growth

Waldemar Wołczyński

Institute of Metallurgy and Materials Science,
Polish Academy of Sciences,
Poland

1. Introduction

Ebeling & Ashby, 1966 were able to prove that the Cu – single crystal can be strengthened by spherical particles of the SiO_2 – phase. Zarubova & Sestak, 1975 studied the analogous phenomenon in the Fe single crystals strengthened by the Si – addition. In the current study, the phenomenon of strengthening is observed in the hexagonal (Zn) – single crystal equipped with regular stripes which contain the Zn-16Ti intermetallic compound.

The studied (Zn) – single crystal was doped by the small amount of titanium and copper. The addition of copper modifies the specific surface free energy at the solid/liquid (s/l) interface. The specific surface free energies are to be determined in the current description for the triple point of the s/l interface to ensure the mechanical equilibrium. Copper does not form an intermetallic compound with the zinc but is localized in the zinc/titanium solid solution, (Zn). The titanium forms, additionally, the intermetallic compound with the zinc, $Zn_{16}Ti$. The (Zn)-Zn_{16}Ti system is the pseudo-binary eutectic system. It exists an opportunity to control the growth of the (Zn) – hexagonal single crystal, and first of all to control the width of the (Zn)-Zn_{16}Ti – stripes which appear cyclically in the single crystal.

Some experiments were performed by means of the *Bridgman* system with the moving thermal field. The system was equipped with the graphite crucible of the sophisticated geometry. It allowed to localize a crystal seed of the desired crystallographic orientation just at the crucible bottom.

The full thermodynamic description of the (Zn) – single crystal growth with periodic formation of the (Zn)-Zn_{16}Ti – stripes requires to consider the model for solute redistribution (Wołczyński, 2000). Next, a steady-state solution to the diffusion equation which yields the solute micro-field in the liquid at the s/l interface is given, (Wołczyński, 2007). The above solution involves a proper localization of the thermodynamic equilibrium at the inter-phase boundary (together with the mechanical equilibrium). This solution also allows for calculating the entropy production in such a system and additionally for describing the transition from lamellar into rod-like structure, (Wołczyński, 2010).

2. Solute redistribution within the (Zn) - Single crystal growing cyclically

Solute redistribution along the single crystal changes significantly due to localization of the *liquidus* and *solidus* lines in the phase diagram. In the case of the (Zn) – single crystal which contains strengthening stripes, Fig. 1a, the Ti – solute redistribution should be measured along the part of the (Zn) – single crystal formed between two neighbouring stripes, Fig. 1b.

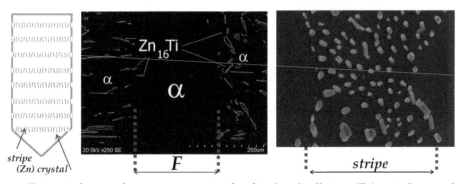

Fig. 1. (Zn) – single crystal structure; a. stripes localized cyclically in a (Zn) - single crystal (scheme), b. the F - distance between stripes; $(Zn) \equiv \alpha$, c. rod-like morphology of a stripe.

2.1 Use of the Zn-Ti phase diagram for the control of the (Zn) - Single crystal growth

According to the Zn-Ti phase diagram the solubility of the Ti in the Zn is equal to 0.000546 at.% at the ambient temperature, Fig. 2. Thus, the strengthening by a solubility is to be neglected. However, an experiment was made to study a strengthening by the eutectic precipitates, for three (I, II, III) solute contents.

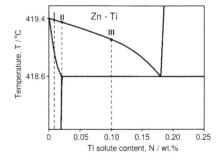

Fig. 2. Zn-Ti phase diagram with the localization of the eutectic point, $N_E = 0.18 \, [wt.\%Ti]$, (Murray, 1990), and all studied nominal solute concentration of the single crystal: I, II, III.

Since the difference between melting point of the Zn and eutectic temperature is small it involves a possibility of the single crystal strengthening by the periodic precipitates of eutectic phase ((Zn) + Zn$_{16}$Ti). The (Zn) – eutectic phase, Fig, 1b, Fig. 1c, is the same as in the bulk single crystal. Therefore, it is to be supposed that the Zn$_{16}$Ti eutectic phase (intermetallic compound) is only responsible for the strengthening of the (Zn) – single crystal.

A peculiar construction of the Zn-Ti phase diagram, Fig. 2, involves a possibility in the control of the precipitates amount. However, a proper model for the solute segregation / redistribution is necessary to predict the amount of the eutectic or intermetallic phase precipitation within a given single crystal.

In the current study the *Bridgman* furnace was working as a closed system and the control of strengthening could be done, to some extent, by an adequate choose of a nominal concentration of titanium in the alloy used for single crystal growth, Fig. 2, and by applying a proper growth rate, v, on which the $\alpha(v)$ back-diffusion parameter depends.

The selection of the nominal solute concentration decides on the stripe thickness as visible in Fig. 2. The nearest is the distance between nominal solute concentration on the *liquidus* line and the eutectic point the wider are the stripes.

Moreover, the applied growth rate decides on the stripes morphology. Three ranges of growth rates were distinguished in the discussed (Zn) – single crystal formation: a. at some low growth rates, the L – shape irregular rod-like structure appears in stripes, b. at some middle growth rates the regular lamellar structure is observed in stripes, c/ at the elevated growth rates the regular rod-like structure exists as an exclusive morphology of stripes.

It is evident that two transformations of morphology were recorded. At the first threshold growth rate the L-shape irregular rods were transformed into the regular lamellae. This transformation cannot occur immediately since the L-shape rod-like into lamellar structure transition is accompanied by the irregular into regular morphology transition. Thus, this transformation occurred continuously within a certain range of rates. However, the lamella into rod transition occurred just at the second threshold rate, immediately.

The control of the stripe thickness can be explained while applying the equilibrium solidification to the *Bridgman* system, Fig. 3.

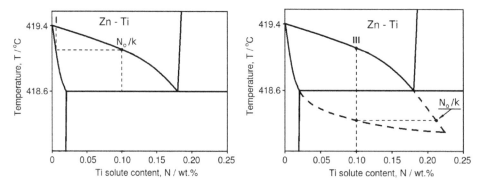

Fig. 3. Solidification paths ($I \rightarrow N_0 / k$) and ($III \rightarrow N_0 / k$) for an equilibrium solidification of the (Zn) - single crystal with: a. $N_0 = I = 0.01 \, [wt.\%Ti]$, and b. $N_0 = III = 0.1 \, [wt.\%Ti]$.

The precipitation is impossible for the equilibrium solidification with the nominal solute concentration: $N_0 = I = 0.01$, since $N_0 / k < N_E$, Fig. 3a. The precipitation is always ensured for the nominal solute concentration: $N_0 = III = 0.1 \, [wt.\%Ti]$, since $N_0 / k > N_E$, Fig. 3b. Moreover, the solidification occurs along the solidification path cyclically, and precipitates are rejected at the end of each cycle to form the stripes.

2.2 Model for the solute microsegregation / redistribution after the back-diffusion

A theory for solute microsegregation accompanied by the back-diffusion which is defined by the - α parameter, has already been delivered, (Brody & Flemings, 1966), with:

$$\alpha = D_S \, t_l \, L^{-2} \tag{1}$$

D_S - diffusion coefficient into the solid, $[m^2 / s]$; t_l - local growth time, $[s]$; L - half the crystal spacing, $[m]$. This theory is not able to describe the solute redistribution since the mass balance is not satisfied, there. Thus, an improved model, (Wołczyński, 2000), based on the α - back-diffusion parameter, Eq. (1) has been delivered. The model describes the solute redistribution after back-diffusion and can be used to calculate an amount of precipitates. The equation which describes the solute redistribution after back-diffusion is as follows:

$$N^B\left(x;X^0,\alpha\right)=\left[k+\beta^{ex}\left(x;X^0\right)\beta^{in}\left(X^0,\alpha\right)\right]N^L\left(x;\alpha\right) \tag{2}$$

k - partition ratio, $[mole\ fr.\ /\ mole\ fr.]$; x - crystal amount, [dimensionless]; $x=X^0$ - crystal amount at arrested growth; β^{ex} - coefficient of redistribution extent, [dimensionless]; β^{in} - coefficient of redistribution intensity, [dimensionless]; N^L - liquid content, $[mole\ fr.]$; Eq. (3).

$$N^L\left(x;\alpha\right)=N_0\left(1+\alpha\,k\,x-x\right)^{(k-1)/(1-\alpha k)} \tag{3}$$

The above equation results from the differential formula:

$$\left[1+\alpha\,k\,x-x\right]dN^L\left(x;\alpha\right)=\left(1-k\right)N^L\left(x;\alpha\right)dx \tag{4}$$

Both, β^{ex} - coefficient of the redistribution extent and β^{in} - coefficient of the redistribution intensity are defined due to the mass balance consideration (Wołczyński, 2000). Their product $\beta\left(x;X^0,\alpha\right)=\beta^{ex}\left(x,X^0\right)\beta^{in}\left(X^0;\alpha\right)$ is equal to zero: $\beta\left(x;X^0,0\right)=0$ for the non-equilibrium solidification (when $\alpha=0$). The product is equal to $(1-k)(1-x)$ for the equilibrium solidification (when $\alpha=1$ with $X^0=1$) that is: $\beta\left(x;1,1\right)=(1-k)(1-x)$. The single crystal growth occurs slowly, thus its growth is performed under condition close to the equilibrium state. The physical limitation for the the α - back-diffusion parameter has also been determined while assuming: $t_D=L^2/D_S$. Then, $\alpha=t_l/t_D$ and finally, $0\leq\alpha\leq1$. with t_D - diffusion time necessary for the full homogeneity of the crystal. Thus, the mass balance is satisfied at each step of the crystal growth. Therefore, Eq. (2) can perfectly fit some measurement points obtained by the EDS technique. On the other side, the solute microsegregation is observable at the moving s/l interface during the experiment. Thus, the redistribution is the only concentration which can be measured after the crystal growth.

Usually, the crystal growth is accompanied by some precipitates. The growth rate, imposed in the *Bridgman* system for a single crystal growth (with a seed) $v>0$, does not follow the equilibrium solidification. Thus, the solidification path is longer than that shown in Fig. 2. Therefore, it was possible to obtain precipitates even for the nominal solute concentrations: $N_0=I=0.01\,[wt.\%Ti]$, $N_0=II=0.02\,[wt.\%Ti]$, when the imposed growth rate, v, involves an elongation of the solidification path beyond the eutectic point ($N_E=0.18\,[wt.\%Ti]$) till the N_K - point on the *liquidus* line. N_K is also the solute content in the liquid at the end of solidification, $[mole\ fr.]$ (N_0, N_E are to be expressed in $[mole\ fr.]$ while calculating N_K). The ratio between an amount of the crystal, x_K, and an amount of precipitate, i_K, depends on the mentioned elongation beyond the eutectic point by the solidification path:

$$x_K\ /\ i_K=\left(N_E-N_0\right)/\left(N_K\left(\alpha,N_0\right)-N_E\right) \tag{5}$$

The end of solidification path is: $N_K\left(\alpha,N_0\right)=N^L\left(x_K,\alpha\right)$, and x_K is defined as follows:

$$x_K(\alpha, N_0) = \frac{1}{1-\alpha k}\left[1 - (N_E / N_0)^{\frac{1-\alpha k}{k-1}}\right] \quad \text{when} \quad 0 \le \alpha \le \alpha_E(N_0) \tag{6a}$$

$$x_K(\alpha, N_0) = 1 \quad \text{when} \quad \alpha_E(N_0) < \alpha \le 1 \tag{6b}$$

with $(\alpha_E k)^{\frac{k-1}{1-\alpha_E k}} = N_E / N_0$, (Wołczyński, 2000).

According to the above considerations, $i_K(\alpha, N_0) = 1 - x_K(\alpha, N_0)$, [dimensionless]. The precipitate is divided into the so-called equilibrium precipitate: $i_E(N_0) = i_K(1, N_0)$ and non-equilibrium precipitate: $i_D(\alpha, N_0) = i_K(\alpha, N_0) - i_E(N_0)$.

The precipitate visible in the crystal reproduces the s/l interface shape which existed "historically" at a given stage of the (Zn) - single crystal growth, Fig. 4.

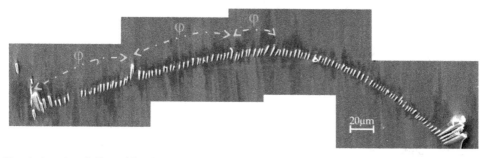

Fig. 4. A stripe deflected by the perturbation wave created at the solid / liquid interface during the (Zn) - single crystal growth; the stripe (precipitate) contains the (Zn) – phase coherent with the bulk single crystal and rods of the $Zn_{16}Ti$ – intermetallic compound; the stripe follows the pattern which seems to be the result of a solitary wave, (φ).

2.3 Measurement / simulation of the Ti - Solute redistribution along the single crystal

The solute redistribution model, discussed above, allows to make some simulations of the Ti – solute redistribution for the (Zn) – single crystal growth. The simulation is developed to fit the measurement points obtained by the EDS technique, Fig. 5. The Ti – solute redistribution was measured along the bulk (Zn) – single crystal just between two neighbouring stripes and for two different samples.

The theoretical solute microsegregation is calculated twice: a. $N^S(x,0) = kN^L(x,0)$ - without the presence of back-diffusion, (Scheil, 1942), b. $N^S(x,\alpha) = kN^L(x,\alpha)$ - with the presence of back-diffusion phenomenon, (Wołczyński, 2000), Fig. 5. The theoretical amount of the bulk crystal, $x_K(\alpha)$ and corresponding amount of the $((Zn) + Zn_{16}Ti)$ – precipitate, $i_K(\alpha)$ (red lines) are also shown. The theoretical solute redistribution is calculated for a selected value of the α - back-diffusion parameter, $(0 \le \alpha \le 1)$, [dimensionless], and additionally for $X^0 = 1$. The partition ratio is applied due to the Zn-Ti phase diagram for the stable equilibrium as equal to $k = 0.11 [wt.\% / wt.\%]$, (Murray, 1990).

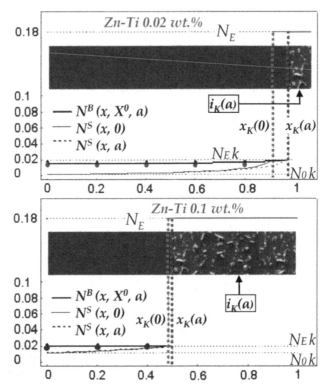

Fig. 5. Ti – solute redistribution as calculated for two alloys of the nominal Ti – solute concentration equal to: 0.02 $wt.\%$ and 0.1 $wt.\%$, respectively; (Boczkal et al., 2010).

3. Solute micro-field in the liquid and the strengthening phase protrusion

The eutectic precipitates are formed under stationary state from the non-homogeneous liquid phase, (Jackson & Hunt, 1966). The solution to the diffusion equation gives a description of the solute concentration micro-field ahead of the s/l interface. The recent solution to the diffusion equation supposes the existence of the mechanical equilibrium at the triple point of the s/l interface and the thermodynamic equilibrium along the inter-phase boundary, (Wołczyński, 2007). The steady-state solution to the diffusion equation is given separately for each eutectic lamella: α - phase lamella, and β - phase lamella. So,
a. for the α - eutectic phase formation (the (Zn) – phase in the Zn – Ti system),

$$\delta C(x,z) = \sum_{n=1}^{\infty} A_{2n-1} \cos\left(\frac{(2n-1)\pi x}{2S_\alpha}\right) \exp\left(-\frac{(2n-1)\pi}{2S_\alpha} z\right) \qquad (7)$$

$$A_{2n-1} = -\frac{4}{(2n-1)\pi} \int_0^{S_\alpha} f_\alpha(x)\cos\left(\frac{(2n-1)\pi x}{2S_\alpha}\right) dx \quad n = 1, 2,... \qquad (7a)$$

b. for the β - eutectic phase formation (the $Zn_{16}Ti$ – compound in the Zn – Ti system),

$$\delta C(x,z) = \sum_{n=1}^{\infty} B_{2n-1} \cos\left(\frac{(2n-1)\pi(x-S_\alpha+S_\beta)}{2S_\beta}\right) \exp\left(-\frac{(2n-1)\pi}{2S_\beta}z\right) \tag{8}$$

$$B_{2n-1} = -\frac{4}{(2n-1)\pi} \int_{S_\alpha-S_\beta}^{S_\alpha} f_\beta(x)\cos\left(\frac{(2n-1)\pi(x-S_\alpha+S_\beta)}{2S_\beta}\right) dx \quad n=1,2,... \tag{8a}$$

C - solute concentration within the micro-field formed in the liquid, $[at.\%]$; S_j - half the width of the eutectic phase lamellae, $[m]$, ($j=\alpha,\beta$), respectively; f_j - function used in formulation of the boundary condition for the α and β - eutectic phases formation, $[at.\%]$, ($j=\alpha,\beta$), respectively; x,z - geometrical coordinates, $[m]$.
The total mass balance is satisfied within the solute concentration micro-field, Eq. (9), Fig. 6.
D - diffusion coefficient in the liquid, $\left[m^2/s\right]$; v - crystal growth rate, $[m/s]$, identical to the thermal field movement rate in the Bridgman system; moreover,

$$B_{2n-1} = A_{2n-1}\left(S_\alpha/S_\beta\right)^2, \quad n=1,2,... \tag{9a}$$

The satisfaction of the total mass balance is shown in Fig. 6. The local mass balance is also satisfied, Fig. 7. But, in the case of the local mass balance the phase protrusion is to be considered, Eq. (10).

$$\int_0^{S_\alpha} \delta C(x,0)\,dx + \int_{S_\alpha}^{S_\alpha+S_\beta} \delta C(x,d)\,dx = 0 \tag{10}$$

d - protrusion of the β - leading eutectic phase over the α - wetting eutectic phase, $[m]$.

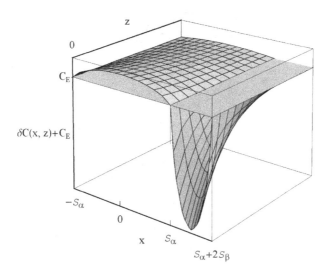

Fig. 6. Total mass balance within the solute concentration field for an eutectic system growing under stationary condition; C_E - eutectic concentration of the solute, $[at.\%]$.

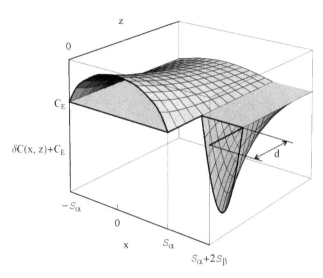

Fig. 7. Local mass balance within the solute concentration field for an eutectic system growing under stationary condition.

The protrusion of the eutectic leading phase, predicted theoretically, Fig. 7, and formerly observed experimentally, (Davies, 1964) has also been revealed within the stripes which contain the $((Zn) + Zn_{16}Ti)$ eutectic, Fig. 8.

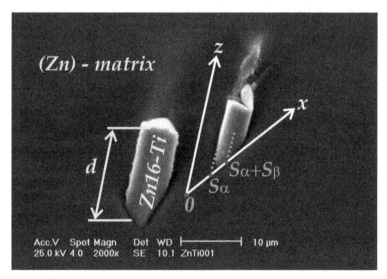

Fig. 8. $((Zn) + Zn_{16}Ti)$ regular rod-like eutectic revealed in the stripes (the EDS observation); the x, z - coordinate system presents the localization of the Ti – solute concentration field; the d - leading phase protrusion is defined due to the (Zn) – single crystal growth arresting.

Since the (Zn) – single crystal growth was arrested and the s/l interface was frozen, it is also possible to reveal the s/l interface shape of the non-faceted (Zn) – eutectic phase, Fig. 9.

Fig. 9. Frozen solid / liquid interface of the ((Zn) + Zn$_{16}$Ti) regular eutectic revealed within the strengthening stripes; a. the EDS observation, b. identification of the eutectic phases: the Zn$_{16}$Ti faceted phase, the (Zn) – non-faceted phase, the frozen liquid, the parabolic envelope (yellow dashed line) superposed onto the s/l interface of the (Zn) – phase.

Since the revealed solid/liquid interface has a parabolic shape, Fig. 9, and no interface destabilization is observed, it is evident that the obtained structure is the regular structure.

4. Thermodynamic selection of the pattern in the growing eutectic structure

(Zn) – single crystal growth occurs under stationary state in the *Bridgman* system with an imposed constant growth rate. Neither, the thermal field does not vary during solidification. Eventually, the Zn$_{16}$Ti eutectic phase (compound) has the same size, Fig. 10, and an inter-phase spacing should not vary during crystal growth.

It should be emphasized that the only condition which defines the stationary state is the criterion of minimum entropy production, (Prigogine, 1968).

Fig. 10. A strengthening stripe ((Zn) + Zn$_{16}$Ti) within the (Zn) – single crystal frozen during its growth in the *Bridgman* system; the Zn$_{16}$Ti - rods are all of the same size, mainly.

Many eutectic systems exhibit either a lamellar or rod-like structure depending on solidification conditions, (Elliott, 1977). It is well visible in the case of directional growth of the Al-Si eutectic alloy, (Toloui & Hellawell, 1976) and (Atasoy, 1984). Moreover, a transition from the lamellar structure into the rod-like structure is observable at a the threshold growth rate typical for a given eutectic alloy, (Cuprys et al., 2000).

Especially, growth rate plays a crucial role in the lamella into rod transformation. Some impurities also involve the transition, (Steen & Hellawell, 1975). The impurities change the specific surface free energies and finally modify a mechanical equilibrium at the triple point of the solid/liquid interface. However, according to the current model assumptions, the mechanical equilibrium varies (rotates) at the s/l interface of the (Zn) – single crystal not only with the copper addition but with solidification conditions (growth rate) as well. An imposed growth rate results in the crystal orientation. Some changes of the orientation from an initial state into a final one give also an effect on the lamella.rod transition.

In the case of the (Zn) – single crystal growth two threshold rates were revealed for the changes in the stripes morphology, as mentioned above. First threshold growth rate, v_1, is connected with the transition from the L – shape irregular (with branches) rod-like structure into the regular lamellar structure. Second threshold growth rate, v_2, is associated with the transition from the regular lamellar structure into the regular rod-like structure.

4.1 Promotion of the rod-like structure or lamellar structure formation

The formerly developed theory (Jackson & Hunt, 1966) has tried to predict the threshold rate at which a structural transition should occur. The theory (Jackson & Hunt, 1966) is based on the description of the s/l interface undercooling with the undercooling defining both lamellar growth and rod-like growth, Eq. (11).

$$\left(\Delta T_L^*\right)^2 = 4vm^2 a^L Q^L ; \quad \left(\Delta T_R^*\right)^2 = 4vm^2 a^R Q^R \tag{11}$$

ΔT_L^* - s/l interface undercooling for the lamellar structure formation, $[K]$; ΔT_R^* - s/l interface undercooling for the rod-like structure formation, $[K]$; v - growth rate, $[m/s]$; $1/m = 1/m_\alpha + 1/m_\beta$, with: m_α - slope of the α - liquidus line, $[K/at.\%]$; m_β - slope of the β - liquidus line, $[K/at.\%]$, $a^L; a^R; Q^L; Q^R$ - parameters associated with the eutectic system capillarity and with a given phase diagram, (Jackson & Hunt, 1966).

This description yields a certain inequality, (Jackson & Hunt, 1966), according to which the rods or lamellae are formed, Fig. 11.

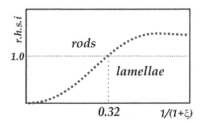

Fig. 11. Prediction of rod-like or lamellar structure formation, (Jackson & Hunt, 1966).

For the isotropic s/l interfacial free energies the r. h. s. of the considered inequality is equal to one, Fig. 11. The discussed inequality, shown in Fig. 11, yields a result, according to which, when the following inequality: $1/(1+\xi) < 0.32$, is satisfied (for the assumed isotropy) then a rod-like structure is stable form, (Jackson & Hunt, 1966). However, this parameter is equal to 0.114, ($1/(1+\xi) = 0.114$), for the Al-Si system. Thus, the rod-like structure should be the stable form. Meanwhile, according to experimental observations a lamellar structure is formed instead of a rod-like structure. Moreover, the lamellar structure transforms into the rod-like structure at a threshold growth rate. It is evident that the inequality, illustrated in Fig. 11, is completely misleading. The discerning analysis shows that the above inequality is able to predict whether an eutectic alloy will manifest lamellar or rod-like structure, only. Thus, the discussed inequality characterizes a given phase diagram. The inequality cannot be applied to describe the lamella into rod transition.

4.2 Thermodynamics of the eutectic s/l interface formation in the pattern selection

Thermodynamics of the eutectic s/l interface formation has already been discussed to some extent, (Wołczyński, 2010). The consideration was focused on the Al-Si eutectic alloy which manifests the irregular structure (with branches) due to directional solidification. It is known that the regular structure areas exist among the generally irregular structure, Fig. 12. It results from the experimental observations of lamellar and rod-like structures formation that the threshold growth rate at which the transformation lamella into rod begins, is equal to 400 $[\mu m / s]$. The transition is completed at the growth rate equal to about 700 $[\mu m / s]$. Both structures coexist within the operating range for the lamella into rod transition, ($400 \mu m / s \div 700 \mu m / s$), as visible in Fig. 13b, but the lamellar structure is obtainable below the threshold solidification rate, exclusively, as it is shown in Fig. 13a.

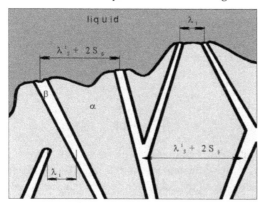

λ_i - inter-lamellar spacing for the regular structure, $[m]$; $\lambda_s^i + 2S_\beta$ - inter-lamellar spacing for the maximal destabilization of the s/l interface of the α - non-faceted phase, $\alpha \equiv (Al)$, $[m]$

Fig. 12. Generally irregular eutectic structure formation; scheme, (Fisher & Kurz, 1980).

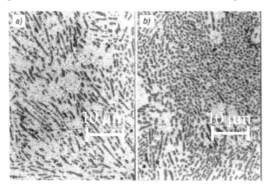

Fig. 13. A cross-sectional morphology of a directionally solidifying Al-Si irregular eutectic;
a. for $v = 370 \mu m / s$, and thermal gradient equal to $G = 100 K / cm$, (lamellae exclusively);
b. for $v = 500 \mu m / s$, and thermal gradient equal to $G = 40 K / cm$, (lamellae + rods).

The thermodynamics of the s/l interface involves the calculation of the *Gibbs'* free energy for the considered eutectic system. The *Gibbs'* free energy formulas which are concerning the s/l interface formation are as follows, for the regular lamellar and regular rod-like structure:

$$\Delta G_L^* = v\lambda Q_{CW}^L - a_{CW}^L \lambda^{-1} \qquad \Delta G_R^* = v R Q_{CW}^R - a_{CW}^R R^{-1} \tag{12}$$

Eq. (12) was developed from Eq. (11) by means of the transformation of the undercooling into the *Gibbs'* free energy.

$\Delta G_L^*; \Delta G_R^*$ - *Gibbs'* free energy for regular lamellar or rod-like growth, respectively, $\left[J / m^3 \right]$; $\lambda \equiv \lambda_i$ - inter-lamellar spacing, $[m]$; R - inter-rod spacing, $[m]$.

$$Q_{CW}^L = \left[m\left(L_\alpha \xi + L_\beta \right)/T_E \right]\left[P^*\left(1+\xi \right)N_0 /\left(\xi D \right) \right] \tag{13a}$$

$$Q_{CW}^R = \left[m\left(L_\alpha \xi + L_\beta \right)/T_E \right]\left[4E N_0 /\left(\xi D \right) \right] \tag{13b}$$

$$a_{CW}^L = \left\{ m(1+\xi)\left[\sigma_\alpha^L \sin\theta_\alpha^L / m_\alpha + \left(\sigma_\beta^L \sin\theta_\beta^L \right)/\left(\xi m_\beta \right) \right] + \sigma_{\alpha\beta}^L \right\} \tag{14a}$$

$$a_{CW}^R = 2\left\{ m\sqrt{1+\xi}\left[\sigma_\alpha^R \sin\theta_\alpha^R / m_\alpha + \left(\sigma_\beta^R \sin\theta_\beta^R \right)/\left(\xi m_\beta \right) \right] + \sigma_{\alpha\beta}^R / \sqrt{1+\xi} \right\} \tag{14b}$$

$$\xi = S_\beta / S_\alpha \quad \text{with} \quad \lambda = 2\left(S_\alpha + S_\beta \right) \tag{15a}$$

$$r_\alpha = 1/\sqrt{1+\xi} \quad \text{with} \quad R = r_\alpha + r_\beta \tag{15b}$$

$L_\alpha; L_\beta$ - heat of fusion per unit volume of a given eutectic phase, $[J / m^3]$; T_E - eutectic melting point, $[K]$; P^*, E - parameters referred to lamellae width, (Jackson & Hunt, 1966); N_0 - difference of the solubility of the B – eutectic element in the A – element and solubility of the A – element in the B – element, according to a given eutectic phase diagram, $[at.\%]$; D - diffusion coefficient in the liquid, $[m^2 / s]$; σ_j^L - specific surface free energy for the lamellar structure, $j = \alpha, \beta$, $[J / m^2]$; σ_j^R - specific surface free energy for the rod-like structure, $j = \alpha, \beta$, $[J / m^2]$; $\sigma_{\alpha\beta}^L$ - α / β phase boundary free energy for lamellar structure, $[J / m^2]$; $\sigma_{\alpha\beta}^R$ - α / β phase boundary free energy for the rod-like structure, $[J / m^2]$.
Geometry for both types of eutectic morphology is shown in Fig. 14.

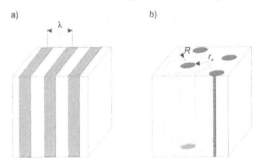

Fig. 14. Geometrical parameters defining both morphologies; a. inter-lamellar spacing, λ in the regular structure; b. rod/(rod + matrix) radiuses in the regular structure, Eq. (15).

Some rearrangements of Eq. (12) allows to formulate the following inequality:

$$\left\{ m\left[a_\alpha^L / m_\alpha + a_\beta^L / \xi m_\beta \right] + \sigma_{\alpha\beta}^L / (1+\xi) \right\} / \left\{ m\left[a_\alpha^R / m_\alpha + a_\beta^R / \xi m_\beta \right] + \sigma_{\alpha\beta}^R / (1+\xi) \right\}$$
$$> 4\left(E / P^* \right) \left(1 / (1+\xi) \right)^{1.5} \tag{16}$$

The obtained inequality was developed in the same manner as the inequality illustrated in Fig. 11. The developed inequality cannot predict the lamella into rod transformation in the eutectic system. Analogously to the mentioned theory, (Jackson & Hunt, 1966), this inequality can be applied, Eq. (16), to some extent, in order to predict, only, whether a given eutectic system (phase diagram) promotes the lamellae or rods formation.
Therefore, the *Gibbs'* free energy, Eq. (12), was calculated with applying the evolution (with the growth rate) of a mechanical equilibrium at the triple point of the s/l interface, Fig. 15.

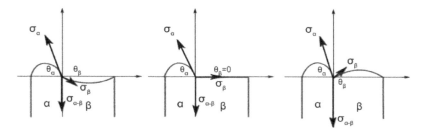

Fig. 15. Evolution (with varying growth rate) of the s/l interface curvature for the regular eutectic growth and adequate mechanical equilibrium situated at the triple point, (scheme).

$\alpha \equiv Si$, and $\beta \equiv (Al)$ - eutectic phases; $\sigma_\alpha \equiv \sigma_\alpha^L$ or $\sigma_\alpha \equiv \sigma_\alpha^R$; $\sigma_\beta \equiv \sigma_\beta^L$ or $\sigma_\beta \equiv \sigma_\beta^R$; $\sigma_{\alpha-\beta} \equiv \sigma_{\alpha\beta}^L$ or $\sigma_{\alpha-\beta} \equiv \sigma_{\alpha\beta}^R$; $\theta_\alpha \equiv \theta_\alpha^L$ or $\theta_\alpha \equiv \theta_\alpha^R$; $\theta_\beta \equiv \theta_\beta^L$ or $\theta_\beta \equiv \theta_\beta^R$, Fig. 15.
The mechanical equilibrium for the both structure types is:

$$\sigma_\alpha^L \sin\theta_\alpha^L \mp \sigma_\beta^L \sin\theta_\beta^L - \sigma_{\alpha\beta}^L = 0 \; ; \; \sigma_\alpha^R \sin\theta_\alpha^R \mp \sigma_\beta^R \sin\theta_\beta^R - \sigma_{\alpha\beta}^R = 0 \tag{17}$$

The *Gibbs'* free energy, Eq. (12) calculated for the threshold growth rate is shown in Fig. 16. The threshold growth rate in calculation equals $v = 400\,\mu m / s$; $\Delta G^* \equiv \Delta G_L^*$ or $\Delta G^* \equiv \Delta G_R^*$, respectively; dot L - denotes localization of the average inter-lamellar spacing, $\bar{\lambda}$ and dot R - denotes localization of the average inter-rod spacing, $\bar{R}$; $\theta_\alpha^L; \theta_\beta^L; \theta_\alpha^R; \theta_\beta^R$ - angles, Fig. 15.
All the minima of the *Gibbs'* free energy for considered growth rates are gathered in Fig. 17. According to the result of the *Gibbs'* free energy calculation this structure is stable form which has its minimum situated lower, Fig. 17. Thus, the calculation allows to determine the threshold growth rate, when minima are at the same level, Fig. 16. The operating range for transition cannot be described, alas. The present simulation, Fig. 17, was possible since the evolution of the s/l interface curvature, Fig. 18, was implemented into the calculation.
The evolution involves some changes of crystallographic orientation of the s/l interface and specific surface free energies together with inter-phase boundary free energy, Fig. 15.

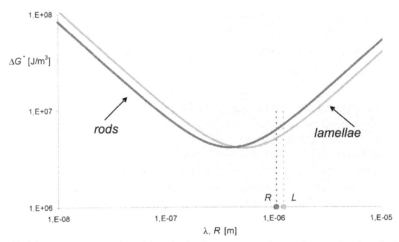

Fig. 16. *Gibbs'* free energy simulated for the both structures formation at the threshold rate.

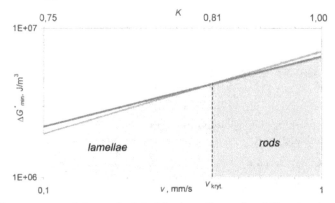

Fig. 17. *Gibbs'* free energy minima calculated for lamellar and rod-like structures formation; the critical growth rate, $v_{kryt.}$, for lamella into rod transition results from the trajectories intersection; $v_{kryt.}$ is equal to the experimental threshold growth rate: $v_{thr.}$.

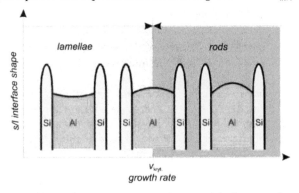

Fig. 18. A model of the s/l interface curvature evolution with the growth rates; $v_{kryt.} \equiv v_{thr.}$.

4.3 Thermodynamics of the whole solidification of regular eutectic structure

Thermodynamics of the whole solidification process supposes a calculation of the entropy production for the regular structure growth, (Lesoult & Turpin, 1969). The regular structure areas can be easily distinguished among generally irregular structure, (Fisher & Kurz, 1980). The regular structure selection under stationary state with the assumption of an isothermal s/l interface is to be described by the criterion of minimum entropy production, (Kjelstrup & Bedeaux, 2008). The entropy production per unit time, P_S^D, $S = R, L$, is as follows:

$$P_S^D = \int_V \sigma \, dV \tag{18}$$

V is the so-called „thermodynamic macroscopic point" (volume) inside of which all essential fluxes are observed, (Glansdorff & Prigogine, 1971). V - volume is shown in Fig. 19.

$$\sigma = R^* \, \varepsilon C^{-1} (1 - C)^{-1} D \nabla^2 C \tag{19}$$

R^* - gas constant, $[J/(mole\,K)]$; ε - thermodynamic factor, [dimensionless]; C - solute content, $[at.\%]$; σ - entropy production per unit time and unit volume, $\left[mole\,fr.^2/(m^3 s)\right]$.

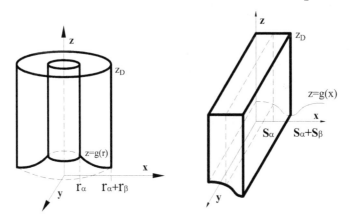

Fig. 19. V - volume, applied in calculation of the entropy production per unit time, Eq. (18); a. for the rod-like structure, $z = g(r)$ - function which describes the s/l interface curvature; z_D - thickness of the diffusion zone, $[m]$; b. for the lamellar structure, $z = g(x)$ - function which describes the s/l interface curvature; z_D - thickness of the diffusion zone, $[m]$.

The solution to Eq. (18) gives the following result:
a. entropy production (associated with the diffusion) for a rod-like structure formation:

$$P_R^D = V_1 \, v \left(r_\alpha + r_\beta\right)^{-1} + V_2 \, v \left(r_\alpha + r_\beta\right)^{-2} + V_3 \, v^2 + V_4 \, v^2 \left(r_\alpha + r_\beta\right) + V_5 \, v^3 \left(r_\alpha + r_\beta\right)^2 \tag{20a}$$

b. entropy production (associated with diffusion) for a lamellar structure formation:

$$P_L^D = W_1 \, v \left(S_\alpha + S_\beta\right)^{-1} + W_2 \, v \left(S_\alpha + S_\beta\right)^{-2} + W_3 \, v^2 + W_4 \, v^2 \left(S_\alpha + S_\beta\right) + W_5 \, v^3 \left(S_\alpha + S_\beta\right)^2 \tag{20b}$$

$V_n; W_n$ $n = 1,...,5$ - constants contain material parameters, (Wołczyński & Billia, 1996).

The calculations of the entropy production applied to the Zn-Ti system are connected with some experimental observations. According to the observations, for:

-   $0 < v < v_1$ irregular L-shape rods (equipped with branches) appear, Fig. 20,

-   $v_1 < v < v_1'$ L-shape rods with disappearing branches and regular lamellae coexist, Fig. 21

-   $v_1' < v < v_2$ regular lamellae are formed exclusively within the range of growth rates, Fig. 22,

-   $v > v_2$ regular rods are stable form within the range of growth rates, Fig. 23.

The characteristic crystal growth rates have the following experimentally determined value:
$v_1 \approx 5\,mm\,/\,h$; $v_1' \approx 5.8\,mm\,/\,h$; $v_2 \approx 10\,mm\,/\,h$.

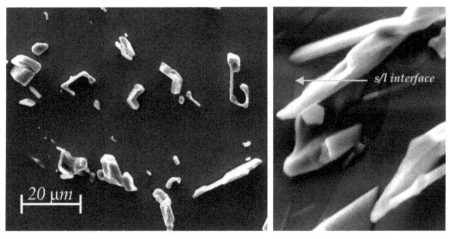

Fig. 20. ((Zn) + Zn$_{16}$Ti) - eutectic morphology in stripes obtained for the growth rate $v < v_1$; a. L-shape irregular rods; b. branching phenomenon with the frozen planar s/l interface.

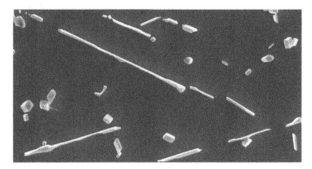

Fig. 21. Coexistence of L-shape rods (disappearing branches) and regular lamellae, $v < v_1'$.

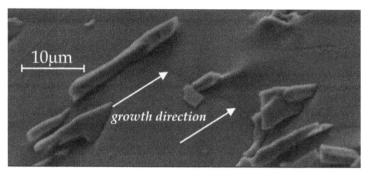

Fig. 22. Lamellar ((Zn) + Zn₁₆Ti) - eutectic morphology for the growth rate $v_1' < v < v_2$.

Arrows juxtaposed onto the morphology, Fig. 22, show the parabolic shape of the s/l interface of the (Zn) – non-faceted phase and emphasize the regularity of eutectic structure.

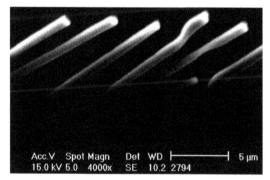

Fig. 23. Regular rods of the Zn₁₆Ti – intermetallic compound obtained for $v > v_2$.

The calculation of the entropy production, P_D, applied to the Zn-Ti system is made with the implementation of the model for rotation of mechanical equilibrium around the triple point of the s/l interface, Fig. 24. The rotation is performed with the increasing growth rate and causes some changes of the s/l interface curvature of the non-faceted (Zn) – phase, Fig. 24.

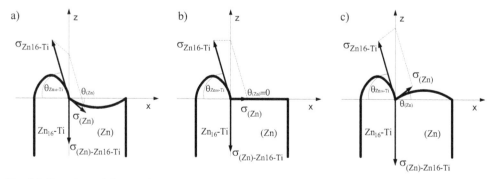

Fig. 24. Rotation of the mechanical equilibrium around the triple point of the s/l interface.

The mechanical equilibrium is established by the parallelogram of the anisotropic specific surface free energies together with the α / β - inter-phase boundary free energy, Fig. 24. As a result of the varying growth rate, the s/l interface curvature changes due to rotating crystallographic orientation. $\sigma_{(Zn)} \equiv \sigma_{(Zn)}^{L}$ and $\sigma_{Zn16-Ti} \equiv \sigma_{Zn16-Ti}^{L}$ - specific surface free energies for (Zn) - phase and $Zn_{16}Ti$ - compound, respectively, $[J / m^2]$; $\sigma_{(Zn)-Zn16Ti}$ - inter-phase boundary free energy, $[J / m^2]$; θ_j - angles, $[^0]$, $j = (Zn), Zn_{16}Ti \equiv \alpha, \beta$

The rotation of the mechanical equilibrium yields some changes of capillarity parameters. This is shown in function of the increasing growth rate, Fig. 25.

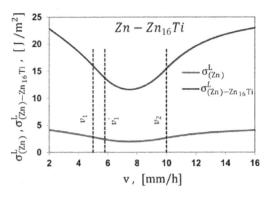

Fig. 25. Changes of the specific surface free energy, $\sigma_{(Zn)}^{L}$, and inter-phase boundary free energy, $\sigma_{(Zn)-Zn_{16}Ti}$; as applied to calculation of the entropy production.

The appearance of the observed structure is accompanied by a proper entropy production, as postulated. The considered solidification occurs under stationary state, therefore, the entropy production, P_D, manifests its minimum. The calculations are made to show the competition between two eutectic regular structures: rod-like and lamellar structure, Fig. 26.

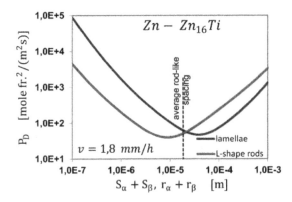

Fig. 26a. Competition between eutectic structures formation: regular rods among irregular L-shape rods and regular lamellae; shown for the growth rate, $v = 1.8 mm / h$; $0 < v < v_1$.

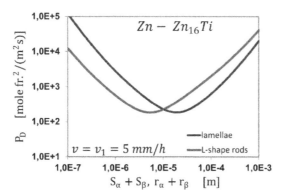

Fig. 26b. Competition between eutectic structures formation: regular rods among irregular L-shape rods and regular lamellae; shown for the first threshold growth rate, $v_1 \approx 5\,mm\,/\,h$.

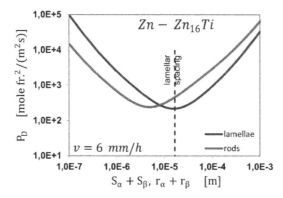

Fig. 26c. Competition between eutectic structures formation: regular lamellae and regular rods; shown for the growth rate, $v = 6\,mm\,/\,h$; $v_1' < v < v_2$.

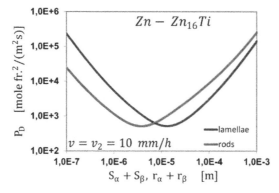

Fig. 26d. Competition between eutectic structures formation: regular lamellae and regular rods; shown for the second threshold growth rate, $v_2 \approx 10\,mm\,/\,h$.

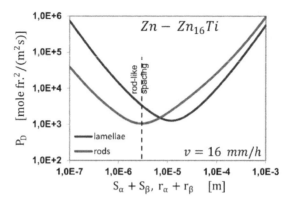

Fig. 26e. Competition between eutectic structures formation: regular lamellae and regular rods; shown for the growth rate; $v = 16\,mm/h$, $v > v_2$.

It is visible that the entropy production calculated for the observed structure formation within a proper range of growth rates, locates its minimum lower than the entropy production determined for a competitive structure, Fig. 26.

Next, the calculated entropy productions were minimized in order to compare their minimal values for all studied growth rate ranges: a. $0 < v < v_1$, where the L-shape rods equipped with branches, (irregular eutectic structure) appear, Fig. 20, b. $v_1 < v < v_1'$, where the L-shape rods with disappearing branches and regular lamellae coexist, Fig. 21, c/ $v_1' < v < v_2$, where the regular lamellae are exclusive form, Fig. 22, and d/ $v > v_2$, where the regular rods are created, Fig. 23. The results of calculation are shown in Fig. 27.

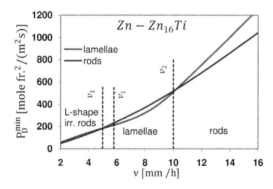

Fig. 27. Comparison of minimal values of the entropy production calculated for both regular rod-like structure formation (formed twice) and regular lamellar structure formation.

The estimation of the average inter-rod spacing for the first rates range: $0 < v < v_1$ requires to apply a definition of the mean value of the investigated spacing.

$$\bar{R} = 0.5R + R_\beta^i + 0.5\lambda_s^i \qquad (21)$$

$R = r_\alpha + r_\beta$ - defined within the areas of the regular structure, Fig. 14b, which are formed among the generally irregular structure, $[m]$; R_β^i - radius of the non-coupled rod which manifests an excess protrusion over the s/l interface, $[m]$, (Wołczyński, 1996).

$$\lambda_s^i = 2\pi\left(\Gamma_{(Zn)} / \left(G - m_{(Zn)} G_C\right)\right) \tag{22}$$

where λ_s^i - a wavelength of perturbation which appears at the s/l interface of the (Zn) – non-faceted phase, (as defined in Fig. 12.) $[m]$, with $\Gamma_{(Zn)}$ - *Gibbs-Thomson* capillarity parameter; G - thermal gradient at the s/l interface, $[K / m]$; $m_{(Zn)}$ - slope of the *liquidus* line of the (Zn) – phase, $[K / at.\%]$; G_C - solute concentration gradient at the s/l interface of the (Zn) – phase, $[at.\% / m]$. In the range: $v_1 < v < v_1'$, the following transformation occurs:

$$\lambda_s^i \to R; \quad R_\beta^i \to 0 \quad \text{and finally} \quad \overline{R} \to R \tag{23}$$

It means that irregular rod-like structure transforms into the regular rod-like structure. This transition leads to the local disappearing of branches. Eventually, the regular rods transform into the regular lamellae within the range, $v_1 < v < v_1'$, Fig. 21.

5. Concluding remarks

A pattern selection among the eutectic morphologies is performed through the competition. In the case of the eutectic morphologies which were created within the stripes strengthening the (Zn) single crystal the L-shape irregular rods, regular lamellae or regular rods were revealed. According to the current model, this eutectic structure appeared which was the winner just in the thermodynamic competition. Therefore, it was necessary to determine the minimum entropy production for the appearance of the all studied structures. The formed structure had its minimum entropy production situated lower than the other participant of the competition. So, it has been proved that the regular L-shape rods (existing among the generally irregular L-shape rods) have their minimum entropy production situated lower than the minimum entropy production determined for the regular lamellae growth (when $0 < v < v_1$). However, the regular lamellae are the winner within the growth rates range $v_1' < v < v_2$. Eventually, the regular rods exist exclusively when the growth rate fulfils the following inequality $v > v_2$ since the minimum entropy production calculated for their formation is localized lower than the minimum entropy production determined for the regular lamellae growth, Fig. 27. Both minima being in competition are at the same level for the threshold growth rates, v_1, and v_2, Fig. 26b, Fig. 26d, respectively, (also Fig. 27).

It has been revealed that the leading eutectic phase (the $Zn_{16}Ti$ intermetallic compound) was growing with the d - protrusion, Fig. 8, Fig. 9 and Fig. 10. The protrusion can be determined based on the solute concentration micro-field in the liquid ahead of the s/l interface, Eq. (7), and Eq. (8), but the local mass balance is to be considered, Eq. (10).

In fact, the theoretical definition for the protrusion results from the analysis of the local mass balance (Wołczyński, 2007) and is formulated for the slow solidification as follows:

$$\sum_{n=1}^{\infty} A_{2n-1} \frac{(-1)^{n-1}}{(2n-1)} \left[1 - \frac{S_\alpha}{S_\beta} \exp\left(-\frac{(2n-1)\pi}{2S_\beta} d \right) \right] = 0 \tag{24}$$

A_{2n-1} - constant in the *Fourier* series. According to the observation of the leading phase protrusion, d, Fig. 8, Fig. 9 and Fig. 10, it is suggested to introduce the revealed protrusion into the structural scheme shown in Fig. 12.

Also, the operating range for the irregular eutectic growth (as defined in Fig. 12) can be justified by some structural observations. These observations are connected with the L-shape rod-like structure revealed for the growth rates range, $0 < v < v_1$. A typical spacing is to be distinguished within the operating range for the irregular eutectic growth:

a. the λ_i - spacing which is associated with the regular lamellar structure formation among generally irregular structure, (or the R - spacing which is associated with the regular rod-like structure formation, Fig. 28a); this spacing is referred to the criterion of the minimum entropy production,

b. the λ_s^i - spacing which corresponds to the maximal destabilization of the s/l interface of the (Zn) - non-faceted phase, Fig. 28b; this spacing (treated as perturbation wave, λ_s^i), is referred to the state of marginal stability which is created at that very moment at the s/l interface.

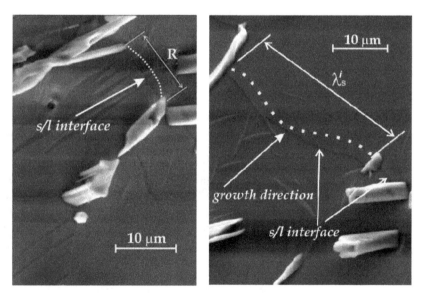

Fig. 28. Selection of the spacing in the irregular eutectic growth; a. λ_i - spacing selected by the minimum entropy production, and b. λ_s^i - spacing selected by the marginal stability.

It is evident that the oscillation of the interlamellar or interrod spacing occurs in the (Zn) – $Zn_{16}Ti$ eutetcic system. The analyzed structural oscillation is shown schematically in Fig. 29.

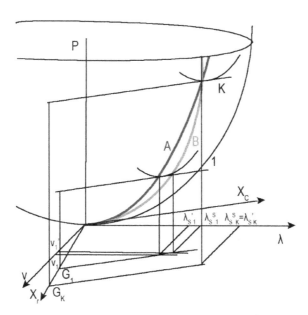

Fig. 29. A paraboloid of the entropy production drawn in function of two thermodynamic forces (X_C, X_r), with the added so-called "technological" coordinate system, (v, λ).

X_C - generalized thermodynamic force associated with the mass transport, $[at.\% / m]$; X_r - generalized thermodynamic force associated with the heat transfer; $[K / m]$; v - crystal growth rate, $[m / s]$; λ - inter-lamellar spacing within the regular eutectic structure, $[m]$

The paraboloid of the entropy production is drawn schematically in the "thermodynamic" coordinate system, that is, in function of two thermodynamic forces. Next, it is assumed (for the simplicity) that the paraboloid does not change its shape when entropy production, P, is calculated for the crystal growth in function of v, λ, Eq. (20b) or v, R, Eq. (20a).

Two trajectories are superposed onto the paraboloid, Fig. 29: a. A - trajectory of local minima of the paraboloid for the regular structure formation, Fig. 28a, /B - trajectory of the marginal stability referred to the maximal destabilization of the s/l interface, Fig. 28b. Both trajectories intersect each other (at K - point) for the critical thermal gradient, G_K, at which an oscillation disappears and irregular structure is completely transformed into regular eutectic structure. However, for a given condition of solidification (crystal growth rate, v_1, and thermal gradient, G_1), the destabilized s/l interface moves slower, with the v_1' - rate, ($v_1' < v_1$), and the perturbation wave, λ_s^i, is equal to λ_{s1}^s. On the other side, the regular structure appears at the v_1 - rate and involves the $\lambda_i = \lambda_{s1}'$ - spacing formation, Fig. 29. Thus, the discussed oscillation occurs between both trajectories as shown for v_1, G_1, Fig. 29.

The oscillation of the eutectic spacing, Fig. 28, is observed in the case of the irregular eutectic structure formation formed within the stripes strengthening the (Zn) - single crystal for the growth rates range equal to $0 < v < v_1$, as mentioned.

The discussed oscillation can be illustrated on the parabola of entropy production. Such a parabola is to be created by an intersection of the paraboloid, Fig. 29, by the plane given for the imposed thermal gradient, $G = const.$, (Prigogine, 1980), as shown in Fig. 30.

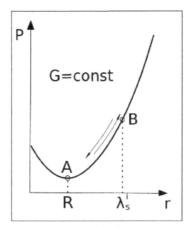

Fig. 30. A parabola of the entropy production for the eutectic rod-like structure formation.

It results from the current model, Fig. 30, that:

a. the regular part of the generally irregular eutectic structure is formed at the minimum entropy production localized at the A - parabola minimum which performs the role of attractor for the eutectic system,

b. the maximal destabilization at the s/l interface of the non-faceted eutectic phase is observed for the B - marginal stability of the eutectic system which corresponds to the branching phenomenon (as visible in Fig. 12) and to maximal deviation from the attractor.

6. Acknowledgement

The financial support from the Polish Ministry of Science and Higher Education (MNiSW) under contract N N 508 480038 is gratefully acknowledged.
Sincere gratitude is expressed to Dr Grzegorz Boczkal. Without his constant support and great patience with experimental observations this chapter would never have been written.

7. References

Atasoy, O. (1984). Effects of Unidirectional Solidification Rate and Composition on Inter-particle Spacing in Al-Si Eutectic Alloy, *Aluminium*, Vol. 60, pp.109-116

Boczkal, G.; Mikułowski, B.: & Wołczyński, W. (2010). Oscillatory Structure of the Zn-Cu-Ti Single Crystals, *Materials Science Forum*, Vol. 649, pp. 113-118

Brody, H.D.; & Flemings, M.C. (1966). Solute Redistribution in Dendritic Solidification, *Transactions of the Metallurgical Society of the AIME*, Vol. 236, pp. 615-624

Cupryś, R.; Major, B.; & Wołczyński, W. (2000). Transition of Flake into Fibre Structure in Eutectic Al-Si, *Materials Science Forum*, Vol. 329/330, pp.161-166

Davies, V.L. (1964). Mechanisms of Crystallization in Binary Eutectic Systems, *Journal of the Institute of Metals*, Vol. 93, pp. 10-14

Ebeling, R.; & Ashby, M.F. (1966). Dispersion Hardening of the Copper Single Crystal, *Philosopher Magazine*, Vol. 13, pp. 805-834

Elliott, R. (1977). Eutectic Solidification, *International Metals Reviews*, Vol. 219, pp. 161-186

Fisher, D.J.; & Kurz, W. (1980). A Theory of Branching Limited Growth of Irregular Eutectics, *Acta Metallurgica*, Vol. 28, pp. 777-794

Glansdorff, P.; & Prigogine, I. (1971). *Thermodynamic Theory of Structure, Stability and Fluctuations*, Wiley – Interscience, 306 pages, John Wiley & Sons, (Ed.), ISBN: 0 471 30280 5, London, UK – New York, USA – Sydney, Australia – Toronto, Canada

Jackson, K.A.; & Hunt, J.D. (1966). Lamellar and Rod Eutectic Growth, *Transactions of the Metallurgical Society of the AIME*, Vol. 236, pp. 1129-1142

Kjelstrup, S.; & Bedeaux, D. (2008). *Non-Equilibrium Thermodynamics of Heterogeneous Systems*, World Scientific Pulishing Co. Ltd., 434 pages, M. Rasetti, (Ed.), ISBN: 13-978-981-277-913-7 and ISBN: 10 981-277-913-2, New Jersey, USA – London, UK – Singapore, – Beijing, China – Shanghai, China – Hong Kong – Taipei, Taiwan– Chennai, India

Lesoult, G.; & Turpin, M. (1969). Etude Theorique sur la Croissance des Eutectiques Lamellaires, *Revue Scientifique de la Revue de Metallurgie*, Vol. 66, pp. 619-631

Murray, J. (1990). Zn-Ti Phase Diagram for the Stable Equilibrium, In: *Binary Alloy Phase Diagrams*, T. Massalski, (Ed.), ASM International, pp. 3500-3502, ISBN: 0-87170-406-4, Park Ohio, USA

Prigogine, I. (1968). *Introduction a la Thermodynamique des Processus Irreversibles*, (Monographies DUNOD), 160 pages, John Wiley & Sons, (Ed.), Paris, France

Prigogine, I. (1980). *From Being to Becoming: Time and Complexity in the Physical Sciences*, The Maple-Vail Book Manufacturing Group, 286 pages, W.H. Freeman and Company, (Ed.), ISBN: 0-7167-1107-9 and ISBN: 0-7167-1108-7, USA

Scheil, E. (1942). Über die Eutektische Kristallisation, *Zeitschrift für Metallkunde*, Vol. 34, pp. 70-80

Steen, H.A.H.; & Hellawell, A. (1975). The Growth of Eutectic Silicon – Contributions to Undercooling", *Acta Metallurgica*, Vol. 23, pp. 529-536

Toloui, B.; & Hellawell, A. (1976), Phase Separtion and Undercooling in Al-Si Eutectic Alloy – The Influence of Freezing Rate and Temperature Gradient, *Acta Metallurgica*, Vol. 24, pp. 565-572

Wołczyński, W. (1996), Thermodynamics of Irregular Eutectic Growth, *Materials Science Forum*, Vol. 215/216, pp. 303-312

Wołczyński, W. (2000). Back-Diffusion Phenomenon during the Crystal Growth by the Bridgman Method, In: *Modelling of Transport Phenomena in Crystal Growth*, J.S. Szmyd & K. Suzuki, (Ed.), pp. 19-59, WIT *PRESS* ISBN: 1-85312-735-3, Ashurst Lodge, Southampton, UK - Boston, USA

Wołczyński, W. (2007). Concentration Micro-Field for Lamellar Eutectic Growth, *Defect and Diffusion Forum*, Vol. 272, pp. 123-138

Wołczyński, W. (2010). Lamella / Rod Transformation as described by the Criterion of the Minimum Entropy Production, *International Journal of Thermodynamics*, Vol. 13, pp. 35-42

Wołczyński, W.; & Billia, B. (1996). Influence of Control and Material Parameters on Regular Eutectic Growth and Inter-Lamellar Spacing Selection, *Materials Science Forum*, Vol. 215/216, pp. 313-322

Zarubova, N.; & Sestak, B. (1975). Plastic Deformation of Fe-3 wt.% Si Single Crystals in the Range from 113 to 473 K, *Physica Status Solidi*, Vol. 30, pp. 365-374.

7

The Influence of Atmosphere on Oxides Crystal Growth

Morteza Asadian
Iranian National Center of Laser Science and Technology,
Tehran,
Iran

1. Introduction

Oxide single crystals are known for wide application in electronics and optical industries, e.g. lasers, substrates, scintillators, nonlinear and passive optical devices (Brandle, 2004). The Czochralski technique has become an important method of choice for the growth and production of many bulk oxide materials owing to the possibility of growing crystals with large size, core free with good optical quality, and a high concentration of dopant, e.g., Nd^{3+}, Yb^{+3} with better homogeneity (Zhaobing et al., 2007; Albrecht et al., 1998).

Numerous researchers (Dupret and Bogaert, 1994; Cockayne et al., 1976; Valentino and Brandle, 1974; Brandle, and Barns, 1974; Jacobs et al., 2007; Mateika et al., 1982; Piekarczyk and Pajaczkowska, 1979) have studied the effect of different parameters on the quality of oxide crystals grown by Czochralski technique. They found that Crystal growth atmosphere composition and melt stoichiometry are strongly influenced on the quality of crystals as well as solid-liquid interface shape.

However, the growth of oxides such as $Gd_3Ga_5O_{12}$ [Brandle et al., 1972], ZnO [Klimm et al., 2008a], $SrLaGaO4$ [Pajaczkowska et al., 2001], single crystals by standard melt growth techniques are difficult due to their high melting point and thermal decomposition upon heating. Otherwise, if the composition of melt would be serious, the chemical composition of these oxides melt will depart from the congruent point and superfluous components would become the harmful impurities for crystal growth.

Due to the volatilization of oxide melts and the necessity of an oxidizing atmosphere, the crucible that contains the melt should consist of a noble metal. The only materials that have been found to tolerate these violent conditions are the platinum group of metals and their alloys (Day, 1963; Cockayne, 1974). Although the growth from platinum crucible can be performed in presence of oxygen, unfortunately, platinum with T_m=2042 K and its alloys cannot use for mostly oxides because of their high melting point (Darling et al., 1970). Very often platinum or alloys of platinum are used only for oxides with low melting point (T<1850 K) (Uitert, 1970). Actually, it is necessary to choose materials of higher melting point, such as the rhodium-platinum alloys, or rhodium or iridium metals (Cockayne, 1968). Owing to a high melting point and high mechanical strength even at T> 2250 K° (Handley,1986), iridium crucible are widely used for growing high melting oxides such as GGG, Al_2O_3 and ZnO. It is important to note that, iridium is more sensitive to oxidation than platinum or rhodium (Weiland et al., 2006), especially in the medium temperature (~870-1000 K°) as the metal would oxidize to IrO_2 (s) (Asadian et al., 2010). Consequently, the addition of oxygen to the furnace chamber reacts with iridium crucible at medium temperatures and it complicates the process of crystallization (Lipinska et al., 2009).

The ceramic oxide crucibles are usually utilized for melting metals (Day, 1963). Since they are all chargeable to chemical react by the molten oxides, unfortunately, they cannot apply as a part to contact directly to the oxides melt (Klimm and Schroder, 1999). The high melting point metal such as molybdenum (T_m=2896 K) and tungsten (T_m= 3680 K) are oxidized rapidly by oxygen at high temperatures and can only be used in reducing or neutral atmospheres (Klimm et al., 2008b).

This chapter is organized as follows. An outline of thermodynamic background is illustrated in section 2. Section 3 contains a detailed discussion of the thermodynamic behaviour of construction materials (crucible, seed holder, after heater). Section 4 describes partial pressures effect of gases atmosphere. The short review of growth atmosphere effect on crystal growth process including two examples is presented in section 5. Section 6 concludes the chapter.

2. Thermodynamic background

Consider the general redox equilibrium reaction

$$aM_xO_y \leftrightarrow bM_zO_w + 1/2O_2 \tag{1}$$

With

$$\text{Log}\left(K_p\right) = \frac{1}{2.30R}(\Delta S^o - \frac{\Delta H^o}{T})(J.K^{-1}.mol^{-1}) \tag{2}$$

The equilibrium constant (K_p) gives the required combination of the activities of the reactants as

$$K_p = (\frac{a_{M_zO_w}^b \times P_{O_2}^{1/2}}{a_{M_xO_y}^a}) .$$

R is the universal gas constant, ΔH^o is the standard enthalpy change and ΔS^o is the standard entropy change for the reaction. The choice of the pure material a standard states makes the activity of material has values of unity in equation (2). In which case for material oxides, if the oxide pure is solid oxides, the activity of solid oxides will be unit. Moreover, when the material oxides would be gas forms the activity of the oxides replace to their pressures (Gaskell, 2003).

For the oxidation of (a)dn mole material oxide from the m valent state to the m- s valent state proceeds under the consumption of (s/2)dn mole oxygen. Herein m=2y/x and s=2/ax. m valent gives a number between 1-8 and always $m \geq s$. The Table 1 indicates four possible conditions of the redox equilibrium reactions which are obtained with simplifying the equation (1).

m	s	The redox equilibrium reaction
even	even	$2MO_{(m/2)} \leftrightarrow 2MO_{(m-s)/2} + \dfrac{s}{2}O_2$
even	odd	$2MO_{(m/2)} \leftrightarrow M_2O_{(m-s)} + \dfrac{s}{2}O_2$
odd	even	$M_2O_m \leftrightarrow M_2O_{(m-s)} + \dfrac{s}{2}O_2$
odd	odd	$M_2O_m \leftrightarrow 2MO_{(m-s)/2} + \dfrac{s}{2}O_2$

Table 1. The different conditions of redox equilibrium reaction.

The reaction equilibrium between pure material M, its oxide and oxygen gas where m=s would take place. In which case $\Delta G°$ is a linear function of temperature and a plot of $\Delta G°$=RT Log (PO$_2$) versus temperature gives the lines for each oxidation reaction on an Ellingham diagram (Elingham, 1944). These lines separate phase regions where one oxidation state overcomes and the whole graph represents a predominance phase diagram for the material M and its oxides.

3. Construction materials

Platinum group metals and their alloys can usually be safely heated for long periods in contact with the more refractory oxides without serious risk of contamination. Due to their excellent chemical stability, oxidation resistance, and resistance to the action of many molten oxides, the platinum group metals: iridium, platinum and rhodium are widely used for high-temperature applications involving simultaneous chemical attack and mechanical strength. Important applications of platinum group metals are as crucibles for oxides crystal growth from melt. Oxides of major optical interest and their most suitable crucible materials are listed in Table 2.

Material	Oxide melting point (K)	Crucible	Material use
Lithium Niobate (LiNbO$_3$)	1523	Platinum or rhodium-platinum	Electro-optic
Calcium Tungstate (CaWO$_4$)	1839	Rhodium	Laser host
Gadolinium Gallium Garnet (Gd$_3$Ga$_5$O$_{12}$)	2023	Iridium	Laser host
Zinc Oxide (ZnO)	2248	Iridium	Green laser-UV light emitters
Sapphire (Al$_2$O$_3$)	2327	Iridium	Laser host

Table 2. Application of oxides single crystal and appropriate crucibles.

3.1 Platinum

The only materials that have been found to withstand high temperature chemical stability, oxidation resistance and resistance to react with many molten oxides are the platinum group of metals and their alloys. Although growth from Platinum crucibles can be performed in air, platinum does not appear to be widely used for the growth of oxide materials, probably due to its relatively low melting point (2042 K) in comparison to that for most refractory oxides or mixtures of oxides. The low melting point of lithium niobate ($LiNbO_3$), T_m=1523 K, (Day, 1963) and the chemical inertness of platinum suggested that this metal might be a useful crucible material, while platinum also offered some advantage in cost over the platinum group metals.

3.2 Rhodium

In a few instances, unalloyed platinum crucibles have been found satisfactory, but generally, it is necessary to choose materials of higher melting point, such as the rhodium platinum alloys, or the metals rhodium or iridium.

In the past, there has been some hesitation in the use of rhodium and iridium crucibles in oxidizing atmospheres at high temperatures through fear of high losses from the formation of volatile rhodium and iridium oxides. Based on this concept, Nassau and Broyer (1962) have used rhodium and iridium crucibles successfully for growing barium titanate single crystals. They were grown at about 1920 K from rhodium crucibles, when 0.36 weight percent of rhodium was detected in the crystals and from iridium crucibles when 0.02 weight percent of iridium was observed.

It is known that Rh reacts with oxygen above 1000 K (Chaston, 1965) forming Rh_2O_3 (s). Fig. 1 shows the stability regions phase in the Rh-O system from available thermodynamic data (Binnewies and Milke, 2008). As illustrated in Fig. 1, Rh_2O_3 (s) is stable up to 1412 K where the oxygen pressure reaches 1.0 atm. Above this temperature, Rh_2O_3 (s) decomposes to gaseous phase according to the equation

$$Rh_2O_3 \ (s) = RhO_2 \ (g) + 1/2O_2 \tag{3}$$

As pointed out, the dissociation of Rh_2O_3 at very low oxygen pressure $PO_2 < 10^{-35}$ (atm) and at room temperature, it maybe possible to take place this reaction

$$Rh_2O_3 \ (s) = Rh \ (s) + 3/2O_2 \tag{4}$$

From Fig. 1, it is limited to apply rhodium at high oxygen pressure as crucible because of evaporate to RhO_2. Therefore, the oxygen partial pressure must be low to avoid oxidation of crucible especially at low temperature.

3.3 Iridium

Owing to have a high melting point and high mechanical strength even at T> 2250 K°, iridium is a particularly suitable material for applications such as the stress-rupture strength, creep behavior and thermal shock which preclude the use of platinum alloys or rhodium (Weiland et al., 2006). Important applications of iridium is as crucible for pulling refractory oxides crystals such as GGG (T_m=2023 K) and ZnO (T_m=2248 K).

It is noticeable that iridium crucible are sensitive to oxygen, especially at the medium temperature (~870-1000 K°) as the metal would oxidize to solid iridium oxide. IrO_2 (s) decomposes to iridium metal at temperature higher than 1370 K° according to the equation

$$IrO_2 (s) = Ir (s) + O_2 \tag{5}$$

As illustrated in Fig. 2, IrO_2 (s) is stable at the temperature less than 1370 K° at standard condition. If oxygen were applied at higher temperature than 1370 K°, iridium parts would not be oxidized.

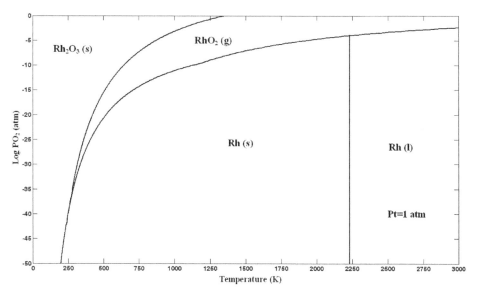

Fig. 1. Predominance diagram for Rh-O_2 system.

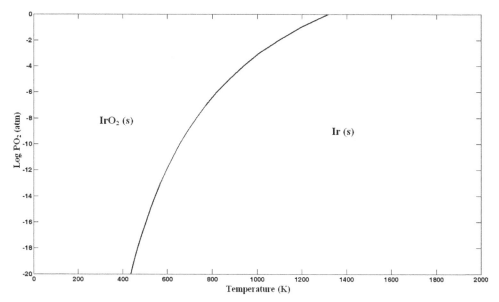

Fig. 2. Predominance diagram for Ir and IrO_2 (s) in the temperature- PO2 plane.

Several studies (Cordfunke and Mayer, 1962; Schafer and Heitland, 1960) have been shown that the oxidation of iridium at high oxygen pressure forms IrO_3 (g). As mentioned before, the decomposition temperature of IrO_2 (s) in oxygen at 1 atm is about 1370 K. Hence, above this temperature the equilibrium forming IrO_3 (g) follows

$$2Ir \ (s) + 3O_2 = 2IrO_3 \ (g) \tag{6}$$

Below 1370 K the volatile oxide dissociates to IrO_2 (s) and oxygen by the reaction (Cordfunke and Meyer, 1962)

$$2IrO_3 \ (g) = 2IrO_2 \ (s) + O_2 \tag{7}$$

The major volatile species in the iridium-oxygen system is IrO_2 (g) (Chandrasekharaiah et al., 1981; Carpenter, 1989). For the reaction

$$Ir \ (s) + O_2 = IrO_2 \ (g) \tag{8}$$

The results of these thermodynamics calculation are plotted in Fig. 3 as Log PO_2 against temperature. It can be seen that IrO_3 (g) is identified as the major gaseous species at low temperatures (800 K- 1700 K). At 2033 K, Norman et al. (1965) determined the IrO_2 (g) pressure to be 1.9×10^{-8} atm and the IrO_3 (g) pressure to be 3.1×10^{-9} atm. This is indicated that at elevated temperatures IrO_2 (g) is predominant gaseous species (seen Fig. 3).

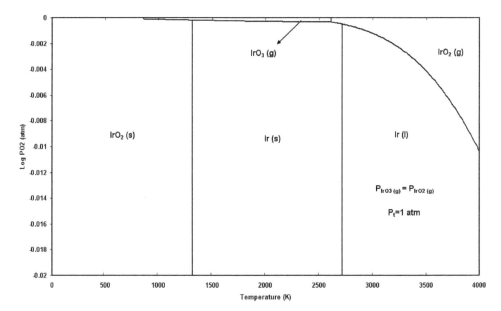

Fig. 3. Predominance diagram for $Ir-O_2$ system.

To sum up, although iridium is more sensitive to oxidation than platinum (E= -0.1474 V for Ir and E= -0.4422 V for Pt (Klimm et al., 2008c), it is the most chemically resistant of all metals. The high melting point of iridium (2716 K) and its resistance to attack by stable oxide melts is maintained up to temperatures above 2250 K make it a particularly suitable material for oxides crystal growth from hot melt which preclude the use of platinum alloys or rhodium.

4. Growth atmosphere

The oxides crystal growth atmosphere should have two conditions
1) Consider a type of oxide material Me, in Table 1, i. e.,

$$Me_2O_m(s,l) \leftrightarrow Me_2O_{(m-s)}(s,l,g) + \frac{s}{2}O_2 \tag{9}$$

The oxygen partial pressure must be in that range where the favored valency of the oxide material (Me_2O_m) is stable. Always PO_2 system should be more than $(PO_2)_{min}$ to suppress the decomposition of oxide. Herein $(PO_2)_{min}$ is the minimum partial pressure of O_2 where the reaction (9) moves to the left side for the given growth temperature $((PO_2)_{eq} > (PO_2)_{min})$.
2) For the reaction equilibrium between a pure solid construction material M, its pure oxide and oxygen gas

$$2M(s) + \frac{m}{2}O_2 \leftrightarrow 2MO_{(m/2)}(s, g) \tag{10}$$

The oxygen partial pressure should be low enough to avoid oxidation of construction parts (crucible, seed holder) being in contact with the melt. Herein PO_2 system should be less than $(PO_2)_{max}$ with $(PO_2)_{max}$ - the maximum oxygen partial pressure where the construction material is equilibrium with its oxide in the m valency state $((PO_2)_{max} > (PO_2)_{eq})$. Because the oxide construction parts are often stable at lower temperatures and oxide crystal almost decomposed at higher temperatures, both conditions are accomplished, if $(PO_2)_{min} < (PO_2)_{eq} < (PO_2)_{max}$. Commonly, amounts of oxygen, e.g. for the growth of $Gd_3Ga_5O_{12}$ (GGG) is about 1-2 vol% (Ganschow, 2010), add to the growth atmosphere to prevent volatile oxide. However, this value (PO_2) often is more than $(PO_2)_{max}$ at lower temperatures. Therefore, it causes that the construction material oxidized.
A solution to this problem is that mixtures containing an oxygen bearing gas like CO_2 or H_2O would be utilized in the growth atmosphere. They produce a temperature dependent oxygen partial pressure in this manner that the O_2 partial pressure increases with temperature. Hence, oxidation of construction parts decrease.
Another solution is that the protective gas (N_2, Ar) atmosphere would be charged into the furnace chamber in ambient temperature before heating the raw material and beginning the crystal growth process. The atmospheric pressure of the chamber should be more than ambient pressure. To prevent the evaporation of oxide, amount of oxygen should add into the during the crystal growth process at temperatures higher than temperature $MO_{m/2}$ decomposition. Growth experiments for Nd:GGG (Asadian, 2011) prove that the iridium crucible is not oxidized when the required oxygen was charged at T>1370 K°.

4.1 CO₂-CO system
CO_2 decomposes into CO and oxygen with increasing temperature according to

$$2CO_2 = 2CO + O_2 \tag{11}$$

If a moles of CO and 1 mole of CO_2 are mixed ($a=CO/CO_2$), then from stochiometry of Eq. (11), 2x moles of CO_2 would decompose to form 2x moles of CO and x moles of O_2 such that, at total pressure P_t, the PO_2 in CO-CO_2 system in various temperatures would be

$$Log \frac{x(a+2x)^2}{(1+a+x)(1-2x)^2} P_t = 9.04 - \frac{29613.35}{T} \tag{12}$$

Which,

$$PO_2 = \frac{x}{1+a+x} P_t \tag{13}$$

The partial pressure of oxygen can be "automatically" increased in the heating-up phase of the growth process by the thermal decomposition of carbon dioxide Eq. (11). Temperature dependence of PO_2 with comparing different gases for the above reaction is shown in Fig. 4.

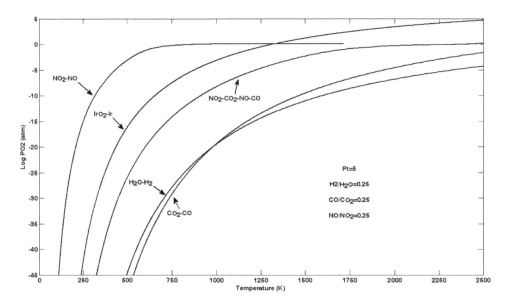

Fig. 4. Temperature dependence of PO_2 within different gases and gas mixtures comparing with stability regions for iridium (iridium metal and its solid oxide).

Thermodynamic calculations (Tomm et al.,2000) of the chemical equilibria between Ga_2O_3 (solid and liquid phase) and the surrounding gas-phase showed that an Ar+10% CO_2 atmosphere delivers an oxygen partial pressure that increases in such a manner with temperature, that Ga_2O_3 crystals can be grown from the melt. This amount of liberated oxygen is just sufficient to decrease the evaporation of molten Ga_2O_3 as well as low enough to allow the use of iridium crucibles, i.e., allowing Czochralski growth of Ga_2O_3.

4.2 H₂O-H₂ system

H_2O-H_2 system are used when it is required that the partial pressure of oxygen in a gas phase be fixed at a very low value at low temperatures. For example, if it were required to have a gaseous atmosphere containing an oxygen partial pressure of 10^{-30} atm at 705 K, then such an oxygen potential pressure can be obtained with simple relation by establishing the equilibrium

$$2H_2O = 2H_2 + O_2 \tag{14}$$

The partial pressure of oxygen in H_2-H_2O gaseous atmosphere in various temperatures from Eq. (14)

$$Log\frac{x(c+2x)^2}{(1+c+x)(1-2x)^2}P_t = 4.56 - \frac{25302.15}{T} \tag{15}$$

Which,

$$C = H_2/H_2O$$

$$PO_2 = \frac{x}{1+c+x}P_t \tag{16}$$

The oxygen pressure of H_2-H_2O is shown in Fig. 4. Comparing the H_2-H_2O system with CO-CO_2 indicate that H_2-H_2O system products more oxygen at lower temperatures. However, at higher temperatures CO-CO_2 system is the more efficient oxidant agent.

4.3 NO₂-NO system

Similarly, the partial pressure of oxygen in NO_2-NO system can be determined by establishing the reaction

$$2NO_2 = 2NO + O_2 \tag{17}$$

Thus

$$Log\frac{x(b+2x)^2}{(1+b+x)(1-2x)^2}P_t = 7.67 - \frac{5985.45}{T} \tag{18}$$

Which,

$$b = NO/NO_2$$

$$PO_2 = \frac{x}{1+b+x}P_t \tag{19}$$

The oxygen partial pressure in NO_2-NO gaseous atmosphere in various temperatures is drawn in Fig. 4. Comparing with others gaseous components, at lower temperatures the released oxygen is too much in NO_2-NO system. Therefore, the iridium crucible indeed would be oxidized.

4.4 NO₂-CO₂-CO-NO system

Consider the reaction CO_2 and NO_2 are mixed in the molar ratio 1:1 to form NO, CO and O_2 according to

$$CO_2 + NO_2 = CO + NO + O_2 \tag{20}$$

To obtain this gas mixture, CO_2 and CO, NO_2 and NO at total pressure P_t, would be mixed in the molar ratio $CO/CO_2 = a$, $NO/NO_2 = b$ and allowed to equilibrate. Which,

$$Log \frac{x(a+x)(b+x)}{(2+a+b+x)(1-x)^2} P_t = 8.36 - \frac{17799.40}{T} \tag{21}$$

$$PO_2 = \frac{x}{2+a+b+x} P_t \tag{22}$$

According Fig. 4, NO_2-NO and CO_2-CO gas mixtures can be used when it is required that the partial pressure of oxygen very low at lowers temperature to suppress oxidation of iridium crucible. Also they can be utilized when it is required that the high partial pressure of oxygen at elevated temperature. In this case, ZnO crystal growth is a good example. For the growth of ZnO in iridium crucible by Czochralski method, the oxygen partial pressure of chamber at ZnO melting point (2248 K) should be more than $(PO_2)_{min} = 0.35$ atm (Klimm et al. , 2009) to have ZnO melt. In order to suppress the oxidation of iridium crucible, the PO_2 of system should be less than $(PO_2)_{max}$. According Table 3, NO_2-NO and CO_2-CO gas mixtures can be used for ZnO crystal growth.

PO₂ (atm) Temperature	$(PO_2)_{min}$ (Ga_2O_3)	$(PO_2)_{min}$ (ZnO)	H_2-H_2O $(PO_2)_{eq}$	CO-CO₂ $(PO_2)_{eq}$	NO-NO₂ $(PO_2)_{eq}$	CO₂-NO₂-CO-NO $(PO_2)_{eq}$	$(PO_2)_{max}$ (iridium)
900 K	7.94×10⁻³⁸	9.52×10⁻¹⁴	5.6×10⁻²³	2.0×10⁻²³	1.06	6.3×10⁻¹¹	2.57×10⁻⁵
$(T_m(G_2O_3))$ 2080 K	1.12×10⁻⁵	1.53×10⁻²	4.89×10⁻⁷	1.01×10⁻⁴	1.33	0.63	10² <
$(T_m(ZnO))$ 2248 K	-------	7.76×10⁻²	4.0×10⁻⁶	1.20×10⁻³	1.41	0.89	10² <

Table 3. Maximum, minimum and required PO_2 that is supplied by different gas mixtures at different temperatures.

5. Oxides crystal growth

Oxides crystal growth are often performed from melts contained in crucibles, e.g. by Czochralski or Bridgman technique. For oxide components, oxygen partial pressure during growth is one of the most important parameters to decide about success or failure of crystal growth process. Many oxides may be easily decomposed at high temperatures and low oxygen partial pressure. Therefore, oxides crystal growth is often carried out in a protective gas atmosphere such as Ar and N_2 to avoid the oxidization of crucibles and to minimize dissociation of oxide. An amount of oxygen partial pressure is always necessary to remain the stability of the oxide.

5.1 Ga_2O_3
The growth of GGG single crystal by the Czochralski method, in high-quality and large-sized is hard because of the dissociation of Ga_2O_3 (Luo et al. ,2005; Li et al. , 2007). At the T> 1500 K° (Klimm et al., 2008), reduction to evaporate Ga_2O may occur, whereas, the serious evaporation of Ga_2O_3 takes place at high temperatures. Fig. 5 represents a schematic of the experimental setup employed for the Nd:GGG crystal growth.

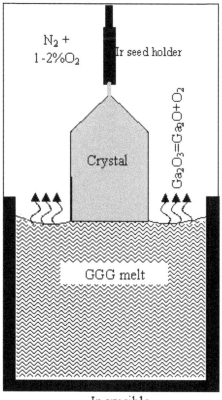

Fig. 5. Sketch diagram of GGG crystal growth setup.

Based on thermodynamic analysis, Ga_2O (g) and Ga (l) is main decomposed constituent via heating Ga_2O_3. During single crystal growth process by the Czochralski method, Ga_2O_3 sublimes by dissociation:

$$Ga_2O_3 \text{ (s,l)} = Ga_2O \text{ (g)} + O_2 \tag{23}$$

The diagram Log P_{O_2} (T) for gallium oxides is shown in Fig. 6. At above 1500 K°, the evaporation of Ga_2O_3 may take place but the most serious Ga_2O_3 volatile will be happened at temperatures than higher 1950 K°. According to the chemical equilibria law, charging amount of oxygen into system decreases the volatilization of Ga_2O_3. For GGG crystal growth by Czochralski method, the pressure of oxygen (PO_2) should be more than 10^{-4} atm to preserve stability of growth process.

Additionally, it would be extracted from Fig. 6, at lower pressures ($PO_2 < 10^{-11}$ atm), the decomposition of Ga_2O_3 to liquid gallium may occur.

$$Ga_2O_3 \text{ (s)} = 2Ga \text{ (l)} + 1.5O_2 \tag{24}$$

The liquid gallium will rapidly make to alloy with iridium crucible. Unfortunately, the formation of iridium-gallium alloys can definitely destroy the iridium crucible. Moreover, the Ga^+ ions decomposed from the Ga_2O_3 enter the lattice of Nd:GGG crystal ultimately cause the spiral growth.

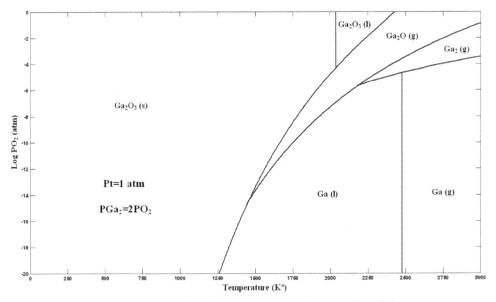

Fig. 6. Predominance diagram for Ga-O_2 system in dependence on T and the oxygen pressure PO_2 at total pressure 1 atm.

A solution to this problem is that the addition of oxygen would be replaced by carbon oxide (CO_2) in the growth atmosphere. The O_2 partial pressure increases with temperature; hence, oxidation of iridium crucible decreases (see Fig. 6). Matika et al. (1982) used a gas mixture of 50% CO_2 and 50% N_2 to reduce the dissociation of Ga_2O_3 and to suppress the formation of iridium within crystals. They obtained the densities of dislocations and inclusions for single crystals were below 5 cm^{-2}. Tomm et al. (2000) reported the first successful growth of Ga_2O_3 single crystal by Czochralski method. They have decreased evaporation of molten Ga_2O_3 by means of Ar plus 10% CO_2 instead of O_2 in the atmosphere.

5.2 ZnO

Zinc oxide is a group II-VI semiconductor that most commonly exhibits wurtzite structure in its unstrained phase (Park et al., 2006). Wurtzite crystals can be cleaved cleanly along several different planes. The c-plane (0001) intercepts only the z axis and in the case of zinc oxide produces polar surfaces as a result of the lack of inversion symmetry along the c axis (Miller, 2008). Furthermore, because of its wurtzite crystal structure and lattice parameter (a=3.25A$^{\circ}$ in the a-direction) it is isomorphous with GaN and has potential as a substrate material for GaN-based epitaxial devices (Pearton, et al., 2005). It has a large exciton binding energy of 60 meV and, as a result, stimulated emission and lasing have been observed at room temperature (Bagnall et al., 1998)]. This means that ZnO excitons will remain bound at much higher temperatures than GaN and in doing so, offer the potential for greatly improved efficiency over GaN. This indicates that ZnO can be the preferred choice for future opto-electronic devices.

Attempts are made to grow ZnO a large number of different methods. Bulk ZnO crystals can be grown using melt (Klimm et al.,2008a), hydrothermal (Suscavage et al.,1999) and vapor phase growth techniques (Look et al., 2002).

Unfortunately, traces of the solvent are always incorporated in such hydrothermal crystals and hydrogen and lithium are found typically in concentrations of several 10^{18} cm^{-3}. In addition, the suitability of ZnO crystals for epitaxy is highly dependent on surface preparation and subsequent characterization.Therefore; Attempts have recently been made to grow single crystal bulk from the melt.

Single crystals of ZnO can be grown from the melt at high oxygen pressure and high temperature. The Czochralski (Klimm et al., 2008a) and Bridgman (Jacobs et al., 2009) methods are the most common melt-growth techniques used for production of the bulk single-crystal. The advantages of the Czochralski and Bridgman growth methods are relatively high growth rates (in the range of several millimeters per hour) and the nearly thermodynamically equilibrium conditions are capable to generate bulk crystals of high structural perfection. In addition, the doping techniques are well established in these two methods.

There are technical obstacles to the growth of single crystal ZnO from the melt. The triple point of ZnO (the temperature and pressure at which all three phases of that substance, gas, liquid and solid phases, coexist in thermodynamic equilibrium) is 2248 K under the total (Zn and O_2) vapor pressure of 1.06 atm.

Figs. 7 and 8 show the predominance diagram in the relevant temperature range of the Zn-O system at total pressures 1 and 5 atm, respectively. As it can be seen from Fig. 7, before the melting point, ZnO evaporates under dissociation

$$ZnO \ (s) = Zn \ (g) + 1/2O_2 \tag{25}$$

It has strongly impeded the development of bulk crystal growth from the melt. To maintain the ZnO melt stable, the total pressure of oxygen-containing atmosphere in the growth chamber must be considerably larger than 1.06 atm.

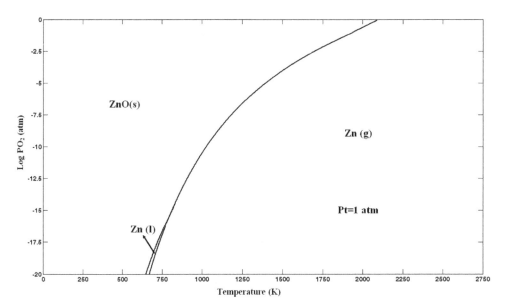

Fig. 7. Predominance diagram for Zn-O_2 system in dependence on temperature and the oxygen pressure PO_2 at total pressure 1 atm.

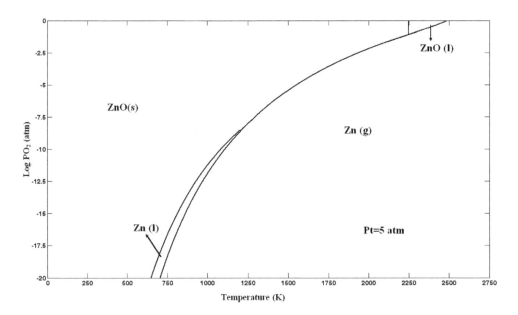

Fig. 8. Predominance diagram for Zn-O_2 system in dependence on temperature and the oxygen pressure PO_2 at total pressure 5 atm.

However, an essential difficulty to overcome is the lack of a perfect crucible material. The only metal withstanding high temperatures and an oxidizing atmosphere appears to be iridium. The design of a crucible containing molten ZnO capable of withstanding highly oxidizing conditions (temperatures of about 2300 K in oxygen containing atmosphere) is a serious challenge. The growth of ZnO crystals from a melt contained in an iridium crucible is continuously investigated by several references (Klimm et al., 2008a; Jacobs et al., 2009).

Indeed, it is well known that an atmosphere where iridium parts are to be heated must not contain more than 1-2% O_2 to avoid oxidizing of the metal. A solution could be try to heat the growth set up in the protective gas (N_2, Ar) –like GGG crystal growth- with an oxygen free atmosphere to 1000-1400 K and to add O_2 later, but practically this is not a solution for the problem. Although iridium crucible would not oxidize when the required oxygen was charged at T>1370 K°, crystal growth process would not stable growth because of sublimation of zinc oxide.

Similar problems during the melt growth of Ga_2O_3 were solved by working in a CO/CO_2 gas mixture. CO_2 yields an oxygen partial pressure well inside that corridor. At low temperatures, equilibrium reaction (11) is far on the left side and the oxygen partial pressure is very low. With increasing temperature, the equilibrium is shifted more and more to the right side and the resulting oxygen partial pressure is represented by curve in Fig. 9. Using CO_2 at a total pressure of approximately 10 bars, the authors (Klimm et al., 2008a; Klimm et al., 2008c; Jacobs et al., 2009) have successfully grown ZnO crystals from the melt in a Bridgman-like configuration.

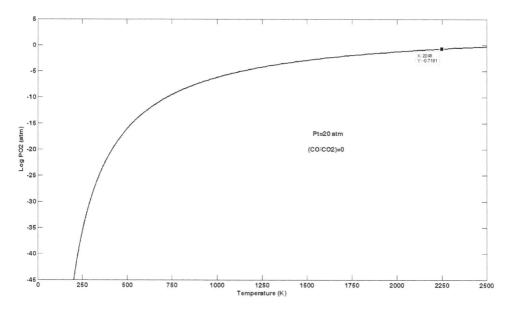

Fig. 9. The Oxygen partial pressure of pure CO_2 is against temperature at total pressure 20 atm.

The researches on growth of ZnO single crystals by the Czochralski method were carried on until now. Unfortunately, all attempts to obtain crystallization of bulk ZnO failed. Only polycrystalline ZnO solidified on the iridium seed rod was reported (Klimm et al., 2008a).

The control of the Czochralski experiments failed as the evaporation rate of ZnO at the melting point is high, even under the pressures up to 20 atm that were used. However, the partial pressure of O_2 at total pressure 20 atm that is supplied by pure CO_2 barely reaches to 1.90×10^{-1} atm at the melting point of ZnO crystal (seen Fig. 9). As shown in Fig. 10, the evaporated material forms white fume laying above the melts surface making optical control of the seeding process almost impossible (Klimm et al., 2008a). Moreover, evaporated ZnO condenses partially on the iridium seed rod, especially where it is lead through the thermal insulation. The sublimate creates mechanical contact between both parts, thus hindering mass control of the crystallizing ZnO by the balance on top of the seed rod that would otherwise allow automatic diameter control of the Czochralski growth process.

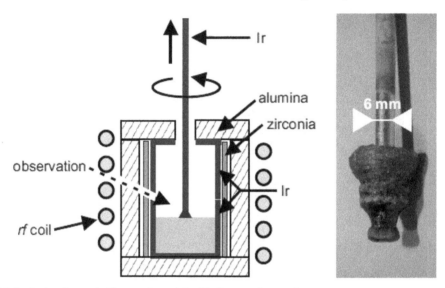

Fig. 10. Left: A schematic illustration of the ZnO crystal growth apparatus,
Right: polycrystalline ZnO solidified on the iridium seed rod. (Klimm et al., 2008a)

The Table 3 shows that at T_m=2248 K (melting point of ZnO), the pressure of oxygen at total pressure 5 atm should be between $(PO_2)_{min}$ and $(PO_2)_{max}$ points ($-1.11 < Log\ PO_2 < 2$) in order to have growth stability and suppress the burning of the construction parts (crucible, seed holder). Based on Table 3, the oxygen partial pressure of gas mixtures containing (CO-NO-CO_2-NO_2) with (CO/CO_2)=0.25 and (NO/NO_2)=0.25 is reached to $Log\ PO_2$=-0.05 atm. Compared with NO-NO_2 system, It is not only placed among of $(PO_2)_{min}$ and $(PO_2)_{max}$ at the melting point of ZnO but also it is less than the partial pressure of oxygen at lower temperatures of 1370 K. Best of all, the ZnO crystal growth can be performed at lower total pressures (< 5 atm) compared with CO-CO_2 system (>20 atm).

6. Conclusion

During the growth of refractory oxides crystal from melt, appropriate partial pressure of O_2 in the chamber is needed to keep the growth process stable. On the other hand, the partial pressure of O_2 should be kept low adequate in lower temperatures to suppress the oxidation of constructive parts specially crucible. The best solution to this problem is that the PO_2 that is supplied by–depends on thermodynamic behavior of oxide- pure CO_2 or by a mixture of CO-CO_2 or NO-CO-CO_2-NO_2 would be utilized in the growth atmosphere. They produce a temperature dependent oxygen partial pressure in this manner that the O_2 partial pressure increases with temperature. This amount of liberated oxygen is just sufficient to decrease the evaporation of molten oxides as well as low enough to allow the use of iridium crucibles, i.e., allowing Czochralski growth of Ga_2O_3 or ZnO.

7. References

Albrecht G.F., Sutton S.B., George E.V., Sooy W.R., Krupke W.F. (1998), *Laser Part. Beams*, vol.16, pp.605

Asadian M., Hajiesmaeilbaigi F., Mirzaei N., Saeedi H., Khodaei Y., Enayati Sh. (2010), *Journal of Crystal Growth*, vol.312 pp.1645

Asadian M., Mirzaei N., Saeedi H., Najafi M., Mashayekhi Asl I. (2011), *Solid State Sciences*, In Press, Corrected Proof

Bagnall D. M., Chen Y. F., Zhu Z., Yao T., Shen M. Y., Goto T. (1998), Appleid Physics Letter vol.73 pp.1038

Binnewies M. and Milke E. (2008), *Thermochemical Data of Elements and Compounds*, Wiley-VCH Verlag GmbH, Weinheim, thierd Edittion

Brandle C. D., Miller D. C., Nielsen J. W. (1972), *Journal of Crystal Growth*, vol.12 pp.195

Brandle C.D. and Barns R.L. (1974), *Journal of Crystal Growth*, vol.26 pp.169

Brandle, C.D. (2004), *Journal of Crystal Growth*, Vol.264, pp. 593-604

Carpenter J. H. (1989), *Journal of the Less Common Metals*, vol.152 pp.35-45

Chandrasekharaiah M. S., Karkhanavala M. D., Tripathi S. N. (1981), *Journal of the Less Common Metals*, vol.80 pp.9-17

Chaston J. C. (1965), *Platinum Metals Review*, vol.9 pp.126

Cockayne B. (1968), *Platinum Metals Review*, vol.12, pp.16

Cockayne B., Lent B., Roslington J.M. (1976), *Journal Material Science*, vol.11 pp.259

Cockaynen B. (1974), *Platinum Metals Review*, vol.18 pp. 86

Cordfunke E. H. P. and Meyer G. (1962), *Rec. Trav. Chimestry*, vol.81 pp.680

Cordfunke E.H.P. and Mayer G. (1962), *Rec. Trav. Chemistry*, vol.81 pp.495

Darling A. S., Selman G. L. Rushforth R. (1970), *Platinum Metals Review*, pp.14, 54-60

Day J. G. (1963), *Platinum Metals Review*, vol. 7 pp.50

Dupret F., Van Den Bogaert N. (1994), *Handbook of Crystal Growth*, Vol. 2, North-Holland, Amsterdam,

Elingham H. J. T. (1944), Reducibility of Oxides and Sulfides in Metallurgical Processes, *Journal Society Chemistry*, vol.63 pp.125

Ganschow S., Schulz D., Klimm D., Bertram R., Uecker R. (2010), *Crystal Research Technology*, vol.45 pp.1219

Gaskell D. R. (2003), *Introduction to the thermodynamics of materials*, 4th Edition, Taylor and Francis, ISBN 1-56032-992-0

Handley J. R. (1986), *Platinum Metals Review*, vol.30, pp.12-13

Jacobs K., Schulz D., Klimm D., Ganschow S. (2009), *Solid State Science*, vol.12 pp.307

Jia Z., Tao X., Dong C., Cheng X., Zhang W., Xu F., Jiang M. (2006), *Journal of Crystal Growth*, vol.292 pp.386

Klimm D. and Schroder W. (1999), *Journal Korean Association of crystal Growth*, vol.9 pp.360

Klimm D., Ganschow S., Schulz D., Bertram R., Uecker R., Reiche P., Fornari R. (2008b), Preprint submitted to Journal of Crystal Growth, for CGCT4 Sendai

Klimm D., Ganschow S., Schulz D., Fornari R. (2008a), *Journal of Crystal Growth*, vol.310 pp.3009

Klimm D., Ganschow S., Schulz D., Fornari R. (2008c), Preprint submitted to Journal of Crystal Growth, for CGCT4 Sendai;

Klimm D., Ganschow S., Schulz D., Fornari R. (2009), *Journal of Crystal Growth*, vol.310 pp.534– 536

Li X., Hu Z. G., Li J. (2007), *Optical Materials,* vol.29 pp.854

Lipinska L., Ryba-Romanowski W., Rzepka A., Ganschow S., Lisiecki R., Diduszko R., Pajaczkowska A. (2009), *Crystal Research and Technology*, vol.44 pp.477

Look D. C., Reynolds D. C., Litton C. W., Jones R. L., Eason D.B., Cantwell G. (2002), *Applied Physics Letters*, vol. 81pp.1830

Luo Z., Lu M., Bao J., Liu W., Gao Ch., *Materials Letters*, vol.59 pp.1188

Mateika D., Laurien R., Rusche Ch. (1982), *Journal of Crystal Growth*, vol.56 pp.677

Miller P. (2008), Zinc Oxide: *a sperctroscopic Investigation of Bulk Crystals and Thin Films*, Degree of Doctor of Philosophy in Physics, University of Canterbury New Zealand

Nassau K., Broyer A. M. (1962), *Journal American Ceramic Society*, vol.45 pp.474

Norman J. H., Staley H. G. Bell W. E (1965), *Journal Chemistry Physics*, vol.42 pp.1123

Pajaczkowska A., Novosselov A. V., Zimina G. V. (2001), *Journal of Crystal Growth*, vol.223 pp.169

Park J. H., Jang S. J., Kim S., Lee T. (2006), *Applied Physics Letters*, vol.89 pp.121108

Pearton S. J., Norton D.P., Ip K., Heo Y.W., Steiner T. (2005), Applied Physics Letter, vol.50 pp. 293.

Piekarczyk W., Pajaczkowska A. (1979), *Journal of Crystal Growth*, vol.46 pp.483

Schafer H. and Heitland H. J. (1960), *ZAAC*, vol.304 pp.249-265

Suscavage M., Harris M., Bliss D., Yip P., Wang S. Q., Schwall D., Bouthillette L., Bailey J. , Callahan M., Look D.C., Reynolds D.C., Jones R.L., Litton C.W. (1999), *Journal Nitride Semiconductor Research*, vol. 4 pp.40

Tomm Y., Reiche P., Klimm D., Fukuda T. (2000), *Journal of Crystal Growth*, vol.220 pp.510;

Valentino A.J. and Brandle C.D. (1974), *Journal of Crystal Growth*, vol.26 pp.1

Van Uitert L. G. (1970*), Platinum Metals Review*, vol.14 pp.118

Weiland R., Lupton D. F., Fischer B., Merker J. Scheckenbach C., Witte J. (2006), *Platinum Metals Review*, vol.50 pp. 158

Zhaobing W., Qingli Z., Dunlu S., Shaotang Y. (2007), *Journal Rare Earths*, vol. 25, pp. 244

8

Controlling the Morphology and Distribution of an Intermetallic $Zn_{16}Ti$ Phase in Single Crystals of Zn-Ti-Cu

Grzegorz Boczkal

AGH-University of Science and Technology, Faculty of Non-Ferrous Metals, Cracow,
Poland

1. Introduction

The work discusses the relatively poorly investigated area of various phenomena accompanying monocrystallization of hexagonal metal alloys, containing second phase inclusions in the structure. Metals of hcp structure form a, not very numerous but having a fundamental industrial importance, group.

Altogether about 27 metals of hcp structure are known, including 11 actinides [1,2,3]. The industrial applications of hexagonal metals cover many different sectors, from zinc sheets used for roofing up to light titanium-based alloys for parts of planes. Single crystal layers based on zinc are also used in electronics [4].

Hexagonal metals are characterised by features unprecedented for other metals. These include the value of c/a ratio (Table 1), which largely determines the metal properties [1,2], as well as very low mutual solubility observed in alloys composed of two metals with hexagonal structure each (Zn-Ti, Cd, Zn, Mg-Zn , Mg-Zr, Mg-Sc and others) [5,6].

The value of the c/a ratio and the related disorder in an ideal structural model are inherently related with the physics of interatomic bonds. In the case of metals with regular face centred lattice and regular body centred lattice, the bonding is of a purely metallic (non-directional) character and there is no phenomenon of interpenetration of the wave functions originating from lower orbitals. In hexagonal metals, the metallic bond predominates, but there is also a significant share of atomic bonds originating from an interaction that is said to exist between the wave functions of the neighbouring atoms [1,2,7,8].

Metals such as Zn, Cd, Tl were classified by Wyatt as a class of intermediate elements, i.e. the elements of a coordination number low enough to enable the occurrence of covalent interactions [1]. The directionality typical of an atomic bonding disturbs the perfect sequence of atomic layers and, consequently, the c/a ratio assumes values different than 1.633, which is perfect quantity when a rigid sphere model is used. Since the interatomic interaction in metals with c/a <> 1.633 is a combination of metallic and atomic bonds, any change in temperature must significantly affect the properties of metals with a hexagonal structure. Lowering the temperature reduces the distance between the atomic cores and, consequently, increases penetration of the lower orbital wave functions, which leads to an increase in the share of atomic bonds. Additionally, the structure where the atoms do not fill in an ideal space is characterised by low energy required to produce and promote migration of point defects (vacancies), and also by strong variation of diffusion coefficients for different crystallographic directions [7,8,9]. Both the temperature and the degree of deformation also strongly influence the number of vacancies in the material. These phenomena are responsible for the specific properties of hexagonal metals at low temperatures, such as the anomalies of a hardening coefficient and yield strength [9].

Metal	a [Å]	c [Å]	c/a
Na-α	3.767	6.154	1.6336
Be-α	2.287	3.583	1.5666
Mg	3.209	5.210	1.6235
Sc-α	3.309	5.273	1.5935
Y-α	3.647	5.731	1.5714
La-α	3.770	12.159	3.2251
Pr-α	3.673	11.835	3.2221
Nd-α	3.658	11.779	3.2200
Ti-α	2.951	4.684	1.5872
Zr-α	3.232	5.148	1.5928
Hf-α	3.195	5.051	1.5809
Tc	2.743	4.400	1.6040
Re	2.761	4.458	1.6146
Ru	2.706	4.281	1.5820
Os	2.735	4.319	1.5791
Zn	2.664	4.947	1.8569
Cd	2.979	5.619	1.8862
Tl-α	3.456	5.525	1.5986

Table 1. Some of the hexagonal metals and their c/a coefficients [2,3].

Hexagonal metals are characterised by large stress variations in the individual slip systems. With proper orientation respective of the stress applied, these metals can be deformed in one system, operating as a primary system, obtaining a wide range of the deformation values [10-14]. It is typical, in particular, of metals having the c/a ratio > 1.633, such as zinc and cadmium. Zinc single crystals of "soft" orientation are deformed to more than 100% (elongation) within the range of an easy slip in a (0001) <11-20> system. Only a very serious change of orientation caused by deformation and the strong strain hardening effect in the basal system of (0001) <11-20> are capable of activating a different slip arrangement [15].

Conducting research on the phenomena and processes occurring in metallic materials requires samples with clearly defined structure of both matrix and lattice obstacles in the form of phases precipitated or introduced from the outside. Metals and alloys in the polycrystalline form are not suitable for studies aiming at the identification of the mechanism of deformation because of the need to examine the phenomena which occur in different and separated areas characterised by different crystal orientations (grains). The use of materials with single crystal structure of the matrix eliminates this problem, as it becomes possible to analyse various phenomena within the whole sample volume, in a well-defined research environment. The ability to control the structure of the lattice obstacles while maintaining the same single crystal matrix enables, moreover, designing of technological processes by means of which products of the desired structure can be made in a reduced number of the necessary technological steps.

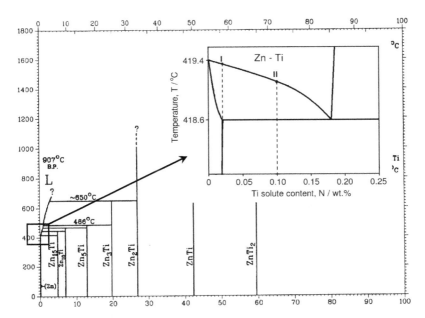

Fig. 1. The Zn-Ti phase diagram [5,6,7].

The specific crystallography of metals with a hexagonal structure, combined with the strong anisotropy of diffusion coefficients and elasticity constants has a strong influence on morphology of the obtained structures.

For example the elastic constants c of the pure zinc are property [16,17]:

- $c_{||}[0001]$ = 13.1 cm^2 dyne^{-1} x 10^{-13}
- $c_{\perp}[0001]$ = 1.93 cm^2 dyne^{-1} x 10^{-13}

The results of other authors suggest moreover, that elasticity of zinc is very sensitive to small variations in amount and kinds of impurities [15,17].

The additional special feature is also a strong influence of even small amounts of the alloying elements on the surface tension of molten zinc, which is a critical parameter for the process of monocrystallization [16,23,24,26].

Another important factor is the existence of a limited number of directions characterised by a high coefficient of diffusion and self-diffusion. At a temperature of 293K, the zinc coefficients of self-diffusion D_0 are, respectively [9]:

- $D_{0\ [0001]}$ = 0.13 x 10^{-4} m^{2}s^{-1}
- $D_{0\ <11-20>}$ = 0.58 x 10^{-4} m^{2}s^{-1}

In hexagonal metals there are three directions preferred by the diffusion; these are the directions of the most dense packing <11-20> located in a hexagonal column on the plane (0001). These are also the privileged directions for the growth of a single crystal structure. This tendency is the stronger, the higher is the imposed from the outside speed of crystallisation and the higher is the content of alloying elements. In the case of single crystals grown without a nucleus, this means a constant axial orientation consistent with the direction [11-20]. This effect is so strong that all attempts at obtaining an orientation different than the one preferred by the growth mechanism require, beside the presence of a nucleus of the desired orientation, also the crucible of special design [18].

Alloys based on hexagonal metals offer low solubility to other metals in the solid state and a tendency to create numerous intermetallic phases [5,6,19-22].

The structural anisotropy strongly influences the mode of nucleation and growth of the secondary phases in single crystals with a hexagonal structure. The large difference in elastic constants at different crystallographic directions of the hcp lattice [16,17] enforces some permanent relationships between the crystal lattice of the newly emerging intermetallic phases and matrix.

2. Methodology

Single crystals of Zn-Ti and Zn-Cu-Ti used in this study were grown by Bridgman method with sliding temperature gradient.

Bridgman's original method [23], which consists in lowering a crucible with the charge inside through the zone of strong temperature gradient, was modified by application of the sliding motion of the furnace, while the nucleus and the charge were left immobile. The growth process was carried out in crucibles made of spectrally pure graphite. To prevent oxidation and also a reaction between the charge and the crucible material, an argon protective atmosphere was used. Permanent purging of the furnace chamber allowed the removal of oxygen and unwanted gaseous products formed during melting of charge. Compositions of the investigated single crystals and crystallisation speeds are summarised in Table 2. The resulting single crystals were oriented with a Bruker D8 Advance X-ray diffractometer, and were cut next into specimens with orientation of the observation planes (0001) and (11-20), respectively. The surfaces of the specimens were pre-polished with abrasive papers and diamond paste, first, and etched next with a chromium reagent. To reveal the shape of precipitates and determine the crystallographic relationship with the matrix, deep selective etching was used. Structural examinations were performed under Hitachi 3300 scanning microscopes with an EDS and EBSD attachments, and under the TESLA-302 microscope.

Single crystal chemical composition	Growth rate [mm/h]
Zn-Ti0.02 wt.%	3
Zn-Ti0.10 wt.%	3
Zn-Ti0.10 wt.% - Cu0.1 wt.%	1.8
	6
	10
	16
Zn-Ti0.2 wt.% - Cu0.15 wt.%	1.8

Table 2. The investigated single crystals and growth rates.

3. Characteristics of phases present in the examined Zn-Ti and Zn-Ti-Cu single crystals

3.1 Types of phase lattice

In the case of Zn-Ti and Zn-Ti-Cu single crystals analysed in this study, the only stable intermetallic phase observed in an around-eutectic range (titanium content in zinc of about 0.2 wt.%) is the $Zn_{16}Ti$ phase [10-14,25]. It is a tetragonal phase of Cmcm structure containing 6.33 at.% of titanium. A single cell of this phase is built of 51 atoms. Each Ti atom is surrounded by 15 Zn atoms, with the additional 16th atom located in a space between the "chain" elements. The $Zn_{16}Ti$ phase has the following lattice parameters [25]:
a.   772.0 pm
b.   1144.9 pm
c.   1177.5 pm
For pure zinc elemental cell has a dimensions [2,3]:
a.   266.49 pm
b.   266.49 pm
c.   494.68 pm
Interplanar distances for the planes (0001) normal to the 'c' Zn direction are $d_{(0001)}=\frac{1}{2}c$ = 247.34 pm. On the other hand, the 'a' parameter of the $Zn_{16}Ti$ phase is 772 pm which is the value roughly three times higher (3 x 247.34 pm = 742.02 pm). Hence, the edge length ratio $a_{Ti}/\frac{1}{2}c_{Zn}$ is 772pm/247.34pm = 3.12.
Other combinations of ratios between the lattice constants of the $Zn_{16}Ti$ phase and the 'a' constant of a unit cell of zinc are:
a. $Zn_{16}Ti$ / a_{Zn} ~ 2.9
b. $Zn_{16}Ti$ / a_{Zn} ~ 4.3
c. $Zn_{16}Ti$ / a_{Zn} ~ 4.42
The edge ratio of a_{Zn16Ti}/a_{Zn}~2.9 is also close to an ideal multiple, but creating an interface of this configuration is energetically less favourable, as it does not affect the planes with the highest packed coefficient.
In this situation, one should expect a strong tendency to the formation of a crystallographic configuration, in which a_{Zn16Ti} edges of the phase cells and $c_{\alpha-Zn}$ edges of the matrix will be mutually parallel to each other.

3.2 Chemical composition and hardness of phases

Hardness of the $Zn_{16}Ti$ phase and of the α-Zn matrix was measured on the (0001) plane of a Zn-Ti0.2-Cu0.15 single crystal.
The α-Zn matrix of these single crystals contains 0.15 wt.% Cu and ~ 0.02 wt.% Ti. The measurements were taken using STRUERS microhardness tester, applying a load of 0.1 kg. To eliminate the error due to possible inhomogeneity of properties and small dimensions of the measured object, mapping was performed on the surface of 1.6 x 1.8 mm in the region containing a single exposed particle of $Zn_{16}Ti$ phase and pure matrix. Altogether, 80 measurements were taken. The results are shown in Figure 2 in the form of a 3D chart.
The obtained results showed that an average hardness of the $Zn_{16}Ti$ phase reaches 296 μHV as compared to 58 μHV of the matrix, with standard deviation not exceeding 8%. Previous studies performed on Zn-Ti and Zn-Ti-Cu alloys have proved that the $Zn_{16}Ti$ phase has little influence on hardening at the initial stage of deformation [11]. What predominates at that

stage is the effect of precipitation hardening derived from copper, which all enters into the solution, and from small amounts of titanium (at the level of hundredths of a weight percent). Its main role in the hardening effect, the $Zn_{16}Ti$ phase starts playing at higher deformations when, owing to their shape and dimensions, the needle-like particles of this phase are acting in a way analogous to fibres in composite materials. Partially coherent combination of the $Zn_{16}Ti$ phase with matrix confirmed by the results of EBSD [10,11,12] suggests good transfer of stresses from the matrix, while high hardness of this phase will increase the properties of the material taken as a whole.

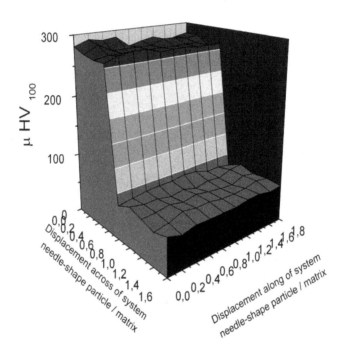

Fig. 2. Microhardness results for system $Zn_{16}Ti$ / α_{Zn}. Displacement in [mm].

4. Crystallographic relations in a matrix/precipitate system

The results of tests performed on Zn-Ti and Zn-Ti-Cu single crystals by TEM and EBSD techniques confirmed the existence of a close crystallographic relationship between the hexagonal crystal lattice of the matrix and the lattice of an intermetallic $Zn_{16}Ti$ phase. The a_{Zn16Ti} edges of the cells of the $Zn_{16}Ti$ phase and the $c_{\alpha-Zn}$ edges of the matrix are parallel to each other. At the same time, it has been observed that the direction of the b_{Zn16T} edges of the crystal cells in the $Zn_{16}Ti$ phase is in the majority of cases parallel to the $a_{\alpha-Zn}$ <11-20> direction of the hexagonal lattice of the matrix. This type of relationship proves an interrelation that exists between the phase lattice and the matrix lattice and, consequently, a partial coherence between the precipitates and the matrix.

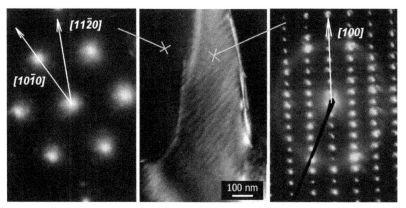

Fig. 3. TEM diffraction results from matrix and the Zn$_{16}$Ti particle. Plane (0001) of the Zn-Ti0.1wt.% single crystal [11].

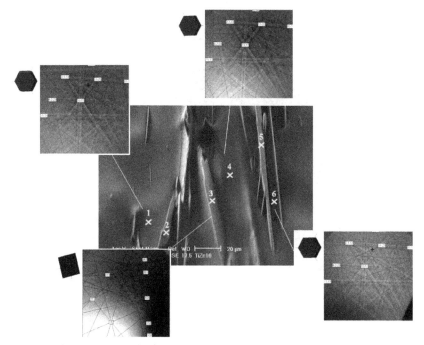

Fig. 4. EBSD results. Plane (0001) of the Zn-Ti0.1wt.% single crystal [26].

5. Control of the Zn16Ti phase morphology

The intermetallic Zn$_{16}$Ti phase commonly occurs also in industrial zinc alloys. Due to its chemical composition, even alloys with a low content of titanium contain a large volume amount of this phase. This applies, first of all, to the hypoeutectic alloys with titanium content below 0.1 wt,%, which are used as a covering sheet metal in building constructions [26].

Previous studies have shown considerable potential for controlling the $Zn_{16}Ti$ phase morphology through the choice of alloy composition and crystallisation speed. In the case of hypoeutectic binary Zn-Ti alloys, the capabilities of forming different structures are limited by the low stability of the crystallisation front which, in turn, translates into intensive nucleation and growth of intermetallic phases in the form of rods of complex cross-sections. This applies to single crystals obtained by Bridgman method, where the limit growth rate for Zn-Ti0.02wt.% alloys and Zn-Ti0.1 wt% alloys with single crystal matrix does not exceed 3 mm/h [26]. With the rate so low, the factor deciding about the morphology of the $Zn_{16}Ti$ phase in a single crystal is the chemical composition. In the case of Zn-Ti0.02wt.% single crystals, the formation of needle-like crystallites is observed. The reason for this is a relatively low concentration of titanium, which forces predominant growth in the directions defined by diffusion and elastic constants of the crystal matrix. The result is an acicular form of precipitates of the $Zn_{16}Ti$ phase.

Much higher content of titanium in Zn-Ti0.1wt.% single crystals enables the growth of crystallites of an intermetallic phase with larger and more complex cross-sections than in the previous case. The low growth rate combined with a high concentration of titanium allows for efficient growth, also on less preferred planes. The result is the presence of the intermetallic phase particles in the form of rods of irregular cross-sections (Figure 5).

Fig. 5. The needle-shape form of the intermetallic phase $Zn_{16}Ti$ observed on (0001) plane in Zn-Ti0.1wt.% single crystals obtained at growth rate of 3 mm/h.

Enhancing the capability of generating the particles with different morphologies is possible through stabilisation of the crystallisation front with an alloying addition increasing the surface tension of the front and thereby significantly limiting the amplitude of thermal fluctuations on its surface (Figure 6). In zinc alloys, very effective has proved to be copper, used in an amount of 0.075wt.% .... 0.15wt.% as a component of industrial Zn-Ti-Cu alloys. Copper addition to Zn-Ti0.1wt.%-Cu0.1wt.% single crystals allows increasing the crystallisation rate up to 16mm/h, which brings a wide range of changes to the morphology of particles produced in an intermetallic phase precipitating at the crystallisation front (morphology changing from columnar through lamellar to acicular) [18].

For a low rate of single crystal growth in the Zn-Ti0.1wt.%-Cu0.1wt.% alloy, i.e. 1.8mm/h, the intermetallic phase occurs in the form of rods of irregular cross-sections (Figure 7). This situation is analogous to the previously discussed single crystals of binary Zn-Ti alloys. The morphology of this phase is shown in Figure 5. Increasing the growth rate of single crystals in Zn-Ti0.1wt.%-Cu0.1wt.% alloy up to 6mm/h changes the Zn$_{16}$Ti phase morphology from columnar to lamellar (Figure 8).

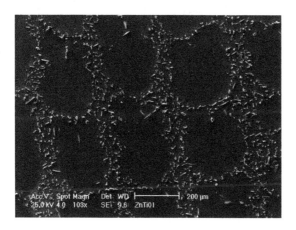

Fig. 6. Structure observed on (11-20) plane in Zn-Ti0.1wt.% single crystals obtained at growth rate of 3 mm/h.

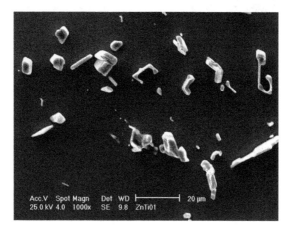

Fig. 7. The Zn$_{16}$Ti intermetallic phase in a form of rods with irregular crossections, elongated on the [11-20] direction, observed in Zn-Ti0.1wt.%-Cu0.1wt.% single crystals obtained at growth rate of 1.8 mm/h.

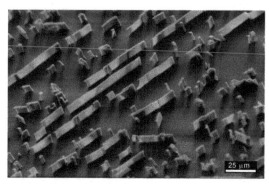

Fig. 8. The lamellar form of the $Zn_{16}Ti$ intermetallic phase observed in Zn-Ti0.1wt.%-Cu0.1wt.% single crystals obtained at growth rate of 6 mm/h. The (11-20) plane [18].

The resulting structure is characterised by a high degree of ordering. Constant crystallographic relationships have been observed between the arrangement of the $Zn_{16}Ti$ phase lamellae and matrix orientation. The precipitates of the $Zn_{16}Ti$ phase in the form of lamellae prefer for growth the pairs of planes from the family {10-11} which, in the case of zinc crystal lattice with the ratio of c/a = 1.856, are oriented at angles of 51° (Figure 9). The choice of a specific pair of planes is determined by the direction of growth of a single crystal belonging to the <11-20> family, which also forms an axis for the band of the selected pair of planes (Figure 10).

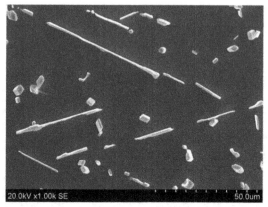

Fig. 9. The lamellar form of the $Zn_{16}Ti$ intermetallic phase observed in Zn-Ti0.1wt.%-Cu0.1wt.% single crystals obtained at growth rate of 6 mm/h. The (11-20) plane [14].

Another change in the morphology of the $Zn_{16}Ti$ intermetallic phase precipitates in single crystals of the Zn-Ti0.1wt.%-Cu0.1wt.% alloy is observed at the growth rate of 10mm/h. In the case of the examined alloy, this is the speed limit at which the crystallisation front is no longer a smooth surface because of the appearance of thermal fluctuations near the axis of growth. A consequence of this phenomenon is the locally varied growth rate of the single crystal matrix structure, changing the kinetics of growth of the intermetallic $Zn_{16}Ti$ phase precipitates at the crystallisation front. This situation is illustrated in Figure 11. The structure visualised here was observed on the plane (11-20) perpendicular to the axis of growth. A clear difference is observed between the morphology of precipitates in the central part of the single crystal cross-section and outer zone close to the crystal faces. The Zn$_{16}$Ti phase in the outer zone has a lamellar morphology, qualitatively identical with that occurring in single crystals at a growth speed of 6mm/h, while central zone is characterised by a morphology close to the acicular one. The reason accounting for this phenomenon is the structure growth velocity locally increased due to thermal fluctuations in the 'empty' oval areas free from the precipitates, and a small radius of the front curvature in these areas. Under such conditions, the nucleation and growth of an intermetallic phase takes place at the inflection points, which best serve this purpose because of the energy expenditure needed to create a nucleus. This situation is shown in Figure 12.

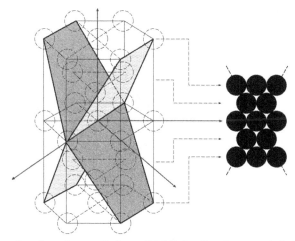

Fig. 10. The planes of preferred growth from {10-11} family, characteristic for lamellar shape of the Zn$_{16}$Ti intermetallic phase.

With the speed of single crystals growth in the Zn-Ti0.1wt.%-Cu0.1wt.% alloy equal to 16 mm/h, the thermal fluctuations now cover the entire surface of the crystallisation front. This is shown in Figure 13. Within the whole examined area, the precipitates of purely acicular morphology arise. Compared to the previously discussed variant of a lower growth rate, the needles of the Zn$_{16}$Ti phase now have round cross-sections, devoid of branches. It has also been observed that the oval areas devoid of precipitates show some degree of elongation on direction which is a trace of the plane (0001), characterised by the closest packing in a hexagonal structure (Figure 11, 12).

In the case of the Zn-Ti0.1wt.%-Cu0.1wt.% alloy, the speed of 16mm/h was the highest one for which a single crystal structure could be obtained (Figure 13,14). Therefore the next change to the morphology of precipitates was initiated by modification of the chemical composition. Considering the amount of the Zn$_{16}$Ti phase formed in alloys with 0.1wt.% of titanium, single crystals of the Zn-Ti0.2wt.%-Cu0.15wt.% alloy were obtained. It is a composition similar to the eutectic point which, on account of the Zn-Ti alloys belonging to the group of "off-eutectic" alloys, has not been at this rate accurately determined.

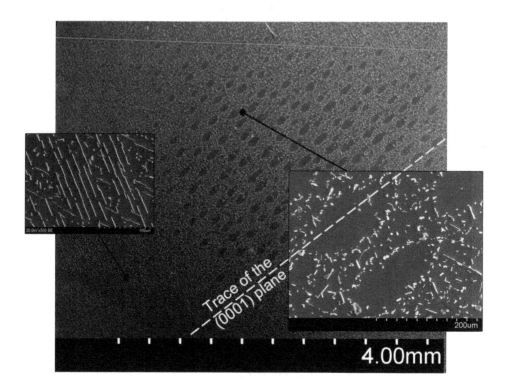

Fig. 11. Structure observed on (11-20) plane in Zn-Ti0.1wt.%-Cu0.1wt.% single crystals obtained at growth rate of 10 mm/h. Morphology of the precipitates in a core of crystals is different from morphology observed near outside walls of the crystal.

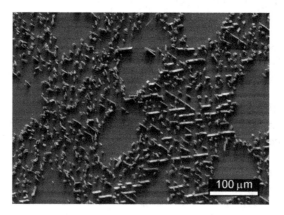

Fig. 12. The beginning of lamellar → fibre transformation observed in a core of the crystal.

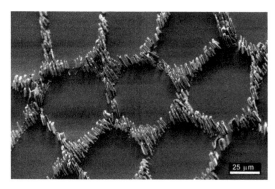

Fig. 13. Structure observed on (11-20) plane in Zn-Ti0.1wt.%-Cu0.1wt.% single crystals obtained at growth rate of 16 mm/h [18].

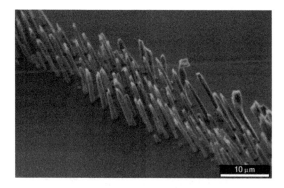

Fig. 14. Structure observed on (11-20) plane in Zn-Ti0.1wt.%-Cu0.1wt.% single crystals obtained at growth rate of 16 mm/h.

Single crystals in the Zn-Ti0.2wt.%-Cu0.15wt.% alloy were obtained at a speed of 1.8mm/h. Structural analysis revealed in the structure the presence of particles of very complex morphology (Figure 15). Observations revealed the existence of precipitates of an acicular morphology and macroparticles shaped like discs and polygons. The performed chemical analysis has shown that all the observed types of precipitates are composed of one intermetallic phase only, i.e. Zn$_{16}$Ti [21]. Until now it has not been possible to identify clearly the reasons which have made one phase assume so many different types of morphologies. One of the reasons can be different mechanism of nucleation of the individual types of particles. The constitution of Zn-Ti0.2wt.%-Cu0.15wt.% macroparticles observed in the structure of Zn single crystals significantly deviates from previous observations of the Zn$_{16}$Ti phase. In the case of macroparticles (Figure 16), the precipitates take the form of a eutectic system, in which there are alternately arranged lamellae of Zn$_{16}$Ti / α / Zn$_{16}$Ti / α / .../ α /. The nucleation of such a system is energetically privileged, compared to the process by which isolated particles of one phase only are expected to precipitate from the solution. This follows from the mechanism of the coupled growth of eutectic phases [27].

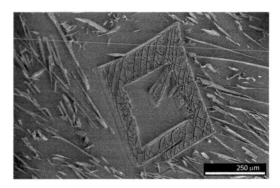

Fig. 15. Multishape morphology of the $Zn_{16}Ti$ intermetallic phase observed on (0001) plane in Zn-Ti0.2wt.%-Cu0.15wt.% single crystals obtained at growth rate of 1.8 mm/h.

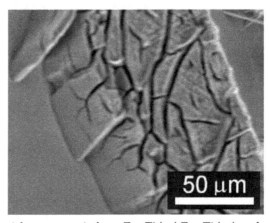

Fig. 16. The macroparticleas a eutectic form $Zn_{16}Ti/α/$ $Zn_{16}Ti/α/...$ observed on (0001) plane in Zn-Ti0.2wt.%-Cu0.15wt.% single crystals obtained at growth rate of 1.8 mm/h.

6. The distribution of $Zn_{16}Ti$ phase in function of growth conditions

6.1 Oscillatory structures

The conditions of growth and chemical composition affect not only the morphology of the $Zn_{16}Ti$ phase precipitates, but also distribution of this phase. A special case of the structure formed at low growth rates of the Zn-Ti and Zn-Ti-Cu single crystals is oscillatory structure [18,26,28]. It is characterised by the cyclic occurrence of typical elements, such as areas rich in precipitates interwoven with areas of pure matrix totally devoid of precipitates.

Oscillatory structures were observed for the first time in the peritectic Sn-Cd alloys [28]. They are formed as a result of changes in the concentration of an alloying constituent ahead of the moving crystallisation front, which affects the conditions necessary for nucleation of individual phases. With proper velocity of the crystallisation front movement, the preferred growth is alternately exhibited by phases a and b (Figure 17).

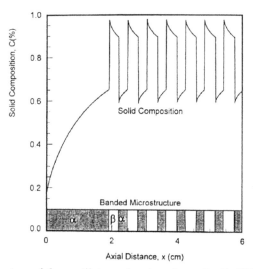

Fig. 17. Growth mechanism of the oscillatory structure in peritectic [28].

Massive macroparticles use in building of their structure a significant amount of titanium, thus reducing the concentration of this element in solution. Then, from the rest of the titanium-depleted solution, the precipitation of fine-grained phases occurs; on account of the low concentration of titanium, these phases do not form a eutectic system in the solution.

A similar phenomenon as described previously for the peritectic Sn-Cd alloys was also obtained in the eutectic Zn-Ti and Zn-Ti-Cu alloys. Studies showed that the condition necessary for the formation of such structures in single crystals based on zinc is low travel speed of the cooling zone, and hence low rate of the crystal growth. In cases under discussion, it did not exceed 6 mm/h. Additional important factors are the chemical composition of the alloy and differences in temperature between the melting point of pure metal (in this case zinc) and the eutectic point. The lower is the difference in temperatures, the stronger tendency the alloy shows to the formation of an oscillatory structure. Depending on the degree to which these conditions are satisfied, the following types of oscillatory structures can be obtained:

a.   purely oscillatory, i.e. a sequence of alternately occurring regions rich in inclusions of the second phase and layers of pure matrix (alpha phase) free from the precipitates [18],

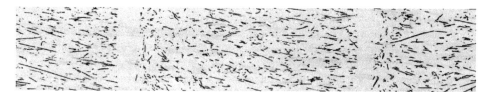

Fig. 18. Oscillatory structure of "a-type" observed in Zn-Ti-Cu single crystals.

b.   oscillatory, in the case of which areas totally free from the precipitates are not observed any longer, while the degree of filling the alternately occurring layers with inclusions of the intermetallic phases is periodically changing.

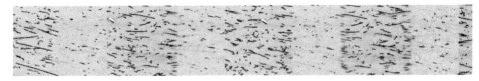

Fig. 19. Model of oscillatory structure of "b-type" observed in eutectic Zn-Ti-Cu single crystals.

Case (a) occurs in titanium alloys with 0.1 wt.% Ti and less. In alloys with the titanium content so low, the zonal segregation occurring during single crystal growth by Bridgman method is sufficiently strong to form areas with the concentration of alloying elements so low that the process of the second phase precipitation is inhibited.

Another phenomenon observed in the case of the oscillatory structures of type (a) is the, linearly changing along the growth direction, density of the inclusions of the second phase in areas rich in precipitates. This phenomenon is particularly evident in single crystals growing at the lowest rates (1.8mm/h). This is presented in Figure 20a. Increasing the growth rate to 6 mm/h makes the structure preserve its oscillatory character but with a uniform distribution of precipitates within the layer (Figure 20b). The reason for this is lower concentration of titanium ahead of the crystallisation front caused by a higher rate of the cooling zone transfer, which weakens the effect of zonal segregations, and consequently changes the conditions for nucleation of an intermetallic phase.

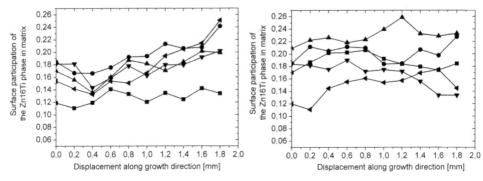

Fig. 20. The $Zn_{16}Ti$ phase distribution along of a precipitation-rich layer: a) oscillatory structure obtained at 1.8 mm/h growth rate; b) oscillatory structure obtained at 6 mm/h growth rate.

On the other hand, structures of type (b) have been observed in Zn-Ti-Cu alloys with around-eutectic titanium content. In this case, the concentration of titanium in the entire volume of the solution is so high that, regardless of the speed of crystal growth, there are conditions for the nucleation and growth of the $Zn_{16}Ti$ phase. What changes is only the intensity of this phenomenon, which manifests itself in an alternate occurrence in the structure of the layers of higher and lower concentration of the precipitates.

6.2 Model analysis of oscillatory structure

To better understand the growth mechanism of oscillatory structures, a mathematical model developed by Wolczynski has been used [29]. It allows us to estimate the impact of growth conditions and chemical composition on the forming sequence of layers rich in an intermetallic phase and free from the precipitates [11,18,26,29,30]. This model has been developed for binary alloys and is based on the diffusion-related changes in alloying constituent concentration. It includes the three major processes occurring during alloy crystallisation:

1. changes in the concentration of alloying element depleting the liquid phase with progressing crystallisation along the liquidus line,

$$N^L(x;\alpha) = N_0 \left(1 + \alpha\, k\, x - x\right)^{(k-1)/(1-\alpha k)}$$

2. microsegregations at the crystallisation front, resulting from the constituent segregation,

$$N^S(x;\alpha) = k N_0 \left(1 + \alpha\, k\, x - x\right)^{(k-1)/(1-\alpha k)}$$

3. redistribution of alloying constituents in grains after the occurrence of reverse diffusion.

$$N^B\left(x; X^0, \alpha\right) = \left[k + \beta^{ex}\left(x; X^0\right) \beta^{in}\left(X^0, \alpha\right)\right] N^L(x;\alpha)$$

where:
- k - partition ratio, x - amount of growing crystal,
- N_0 - nominal concentration of the alloy, α - back-diffusion parameter,
- β^{ex}; β^{in} - coefficients of the redistribution extension and intensity.

$$i_E(N_0) = i_K(1, N_0)$$

$$i_D(\alpha, N_0) = i_K(\alpha, N_0) - i_E(N_0)$$

$$i_K(\alpha, N_0) = 1 - x_K(\alpha, N_0)$$

$$x_K(\alpha, N_0) = \frac{1}{1 - \alpha k}\left[1 - \left(\frac{N_E}{N_0}\right)^p\right]$$

This method allows precise determination of the amount of precipitates of both equilibrium i_E and non-equilibrium i_D character, where x_K is the amount of pure single crystals between the fringes.

$$p = \frac{1 - \alpha k}{k - 1}$$

If α is equal to zero, then the model is diffusionless. Such simplification makes the model unrealistic, although it may serve as a benchmark or reference.

Two binary alloys, i.e. Zn-Ti0.02wt.% and Zn-Ti0.1wt.%, were selected for tests. The structures of single crystals of these alloys are of an oscillatory character, but differ considerably in respect of volume fraction of the $Zn_{16}Ti$ phase and its distribution. Layers rich in precipitates observed in the Zn-Ti0.02wt.% alloy are much smaller in thickness than the layers without precipitates.

Single crystal	Titanium contents [wt.%]	Thickness of precipitation-rich layer [μm]	Thickness of precipitation-free layer [μm]
Zn-Ti0.02wt.%	0.023	10	120.4
Zn-Ti0.1wt.%	0.10	56.6	127.3

Table 3. Structural parameters of the Zn-Ti0.02wt.% and Zn-Ti0.1wt.% single crystals.

For single crystals of Zn-Ti.01wt.% alloy, the layers rich in precipitates are much thicker, while spacing between them is similar to that observed in single crystals of lower titanium content.

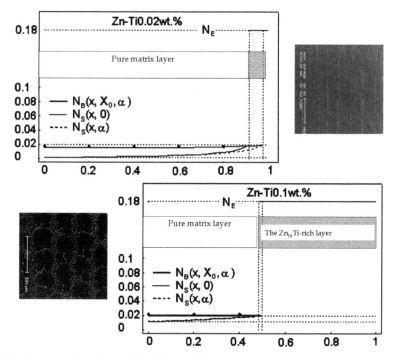

Fig. 21. Calculatin results obtained for the model.

Designations used in Figure 21:
- N_E - eutectic concentration of the Ti-solute [wt.%]
- N_B - solute redistribution after back-diffusion [wt.%]
- N_S - "historical" solute concentration on the solid/liquid interface
- α - back-diffusion parameter [wt.%]

The proposed model based on calculations of the diffusion of the alloying elements fairly well reflects the growth of an oscillatory structure in the binary Zn-Ti alloys. The compliance is particularly strong for the structure of single crystals of Zn-Ti0.02wt.%, where there is clear division between individual elements of the structure. In the case of single crystals of Zn-Ti0.1wt.%, a large amount of the alloying constituent disturbs the arrangement of layers because of "bridges" that are formed in between them. However, in the case of a model based solely on theoretical data, the results obtained can be considered correct. Additionally, Figure 22 presents a hypothetical path of crystallisation for the Zn-Ti0.1wt.% alloy [30].

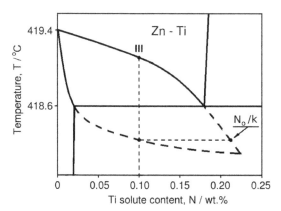

Fig. 22. Solidification $III \rightarrow N_0 / k$ for an equilibrium solidification of the single crystal with $N_0 = III = 0.1 \, [wt.\%Ti]$ [30].

6.3 Continuous and cellular structure in single crystals of Zn-Ti0.1-Cu0.1

Increasing the growth rate of Zn-Ti-Cu single crystals leads to disappearance of oscillatory structure and formation of structures with a uniform distribution of inclusions of the second phase, accompanied by simultaneous change in the morphology of an intermetallic phase.

In the case of an oscillatory structure, the observed phase has the form of rods with heterogeneous cross-sections, while with the crystallisation rates of 6.10 mm/h a lamellar phase appears [18].

The morphological transformation at 10mm/h is a consequence of the crystallisation front changing from flat to cellular. The changing character of the crystallisation front can affect the morphological transformation of the second phase. The phase transformation from lamellar to acicular takes place due to the change of nucleation conditions. According to the thermodynamic rules, the lamellae/rods transformation takes place if and when the minimum free energy for lamellae ΔG_L is equal to the minimum free energy for rods ΔG_R [24,29].

Further increase in the front travel rate brings next change to the distribution and morphology of an intermetallic phase. The distribution of precipitates starts changing from continuous to cellular. This transformation is also accompanied by the appearance of the Zn$_{16}$Ti phase in the form of needles. The beginning of this transformation can be observed in the core of a single crystal, while near the outer walls, the continuous structure with lamellar morphology of the precipitates still prevails. This situation occurs at the rate of growth equal to 10 mm/h.

Thermal fluctuations at the crystallisation front are responsible for different growth rates in microregions, causing the appearance of pseudo-cellular structure. It has the appearance of a lattice with regularly spaced meshes free from the precipitates and surrounded by an acicular phase. The lattice meshes seen on the plane (11-20) are of an oval shape, elongated in the direction indicated by traces of the plane (0001). Observed on the plane (0001), this structure gives a picture of axial sections of the strongly elongated "rods" of pure matrix, enclosed from the outside by acicular precipitates.

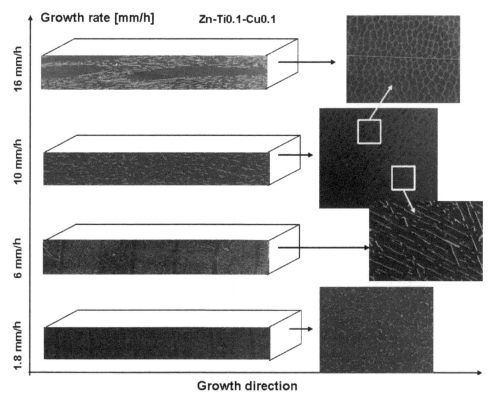

Growth direction

Fig. 23. Comparison of morphology and distribution of the $Zn_{16}Ti$ intermetallic phase in relationship to growth rate of the Zn-Ti0.1wt.%-Cu0.1wt.% single crystals.

An increase in the travel speed of the cooling zone to 16mm/h makes the whole crystal volume assume the form of a pseudo-cellular structure. The observed precipitates assume the form of strongly elongated needles with circular cross-sections, while areas surrounded by them are free from the precipitates and assume an equiaxial shape. Further increase in the crystallisation rate makes the matrix lose its crystallographic continuity and starts the formation of a structure with strongly elongated grains [17].

7. Summary

Regardless of the applied variant of the Zn-Cu-Ti Ti single crystals growth containing up to 0.2wt.% Ti, only one type of the intermetallic $Zn_{16}Ti$ phase was observed to occur. Changing the growth conditions and the chemical composition strongly affected the morphology and distribution of the precipitates of this phase in a single crystal matrix.

It was found that low growth rates of zinc-based single crystals led to the formation of an oscillatory structure. At the same time, in the case of binary Zn-Ti alloys, it was the only available type of structure. Generating other types of structure required higher rates of the single crystals growth. On the other hand, the acceleration of the growth process required stabilised surface of the crystallisation front obtained by reduced thermal fluctuations. This was achieved by adding copper to the Zn-Ti alloy, which resulted in increased surface tension at the crystallisation front and enabled generation of structures characterised by a single crystal matrix at rates in the range of 1.8 - 16 mm/h.

	Shape of the solid/liquid interface	Type of the structure	The intermetallic compound morphology
1.8 mm/h	almost plane	oscillatory	irregular bands
6 mm/h	concave, without instabilities	oscillatory	regular bands
10 mm/h	instabilities appear in the core	continous	bands/needles (fibres)
16 mm/h	instabilities on a whole surface	cellular/pipe	rod-like needles (fibres)

Table 4. Structure types and the intermetallic phase morphology obtained at different growth rates.

Oscillatory structure   Continous structure   Cellular/tube structure

Fig. 24. Change of the intermetallic phase distribution in relationship to growth rate observed in the Zn-Ti0.1wt.%-Cu0.1wt.% single crystals.

Changing the speed of Zn-Ti0.1-Cu0.1 single crystals growth strongly influences the distribution of the Zn$_{16}$Ti phase in a single crystal matrix, as shown in Figures 23 and 24. In the case of oscillatory structures, the following relationships were observed:

a. precipitates-free layers are located at the same distance, λ, for the growth rates 1.8 mm/h and 6 mm/h
b. thickness of these layers, δ, decreases with the growth rate
c. precipitates-rich layer thickness, $\lambda - \delta$, increases with the growth rate
d. oscillatory structure vanishes, cells structure appears in the core of the single crystal, above the growth rate 10 mm/h.

The mathematical model proposed for description of the formation of an oscillatory structure operates well in the case of binary alloys and can be used as a tool helpful in the selection of crystallisation conditions when shaping the desired structure.

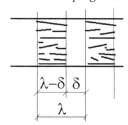

8. Acknowledgment

The authors acknowledge the support of the Polish Committee for Scientific Research, Grant N N508 4800 38 and Grant No. 11.11.180.255

9. References

[1] Oliver H.Wyatt, D.Dew-Hughes: "Metals, Ceramics and Polymers – an introduction to the structure and properties of enginnering materials", Cambridge University Press (1974)

[2] H.Ibach, H.Luth: "Solid-State Physics", Springer-Verlag, (1991)

[3] Landolt-Bornstein, New Series Vol.III, b "structure Data of Elements and Intermetallic Phases", Springer, Berlin, Heidelberg, (1971)

[4] E. Fortunato et al., Materials Science Forum, 514-516, 3, (2006)

[5] J.L.Murray; "Phase Diagrams of Binary Titanium Alloys", J.L.Murray, ed.,ASM International, Metals Park, OH, 336-339 (1987)

[6] T.B.Massalski,: "Binary Alloy Phase Diagrams", editor: H.Okamoto, ASM International (1996)

[7] V. Heine, "Group Theory in Quantum Mechanics", Pergamon, London, (1960)

[8] M. Tinkham, "Group Theory and Quantum Mechanics", McGraw-Hill, New York, (1964)

[9] Gustav E.R. Schulze, Metallphysik, Akademie-Verlag, Berlin (1974)

[10] G.Boczkal, B.Mikulowski, I.Hunsche, C-G.Oertel, W.Skrotzki, Cryst.Res.Technol.,No.2, pp.135-140 (2008)

[11] B.Mikulowski,G.Boczkal, Arch. Metall. Mater.,. 54,: p. 197-203, (2009)

[12] G.Boczkal, B.Mikulowski, I.Hunsche, C-G.Oertel, W.Skrotzki, Cryst. Res. Technol. 45, 111 (2010)

[13] G.Boczkal, B.Mikułowski, Journ.of Alloys and Compound 378, pp.135-139, (2004)

[14] G.Boczkal, Materials Science Forum Vol. 674, pp 245-249,(2011)

[15] B.Mikulowski, Strain Hardening of Zinc Monocrystals with Additions of Silver or Gallium, Metallurgy and Foundry Practice, Scientific Bulletin of Univ. of Mining and Metallurgy 96, Cracow (1982).

[16] G.A.Alers, J.R. Neighbours, The elastic constants of zinc between 4.2° and 670°K ,J.of Phys and Chem of Solids, Vol.7, Iss.1, pp. 58-64, (1958)

[17] C.W.Garland, R.Dalven, Elastic Constants of Zinc from 4.2K to 77.6K, Phys.Rev., vol.111, n.5, Sept. 1, (1958)

[18] G.Boczkal, B.Mikulowski, W.Wolczynski, Materials Science Forum Vol. 649, pp 113-118,(2010)

[19] E.A.Anderson, E.J.Boyle and P.W.Ramsey, Trans. AIME 156, p.278, (1944)

[20] J.A.Spittle, The Effect of Composition and Cooling Rate on the as Cast Microstructure of ZnTi Alloys, Metallography 5, pp.423-447, (1972)

[21] J.A.Spittle, Metallography, Vol.6, pp.115-121 (1973)

[22] Von W.Heine, U.Zwicker, Bd. 53, H.6, (1962)

[23] W. D. Lawson, S. Nielsen: Preparation of Single Crystals. Butterworths Scientific Pub. London. (1958)

[24] B. Chalmers: Principles of Solidification. New York, John Wiley & Sons Inc. (1964)

[25] M.Saillard, G.Develey, C.Becle, J.M.Moreau, D.Paccard, The Structure of ZnTi16, Act.Cryst. 37B, 224-226, (1981)

[26] G.Boczkal, Ph.D. Thesis, AGH University of Science and Technology (2006)

[27] E.Fraś, *Crystallization of metals*, WNT, Warsaw, (2003)

[28] P.Mazumder, R.Trivedi, A.Karma, Metall.and Mat.Trans.A, Vol.31A, pp.1233-1246, (2000)

[29] W. Wołczyński: Modelling of Transport Phenomena in Crystal Growth. Ed.:Szmyd J.S., Suzuki K., Southhampton, Boston, WIT Press (2000)

[30] W.Wolczynski, B.Mikulowski, G.Boczkal, Materials Science Forum Vol. 649, pp 125-130,(2010)

9

High Quality In$_x$Ga$_{1-x}$As (x: 0.08 – 0.13) Crystal Growth for Substrates of λ= 1.3 µm Laser Diodes by the Travelling Liquidus-Zone Method

Kyoichi Kinoshita and Shinichi Yoda
Japan Aerospace Exploration Agency
Japan

1. Introduction

The growth of compositionally uniform alloy crystals is promising for variety of applications because lattice parameters as well as electrical and optical properties can be controlled by composition. Among them, In$_x$Ga$_{1-x}$As bulk crystals are expected as substrates of laser diodes with emitting wavelength of 1.3 µm. High optical gain with small temperature dependence was demonstrated for strained quantum well grown on In$_x$Ga$_{1-x}$As substrates (Ishikawa, 1993; Ishikawa & Suemune, 1994). However, growth of homogeneous In$_x$Ga$_{1-x}$As bulk single crystals is very difficult due to large separation of liquidus and solidus lines in the pseudobinary phase diagram (Bublik & Leikin, 1978). In$_x$Ga$_{1-x}$As bulk crystals were grown by the liquid encapsulated Czochralski (LEC) method with supplying GaAs (Nakajima et al., 1991), zone levelling method (Sell, 1991) and multicomponent zone melting method (Nishijima et al. 2005). InAs mole fraction was limited to 0.2 in LEC method and zone levelling method due to temperature fluctuation in the melt and large separation of liquidus and solidus lines. Multicomponent zone growth (MCZG) using a seed with graded InAs concentration produced a single crystal with InAs mole fraction of 0.3 and length of about 17 mm (Nishijima et al. 2005). However, MCZG requires complicated growth technique and no good reproducible results were obtained. In all of the methods, the most difficult point is to keep the freezing interface temperature constant for growing compositionally uniform crystals since interface travelling rate depends on temperature gradient, mass transport in a melt and solute concentration gradient ahead the interface. As a result, no device quality In$_x$Ga$_{1-x}$As crystals with uniform composition have been obtained so far.

The travelling liquidus-zone (TLZ) method was invented for keeping the interface temperature constant and growing compositionally uniform alloy crystals (Kinoshita et al., 2001, Kinoshita et al., 2004). In the TLZ method, the interface travelling rate can be determined exactly if temperature gradient in the zone is known and then the interface position can be fixed in relation to the heater position by translating a sample device in accordance with the interface travelling rate (Kinoshita et al. 2002). Principle of the TLZ method is proven by the precise measurement of temperature gradient in the melt zone, growth rate and the composition of grown crystals (Nakamura et al. 2003). Since the TLZ method requires diffusion limited mass transport, diameter or thickness of the melt is limited to small values on the ground for suppressing convection in a melt. 2 mm thick platy $In_xGa_{1-x}As$ (x: 0.1 – 0.13) crystals were grown by the TLZ method as substrates of laser diodes with λ = 1.3 µm (Kinoshita et al. 2008). Quality of grown crystals was characterized by electron probe micro analyzer (EPMA), electron back scattering pattern (EBSP), X-ray diffraction (XRD) and by fabricating laser diodes. High temperature stability of output power of laser diodes was proven (Arai et al. 2009).

In this chapter, principle of the TLZ method, growth of high quality $In_xGa_{1-x}As$ platy crystals, fabrication of laser diodes with λ = 1.3 µm and characterization of laser diodes are reviewed.

2. TLZ method

The TLZ method was originally invented for compositionally uniform alloy crystal growth in microgravity. Principle of the TLZ method is studied by growing small diameter $In_xGa_{1-x}As$ crystals on the ground (Kinoshita et al., 2002, 2003, Nakamura et al., 2003, Kinoshita and Yoda, 2010) because convection in a melt can be suppressed in capillary tubes even on the ground. In this section, principle of the TLZ method, one dimensional model and evaluation of a model are reported.

2.1 Principle of a TLZ method

It is well known that compositionally uniform alloy crystals can be grown if diffusion controlled steady-state growth is achieved (Tiller et al., 1953). However, even if such growth is realized during directional solidification, initial transient region is inevitable and long crystal length is required to reach the steady-state. Moreover, growth rate should be sufficiently high because solute in the boundary layer diffuses out and solute should be supplied to keep constant solute concentration at the interface. Segregation at solidification supplies solute and amount of supplied solute is determined by growth rate. This is the reason why sufficiently high growth rate is required for the steady-state growth. High growth rate, however, tends to cause the constitutional supercooling (Tiller et al., 1953). Many investigators tried to grow compositionally uniform alloy crystals by diffusion controlled steady-state growth but none of them obtained satisfactory results. Microgravity experiments are also expected to realize diffusion controlled steady-state growth by suppressing convection in a melt but no one succeeded in growing homogeneous alloy crystals having expected composition. Residual gravity in space crafts on the order of 10^{-4} g hinders diffusion controlled growth. Therefore, a new growth method for obtaining compositionally uniform alloy crystals is anticipated.

When we consider diffusion boundary layer in the steady-state growth, we noticed that zone melting method under temperature gradient is easier to maintain constant solute concentration at the interface because solute is saturated at freezing and dissolving interfaces. Moreover, solute is almost saturated throughout the zone if zone thickness is thin enough. As described later in more detail, in such situation solute concentration gradient in the zone is controlled by temperature gradient and compositionally uniform alloy crystal growth becomes much easier than other growth method. We named this zone melting method under temperature gradient a travelling liquidus-zone (TLZ) method since crystal growth is carried out by travelling almost saturated liquid-zone. Sample configuration, temperature profile, solute profile in a sample and its relation to a phase diagram in the TLZ method is depicted in Fig. 1 as referring to the growth of In$_{0.3}$Ga$_{0.7}$As. It is needless to say that convection in a thin melt zone is suppressed more effectively than in a long melt even in microgravity.

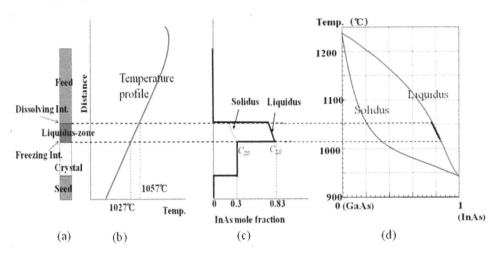

Fig. 1. Schematic drawing of a TLZ method as referring to the growth of In$_{0.3}$Ga$_{0.7}$As. (a) sample configuration, (b) temperature profile, (c) InAs concentration profile, (d) GaAs-InAs pseudobinary phase diagram.

In Fig. 2, solute concentration profile in the diffusion boundary layer in the steady-state growth is compared between directional solidification method and TLZ method (Kinoshita and Yoda, 2011). When temperature gradient is low and zone thickness is thin, solute concentration profile is linearly approximated and it can be related to the liquidus line in the pseudo binary phase diagram as shown in Fig. 1. Liquidus line is also linearly approximated in a narrow temperature range. Then, it should be noticed that solute concentration gradient in a zone can be controlled by temperature gradient in such case as is given by eq. (1), where C_L is solute concentration in the zone, T is temperature, Z is a distance measured from the freezing interface and $\partial C_L/\partial T$ is reciprocal of the slope of the liquidus line in the phase diagram. Then, freezing rate V is given by eq. (3) if we suppose the diffusion controlled steady-state growth is realized as given by eq. (2) and V is determined by temperature gradient in the zone, where D is interdiffusion coefficient between solute and solvent, C_{L0} and C_{S0} are liquidus and solidus concentrations at the freezing interface.

$$\left(\frac{\partial C_L}{\partial Z}\right)_{Z=0} = \left(\frac{\partial C_L}{\partial T}\right)\left(\frac{\partial T}{\partial Z}\right)_{Z=0} \tag{1}$$

$$V(C_{L0} - C_{S0}) = -D\left(\frac{\partial C_L}{\partial Z}\right)_{Z=0} \tag{2}$$

$$V = -\frac{D}{(C_{L0} - C_{S0})}\left(\frac{\partial C_L}{\partial Z}\right)_{Z=0} = -\frac{D}{(C_{L0} - C_{S0})}\left(\frac{\partial C_L}{\partial T}\right)\left(\frac{\partial T}{\partial Z}\right)_{Z=0} \tag{3}$$

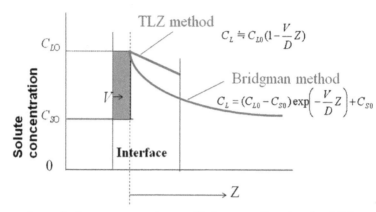

Fig. 2. Comparison of solute concentration profile between directional solidification (Bridgman method) and zone melting method under temperature gradient (TLZ method).

If we know interdiffusion coefficient D and phase diagram date, we can calculate the freezing rate V using eq. (3). Then, it is very easy to fix the interface position as shown in Fig. 3. If a sample device is translated toward opposite direction at the same rate of interface travel rate as calculated by eq. (3), the interface position is fixed and the interface temperature is kept constant, resulting in a compositionally uniform alloy crystals. Thus, compositionally uniform alloy crystals are grown by simply translating a sample device at constant rate in the TLZ method. According to our numerical analysis, 1 to 2 hours are required for establishing an almost saturated solute concentration profile in a zone and spontaneous crystal growth starts.

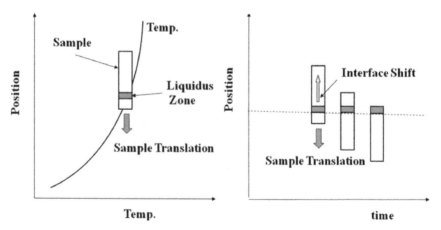

Fig. 3. Schematic view of a liquidus-zone shift in the TLZ method and a way to keep the freezing interface at a fixed position.

2.2 Evaluation of a TLZ growth model

One dimensional TLZ model equation (3) is evaluated by growth experiments. A seed, a zone forming material and a feed were inserted into a boron nitride crucible and the crucible was vacuum sealed in a quartz ampoule at about 1x10^{-4} Pa. In the growth of In$_x$Ga$_{1-x}$As, the zone forming material is InAs or InAs rich solid solution with GaAs. The ampoule was inserted into a temperature gradient furnace and was heated to about 1100℃ at temperature gradients of 10 – 20℃/cm. InAs-GaAs interdiffusion coefficient was measured in microgravity using a sounding rocket (Kinoshita et al. 2000). It was measured to be 1.5±0.2x10^{-8} m^2/s at 1070℃. For In$_{0.3}$Ga$_{0.7}$As growth, C_{L0} is given by 0.83 from phase diagram. $\partial C_L/\partial T$ is also given by phase diagram as -465 K/mol. Then, V is calculated to be 0.22 mm/h for the temperature gradient of 10℃/cm (Nakamura et al., 2003). Crystal growth experiments were performed for crystals with 2 mm diameter in order to suppress convection in a melt on the ground. Concentration profiles were measured by EPMA. Results are shown in Fig. 4. In the figure, crystal growth started at the length of about 8 mm and the sample was quenched at the length of about 32 mm. Therefore, about 24 mm long crystal was grown. It should be noted that expected InAs concentration (15 at%) is realized without initial transient region. This is one unique point typical to the TLZ method when compared with diffusion controlled steady-state growth in the directional solidification. This is because steady-state solute concentration profile can be established without transient region in the case of zone melting method under temperature gradient. Compositional homogeneity is excellent; In concentration 15±0.5 at% is achieved for the length of more than 20 mm. Scattered concentration region was liquidus-zone before quenching. Dendrite crystals were grown during quenching and dendrite growth resulted in concentration scattering.

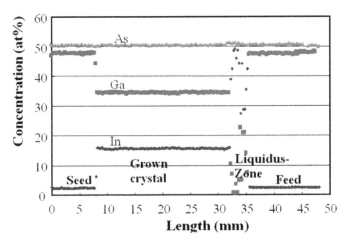

Fig. 4. Axial In, Ga and As concentration profiles for a TLZ-grown 2 mm diameter crystal at a sample translation rate R = 0.22 mm/h.

It is astonishing that excellent compositional homogeneity was achieved at just calculated sample translation rate using eq. (3). Since temperature gradient in the zone was measured precisely by sophisticated method namely by measuring solidus compositions at two interfaces (freezing and dissolving interfaces), model evaluation was possible. Solidus compositions measured at two interfaces are shown in Fig. 5 together with a determined temperature gradient in the zone (Nakamura et al., 2003, Kinoshita et al., 2005). At the freezing interface, In concentration was 15.1 at% and this concentration gives solidus temperature of 1083.1°C from the phase diagram, while In concentration at dissolving interface 12.3 at% gives solidus temperature of 1099.8°C. Distance between two interfaces is 16.8 mm and temperature gradient 10°C/cm was determined. In the temperature gradient measurement, a special sample whose In concentration is gradually decreasing towards a feed and solid-liquid coexistence at the dissolving interface was used. This enabled us to measure solidus composition at the dissolving interface.

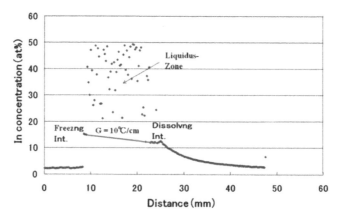

Fig. 5. Measured solidus In concentrations at freezing and dissolving interfaces and determination of temperature gradient (G) in the melt zone.

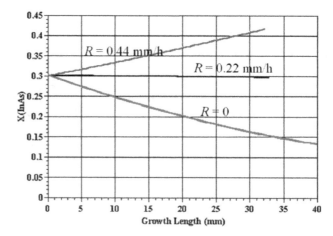

Fig. 6. Results of numerical analysis of axial InAs concentration in mole concentration as a parameter of sample translation rate R.

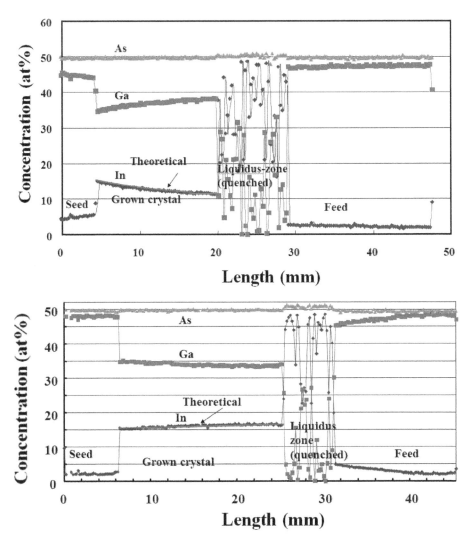

Fig. 7. Axial compositional profiles of TLZ-grown In$_x$Ga$_{1-x}$As crystals at a sample translation rate (a) $R = 0$ and (b) $R = 0.27$ mm/h.

Numerical analyses show that InAs concentration increases when sample translation rate is faster than the predicted one from eq. (3), while it decreases when sample translation rate is slower than the predicted one (Fig. 6). It should be noted that spontaneous growth occurs even if sample is stationary. This is because solute concentration difference is established between the two interfaces and interdiffusion occurs in a melt zone. At the freezing interface, In concentration is higher and In is transported towards the dissolving interface by diffusion. Indium concentration decrease causes solidification at the freezing interface because equilibrium temperature gets higher. However solidification segregates In at the interface. The segregated In is diffused away towards the dissolving interface. At the dissolving interface, a feed is dissolved by transported In because In concentration gets higher than the equilibrium one. Thus, crystal grows at the freezing interface and a feed is dissolved at the dissolving interface; crystal growth proceeds spontaneously towards lower In concentration side (dissolving interface side, which is a higher temperature side). Sample translation towards a lower temperature side plays a role of keeping the interface temperature constant and it is not a driving force for crystal growth. The driving force in the TLZ growth is interdiffusion between solute and solvent in a zone and segregation on solidification. This should be noted clearly as one of features of the TLZ growth. Figures 7 (a) and 7 (b) show experimental results when sample translation rates do not match the freezing rate of 0.22 mm/h at a temperature gradient of 10°C/cm (Kinoshita et al., 2005). Fig. 7 (a) is a case of $R = 0$ (stationary) and Fig. 7 (b) is a case where R is faster than 0.22 mm/h. Both results agree well with the numerical analyses. As described later, higher sample translation rate than 0.3 mm/h caused constitutional supercooling and resulted in fluctuation of composition. Therefore, in Fig. 7 (b) an experimental result at $R = 0.27$ mm/h is compared with a numerical analysis. Thus, TLZ growth model was confirmed by growth experiments at various sample translation rates and comparison of compositional profiles with those obtained by numerical analyses. Latent heat effect on growth rate was also investigated by numerical analysis and the effect is shown to be negligibly small except for end menber region (Adachi et al., 2004b).

2.3 Limitation in the TLZ method

TLZ growth model is established when solute concentration gradient in the melt zone is linearly approximated as shown in Fig. 2. When temperature gradient is low and zone thickness is thin, such approximation is valid. However, freezing rates deviate from those predicted by eq. (3) at high temperature gradient and for long zone length. Amount of solute in a zone also limits the length of a grown crystal. Moreover, TLZ growth model is based on the diffusion controlled steady-state growth and compositional uniformity is affected by convection in a melt. The TLZ method also lies side-by-side with constitutional supercooling since almost saturated melt is utilized. In this section, such limitations will be described in more detail.

2.3.1 Limitation by temperature gradient

TLZ method is classified as diffusion controlled steady-state growth method as well as a zone method. Solute concentration profile in the diffusion boundary layer is expressed as eq. (4) (Tiller et al., 1953). When VZ/D is small, concentration gradient is approximated by linear relation as eq. (5). Such situation has already been described in Fig. 2. Low temperature gradient gives small value of V and gives a small error in growth rate calculation using eq. (3). However, when temperature gradient is high, V gets large and an error in growth rate calculation becomes large (Adachi et al. 2004a). Figure 8 shows deviation of growth rate from linear relation to temperature gradient. From Fig. 8, it can be said that TLZ growth rate defined by eq. (3) is valid within 10% error when temperature gradient is smaller than 40°C/cm for the zone thickness of 20 mm. Therefore, TLZ growth should be carried out at low temperature gradient like 10 - 20°C/cm at zone thickness of 20 mm.

$$C_L = (C_{L0} - C_{S0})\exp\left(-\frac{V}{D}Z\right) + C_{S0} \tag{4}$$

$$C_L = C_{L0}\left(1 - \frac{V}{D}Z\right) \tag{5}$$

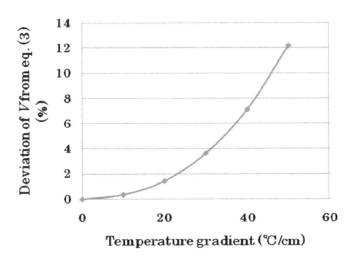

Fig. 8. Deviation of V from eq. (3) as a function of temperature gradient.

2.3.2 Limitation by zone length

Distance Z is also a variable in eq. (4). Therefore, zone length gives similar effect in growth rate calculation. When the temperature gradient is 20°C/cm, an error in the growth rate calculation using eq. (3) is about 7 % for the zone length of 40 mm but it beyonds 12 % when the zone length is 50 mm. If growth rate deviation from linearity is desired less than 10%, zone length should be less than 45 mm. Limitation of zone length is natural since long zone length cannot be distinguished from directional solidification.

2.3.3 Limitation of growth length

In the In$_x$Ga$_{1-x}$As crystal growth, we usually start from a combination of a GaAs seed, an InAs zone forming material and a GaAs feed. GaAs saturated InAs rich zone is formed and this zone travel towards a GaAs feed. Therefore, InAs is consumed according as crystal growth proceeds. If an InAs length is 10 mm, length of a grown crystal is limited to 10/3 mm for an In$_{0.3}$Ga$_{0.7}$As crystal. In order to increase the length of grown crystal longer zone or InAs contained feed should be used. However, zone length is limited from the point of zone growth as described above and in an InAs contained feed InAs content is restricted to low for preventing melting of a feed. In this meaning, the length of a TLZ-grown crystal is limited.

2.3.4 Limitation by convection in a melt

The TLZ method is possible in the diffusion limited regime. If convection occurs in a melt, solute concentration profile ahead the interface is disturbed and the growth rate does not obey eq. (3). We grew 2 mm diameter $In_xGa_{1-x}As$ crystals by suppressing convection in a melt and confirmed the TLZ growth model. However, 2 mm diameter crystals cannot be used for opt-electronic devices and we tried increase of crystal diameter. Result of a growth experiment for 10 mm diameter crystal is shown in Fig. 9. Axial compositional homogeneity is achieved for the length of more than 30 mm but radial concentration inhomogeneity exists in a 10 mm diameter crystal. According to numerical analysis, convective vortices occur at both freezing and dissolving interfaces as shown in Fig. 10. When such convective vortices are taken into consideration, radial concentration difference is understood; In-rich melt is transported towards the centre by convection and InAs concentration increases in the centre of a grown crystal. As for the strength of convective driving force, Grashof number G defined by eq. (6) gives order of magnitude where g is gravity acceleration, β is volume expansion coefficient of a melt, ΔT is a temperature difference between the two ends of a melt, L is characteristic length of a melt and v is kinematic viscosity of a melt. Comparison of convective driving force between 2 and 10 mm diameter melts, convection in 10 mm diameter melt gives 2 orders of magnitude higher than that in 2 mm diameter melt. Result in Fig. 9 shows importance of suppressing convection in a melt for compositionally homogeneous alloy crystals in radial direction as well as in axial direction (Kinoshita et al., 2006).

$$G_r = \frac{g\beta\Delta T L^3}{v^2} \tag{6}$$

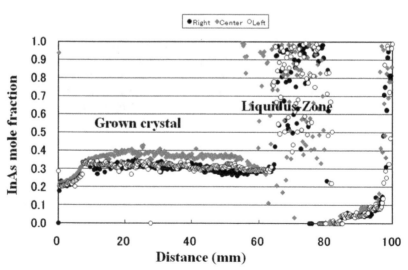

Fig. 9. Axial InAs concentration profiles compared among centre line, right and left peripheral regions.

High Quality In$_x$Ga$_{1-x}$As (x: 0.08 – 0.13) Crystal Growth for Substrates of λ= 1.3 μm Laser Diodes by the Travelling Liquidus-Zone Method

199

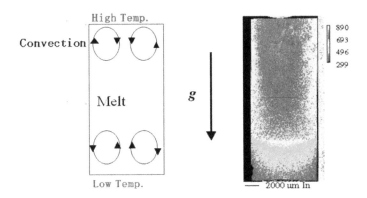

Fig. 10. Convective vortices in a melt and two dimensional In concentration map for a 10 mm diameter crystal.

2.3.5 Limitation by constitutional supercooling

The TLZ growth utilizes almost saturated melt. In this point of view, constitutional supercooling tends to occur. Constitutional supercooling in the directional solidification of alloy crystals in the diffusion limited growth was studied extensively by Tiller et al. (1953). According to them, the growth rate should be high and should fulfil conditions defined by eq. (7) where G is temperature gradient, R is growth rate, m is the slope of the liquidus line. The TLZ method utilizes solute concentration gradient in diffusion boundary layer ahead the freezing interface. Therefore, growth rate should fulfil the conditions proposed by Tiller et al. When we change notation of eq. (3) to similar ones given by Tiller et al. for comparison, equation (8) is given. Equation (8) is further modified into eq. (9). When we compare eq. (7) with eq. (9), we notice that the growth rate for realizing compositional uniformity in the TLZ method is the critical growth rate for avoiding constitutional supercooling. If growth rate is higher than this, constitutional supercooling will occur. From this discussion, it is clearly shown that homogeneous crystal growth in the TLZ method is side-by-side the constitutional supercooling.

$$\frac{G}{R} \geq \frac{m(C_{L0} - C_{S0})}{D} \tag{7}$$

$$R = \frac{D}{(C_{L0} - C_{S0})} mG \tag{8}$$

$$\frac{G}{R} = \frac{m(C_{L0} - C_{S0})}{D} \tag{9}$$

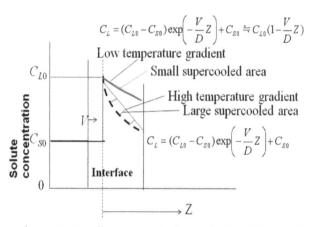

Fig. 11a. Comparison of constitutionally supercooled area during TLZ growth at low and high temperature gradients. At high temperature gradient, non-linearity of solute concentration profile aheat the freezing interface is large and gives larger supercooled area

Constitutional supercooling is inevitable at high temperature gradient (Kinoshita and Yoda, 2011). This may be peculiar from the requirement of high G/R ratio shown by eq. (7) but this requirement comes from constant G/R for compositional uniformity and solute saturation at two interfaces (freezing and dissolving interfaces) in the TLZ method. At high temperature gradient, non linearity of solute concentration profile increases as shown in Fig. 11(a). In the figure, the straight line shows equilibrium concentration at the given temperature gradient and dashed region is constitutionally supercooled region. It should be noted that constitutionally supercooled area is larger at higher temperature gradient. This is unique point in the TLZ method. It is obvious that degree of constitutional supercooling is greater at higher temperature gradient because deviation from linearity in solute concentration profile is greater at higher temperature gradient (Kinoshita et al., 2011). From this point of view, temperature gradient is limited and growth rate in turn is also limited. In the $In_xGa_{1-x}As$ crystal growth experiments, it is shown that temperature gradient that exceeds 30°C/cm resulted in constitutional supercooling. This means that growth rate is limited to 0.66 mm/h. In general, compositional fluctuation increases in constitutionally supercooled region. An example of compositional profile of constitutionally supercooled sample is shown in platy crystal growth in later section (an $In_{0.15}Ga_{0.85}As$ platy crystal grown at a temperature gradient of 37°C/cm showed large compositional fluctuation as shown in 3.2.3).

3. Platy crystal growth

For device application, large surface area is required. However, increase in crystal size increases convection in a melt and compositional uniformity of TLZ-grown crystals is deteriorated. Suppressing convection in a melt and large surface area is fulfilled by the growth of platy crystals if thickness of the platy crystal is sufficiently thin (Kinoshita et al., 2008, 2010). Such situation is schematically depicted in Fig. 11(b). We have experimentally confirmed that convection in a 2 mm diameter melt is suppressed and TLZ mode crystal growth is possible. Then, we determined thickness of platy crystals as 2 mm. In this section, growth and characterization of platy crystals for substrates of laser diodes with emitting wavelength $\lambda = 1.3$ μm are described.

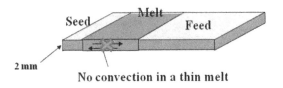

No convection in a thin melt

Fig. 11b. Schematic view of a platy crystal growth.

3.1 Growth procedures and characterization methods

Growth method for platy crystals is similar to that of the cylindrical crystal. A GaAs seed, an InAs zone forming material, and a GaAs feed were cut into plates with 2 mm thickness and were inserted into a boron nitride crucible with a rectangular bore and were sealed in vacuum in a quartz ampoule and then heated in a gradient heating furnace. Growth interface temperature was adjusted around 1100℃ so that In$_x$Ga$_{1-x}$As (x: 0.1 – 0.15) crystals were grown. Temperature gradient was set around 20℃ so that constitutional supercooling at high temperature gradient was avoided. Samples were translated at the rate calculated using eq. (3). Seed orientation was <100> or <110>. Sn was doped as n-type dopant. At the beginning of crystal growth, we thought that In$_{0.3}$Ga$_{0.7}$As composition was required as substrates for λ = 1.3 μm laser diodes. However, it turned out that composition around In$_{0.1}$Ga$_{0.9}$As can be used as substrates since thin film growth technology has been developed and large lattice mismatch between the substrate is conquered. As for crystal size, we started from the growth of 10 mm width and increased crystal width. At the present, we are successful in growing 50 mm wide platy crystals. Grown crystals were polished and crystalline nature was investigated. Compositional profiles of grown crystals were measured by EPMA on polished surfaces. Two-dimensional semi-quantitative mapping analysis was performed for measuring overall compositional distribution. Quantitative analysis was performed along growth axis with precision of 0.1 at% for each constituent element. Crystal quality was evaluated by measuring X-ray rocking curves. Electrical properties of crystals were measured by Hall measurements. Etch pit density was measured after KOH etching.

3.2 Results and discussion

Experimental results for platy crystal growth are summarized and factors that effect on crystal quality are discussed in this section. Factors for affecting crystal quality are common for the TLZ method except for seeding in rectangular crucible.

3.2.1 Single crystalline nature

Typical two examples of roughly polished surface of two platy crystals are shown in Fig. 12. Composition of Fig. 12 (a) is In$_{0.10}$Ga$_{0.90}$As and that of Fig. 12 (b) is In$_{0.13}$Ga$_{0.87}$As. When InAs mole fraction in a grown crystal is less than 0.1, a single crystal that takes over the seed orientation was grown. However, when InAs mole fraction is more than 0.13, single crystal growth that takes over the seed orientation became very difficult. This may be due to increase in lattice mismatch between a GaAs seed and a grown crystal. Difficulty in single crystallization was settled by using a feed having the same orientation of a seed. In Fig. 12 (b), polycrystallization is observed at the seed and grown crystal interface. However, a single crystal was grown again as crystal growth proceeded.

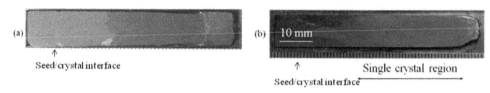

Fig. 12. Roughly polished surface of two platy crystals, (a) $In_{0.10}Ga_{0.90}As$ and (b) $In_{0.13}Ga_{0.87}As$.

The mechanism of single crystal growth is not clear but it may be related to a crucible shape having rectangular bore and nucleation of the same orientation as a feed might occur in a narrow gap between the crucible wall and feed surface. In wider crystal growth too, single crystals were grown by the same mechanism. Reproducibility of single crystal growth was very good and single crystal length more than 50 mm was obtained.

3.2.2 Growth temperature

Freezing interface temperature determines composition of grown crystals. In this point of view, growth temperature is very important. In the TLZ method, solute in a melt is almost saturated and it is very easy to control interface temperature. We set 8 thermocouples around a growth ampoule and one of them is set at seed/zone interface. Only procedure is to monitor temperatures measured by these thermocouples and to change the interface temperature. It is usual that surface temperatures of a quartz ampoule are not equal to inner temperatures of a crucible but difference between the two was less than 5°C and minor adjustment gave target compositions. Figure 13(a) is an example of $In_{0.2}Ga_{0.8}As$ crystal and Fig. 13(b) is an example of $In_{0.13}Ga_{0.87}As$ crystal. In $In_{0.2}Ga_{0.8}As$, interface temperature was set at 1057°C and in $In_{0.13}Ga_{0.87}As$ it was set at 1095°C. In both cases, target compositions were obtained at the seed/crystal interface and no initial transient regions were found as is the case of $In_{0.3}Ga_{0.7}As$ (Fig. 4). Compositional uniformity was excellent for $In_{0.13}Ga_{0.87}As$; InAs mole fraction was 0.13±0.005, but uniformity was degraded for $In_{0.2}Ga_{0.8}As$. Single crystallinity was also better for lower In concentration crystals. Since we succeeded in laser operation at the wavelength of 1.3 μm for laser diodes on $In_{0.13}Ga_{0.87}As$ substrates (Arai et al, 2007), we improved crystal quality for the composition of around $In_{0.13}Ga_{0.87}As$.

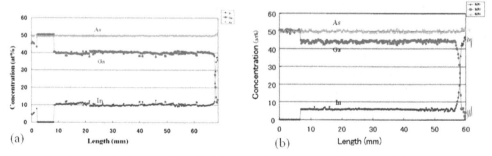

Fig. 13. Axial concentration profiles of platy crystals; (a) for $In_{0.2}Ga_{0.8}As$ and (b) for $In_{0.13}Ga_{0.87}As$.

High Quality In$_x$Ga$_{1-x}$As (x: 0.08 – 0.13) Crystal Growth for Substrates of λ= 1.3 µm Laser Diodes by the Travelling Liquidus-Zone Method

203

3.2.3 Temperature gradient

Temperature gradient determines spontaneous growth rate in the TLZ method as given by eq. (3). So long as the sample translation rate matches this spontaneous growth rate, compositionally uniform alloy crystals can be grown. However, high temperature gradient causes constitutional supercooling and results in compositional fluctuation. Figure 14 compares compositional profiles of crystals grown at different temperature gradient (a) 25°C/m and (b) 37°C/cm. Note that the crystal grown at lower temperature gradient shows lower compositional fluctuation. As described in the section 2.3.5, maximum temperature gradient for suppressing constitutional supercooling may be 30°C/cm. It is concluded that temperature gradient should be as low as possible for suppressing compositional fluctuation but the lowest limit of temperature gradient should be determined from the point of mass productivity of crystals.

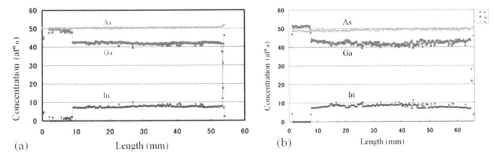

Fig. 14. Comparison of axial concentration profiles of crystals grown at different temperature gradient (a) 25°C/m and (b) 37°C/cm. Average composition of both crystals is about In$_{0.15}$Ga$_{0.85}$As.

3.2.4 Temperature stability

Effect of temperature stability on compositional uniformity was investigated with average composition of In$_{0.13}$Ga$_{0.87}$As (Kinoshita et al., 2008). The most stable temperature was achieved when air flow in a furnace tube was shut. Temperature stability was ±0.1°C as shown in Fig. 15 (a). With air flow conditions, temperature stability got worse to ±0.2°C as shown in Fig. 15 (b). The stability difference was small but this small difference gave a considerable effect on the compositional uniformity as shown in Fig. 16. When the temperature stability was ±0.1°C, InAs mole fraction was 0.13 with σ of 0.0005 where σ is the standard deviation in the distance between 20 and 40mm. When the temperature stability was ±0.2°C, InAs concentration uniformity was 0.13 with σ of 0.006 in the distance between 20 and 40mm. The reason why the temperature fluctuation influence on the compositional uniformity may be related to the crystallization from the almost saturated melt, namely, temperature fluctuation gives rise to the formation of constitutionally supercooled region in a melt and such region crystallizes earlier, then it is not strange that earlier crystallized region has lower In concentration. Improvement in compositional stability resulted in higher crystal quality. X-ray rocking curve measurements showed that full width at half maximum (FWHM) of rocking curve ranges from 0.03 to 0.04 degrees for crystals grown at temperature stability ±0.1°C. Such small FWHM shows good crystallineity which can be used as a substrate of laser diodes.

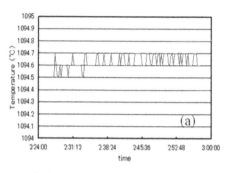

Fig. 15. Comparison of temperature stability.

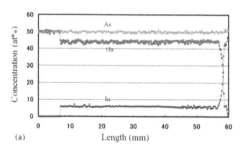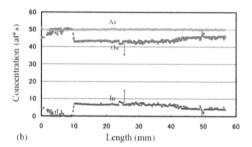

Fig. 16. Comparison of concentration profiles for two crystals grown at different temperature stability (a)±0.1℃ and (b)±0.2℃.

3.2.5 Characterization as substrates of laser diodes

Improvement in compositional stability resulted in higher crystal quality. Figure 17 shows X-ray rocking curves in the distance between 26 mm and 56 mm at an interval of 5mm for the crystal grown at the temperature stability of ±0.1°C. The ω-scan around the (004) diffraction peak was detected by an open detector. Cu $k_{\alpha 1}$ radiation from an X-ray tube (loaded power; 40 kV, 30 mA) was monochromated by a four crystal Ge(220) monochromator and was incident on a crystal. The illuminated area was limited to about 1×1 mm² by a divergence and a scatter slit. Full width at half maximum (FWHM) of rocking curve ranges from 0.03 to 0.04 degrees (from 108 to 144 arc seconds). Such small FWHM shows good crystallineity which can be used as a substrate of laser diodes. High quality region extended to 10 × 30 mm². When InAs mole fraction increases, FWHM increased.

This phenomenon is usual in many alloy crystals. When solute concentration increases, compositional uniformity is deteriorated and strain increases in the grown crystal. The increased strain degrades crystalline quality. Therefore, the lower InAs mole fraction, the higher is the quality of grown crystals. In the course of our study, we found that $In_{0.3}Ga_{0.7}As$ could be used as substrates of λ = 1.3 μm laser diodes instead of $In0.3Ga_{0.7}As$. This is beneficial for ternary crystal growth since higher quality can be expected. Sn was doped as n-type dopant for substrate use. Carrier concentration was measured to be 5 – 8×10¹⁸ cm⁻³. Etch pit densities were in the range between 1×10³ and 3×10⁴ cm⁻², which is sufficiently low for a substrate.

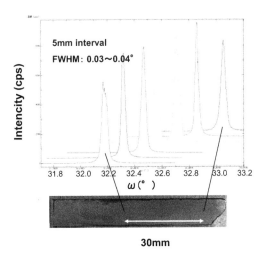

Fig. 17. FWHMs of X-ray rocking curves at seven positions. They range from 0.03 to 0.04° for
the length of 30 mm.

3.3 Increase of a crystal width

Convection in a melt is suppressed in a thin melt. In platy crystal growth, width is
considered to be independent of convection strength when thickness of crystals is limited.
From this point of view, width of platy crystals was increased step by step from 10 mm to
20, 30, and 50 mm (Kinoshita et al., 2010). In all cases, thickness of platy crystals was set to
be 2 mm. The most difficult point was to keep high temperatures in a feed area as growth
ampoule was translated towards low temperature side. When width of platy crystals was
increased, diameter of boron nitride (BN) crucible increased. BN has higher thermal
conductivity than quartz and large diameter BN crucible transferred more heat to lower
temperature side. For keeping high temperatures in a feed area, length of BN crucible was
increased to receive more radiation from a heater.

Figure 18 shows an example of 30 mm wide platy crystal. Outlook of roughly polished
surface is shown in Fig. 18 (a) and InAs mole fractions measured along a centre line and 10
mm away from the centre line are shown in Fig, 18 (b). Single crystalline area larger than
30×30 mm² was obtained. InAs mole fractions were constant and 0.10±0.01 was achieved for
the distance of about 40 mm. Good crystallinity with the FWHM of less than 0.04° in X-ray
rocking curves was obtained. Such excellent compositional uniformity and crystal quality
owe to matching of sample translation rate and freezing rate and resulting in a fixed
freezing interface position relative to heater segments. In this point of view, constant
temperature gradient during translation of a sample was important for obtaining
compositional uniformity. Such increase in surface area in platy crystals without
deteriorating crystal quality also shows that convection in a melt is suppressed by the
limitation of melt thickness (2 mm) and increase in melt width did not cause convection. If
convection occurs in a melt, local inhomogeneity will be resulted due to stirring of a melt by
convection. Good homogeneity shows that convection was suppressed in the course of
crystal growth.

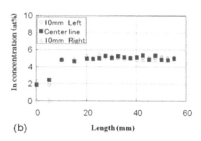

(a)

(b)   Length (mm)

Fig. 18. An example of 30 mm wide platy crystal; (a) roughly polished surface of a crystal and (b) InAs mole fractions along growth axis.

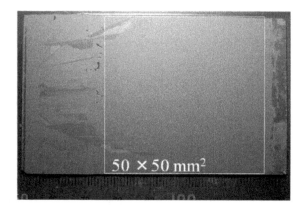

Fig. 19. An example of 50 mm wide crystal (Single crystalline area of almost 50×50 mm^2 is obtained).

Now we have succeeded in growing 50 mm wide platy crystals. Figure 19 shows an example of 50 mm wide crystal. Single crystalline area of almost 50 ×50 mm^2 is obtained. For mass production of laser diodes, large platy crystals with good reproducibility are required. As described earlier, crystals do not take over seed orientation but take over feed orientation. In this case, similar single crystallization mechanism works, too. Reproducibility is now more than 90%.

4. Laser diodes fabrication

Laser diodes with emitting wavelength λ = 1.3 µm were fabricated on In$_x$Ga$_{1-x}$As (x: 0.1 – 0.13) substrates. Ishikawa (1993) demonstrated merits of a ternary substrate for high optical gain with small temperature dependence. In the course of study, we found that In$_x$Ga$_{1-x}$As (x: 0.1 – 0.13) crystals can be used as substrates owing to development of thin film growth technology. Low InAs content crystals are beneficial to substrates because they have better thermal conductivity than high InAs content crystals.

4.1 Fabrication procedures

In$_{0.13}$Ga$_{0.87}$As platy crystals were polished mechano-chemically to 0.5 mm thickness. Surface roughness of mirror polished substrates was measured to be less than several nano-meters. Strained multiple-quantum-wells (MQWs) having the combination of In$_{0.12}$Ga$_{0.88}$As and In$_{0.38}$Ga$_{0.62}$As as shown in Fig. 20 (a) were prepared by metal-organic vapor phase epitaxy (MOVPE) on the cleaned substrate (Arai et al., 2009). The active region was consisted of four In$_{0.38}$Ga$_{0.62}$As wells and five In$_{0.12}$Ga$_{0.88}$As barriers sandwitched by two seperate-confinement heterostructure (SCH) layers. The stripe mesa type laser was fabricated by chemical etching. Cross sectional view of a fabricated mesa stripe is shown in Fig. 20 (b). The stripe mesa was in the [011] direction, forming a reverse mesa. Bottom ridge width of 1.7 μm enabled single mode lasing. Selective wet etching was utilized for such narrow ridge formation.

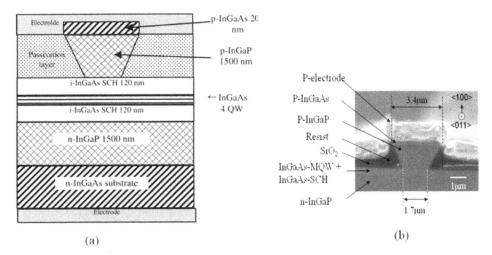

(a)   (b)

Fig. 20. Cross sectional view of a fabricated laser diode, (a) schematic drawing and (b) electron micrograph.

4.2 Characterization of laser diodes

Figure 21 (a) shows the bias current versus output power (I-L) characteristics of a fabricated laser diode The threshold current is 7.2 mA. Figure 21 (b) is the lasing spectrum at a bias of 20 mA. The lasing wavelength is 1.31μm. The laser exhibits a maximum operating temperature of 80 and 150°C under CW and pulsed operation, respectively. Therefore, the limiting factor was self-heating. An effective way of overcoming this thermal problem is to reduce the indium content of the In$_x$Ga$_{1-x}$As substrate and to introduce InAlGaAs barrier layers. We therefore tried to realize 10 Gbps modulation over 85°C by eliminating the thermal problems and we prepared an In$_{0.08}$Ga$_{0.92}$As substrate.

Laser diodes on an In$_{0.08}$Ga$_{0.92}$As substrate showed improved lasing characteristics. Self heating problem due to bad thermal conductivity of In$_{0.13}$Ga$_{0.87}$As substrates was settled by reducing InAs content in the substrate. However, emitting wavelength was 1.26 μm due allowance limit of InAs content in active layers. Figure 22 shows continuous wave (CW) lasing characteristics of this laser diode at various temperatures. CW operation was possible up to 150°C and was much improved. Characteristic temperature of the threshold current density between 25 and 125°C was about 90 K and showed a higher value compared with that of 1.31 μm laser (about 70 K). Such high temperature stability in output power is also expected in 1.31 μm laser diodes by optimizing heat sink structure and heat sink materials.

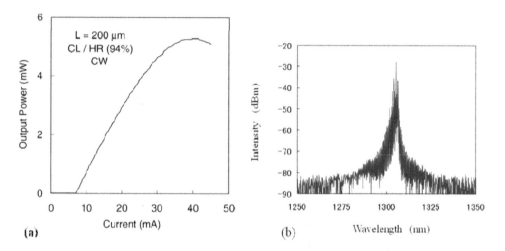

(a)

(b)

Fig. 21. Lasing characteristics of a fabricated laser diode. (b) is the lasing spectrum at a bias of 20 mA.

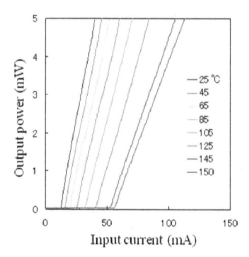

Fig. 22. CW lasing characteristics at various temperatures.

High Quality In$_x$Ga$_{1-x}$As (x: 0.08 – 0.13) Crystal Growth for Substrates of λ= 1.3 μm Laser Diodes by the Travelling Liquidus-Zone Method

209

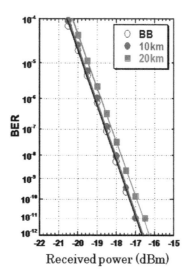

Fig. 23. Bit error rate after 20 km transmission.

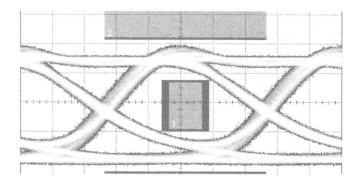

Fig. 24. Eye diagram at the modulation of 10 Gbps at 85°C.

A 10-Gbps direct modulation and transmission tests through a single mode fiber up to 20 km at 25°C were successfully performed using a fabricated laser diode with λ = 1.31 μm as shown in Fig. 23. Bit error rate (BER) for back-to-back configuration after 20 km transmission was less than 10^{-7} at -18 dBm. We also tested 1.26 μm wavelength laser diodes. In this laser diodes, 10 Gbps modulation at 85°C was confirmed by an electrically filtered back-to-back (BB) eye diagram as shown in Fig. 24. The extinction ratio and the mask margin of the synchronous optical network (SONET) mask were 7 dB and 9%, respectively. These results show feasibility of fabricated laser diodes.

5. Conclusions

The TLZ method has been invented for growing compositionally uniform alloy crystals. The solute concentration gradient is controlled by applied temperature gradient utilizing saturation of solute in a zone. Owing to this merit, the spontaneous growth rate is calculated from the diffusion controlled steady-state growth conditions. Compositionally uniform alloy crystals are obtained merely by translating samples at the calculated growth rate relatively to a heater. Principle of the TLZ method was proved by the growth of 2 mm diameter crystals since convection in a melt is suppressed and diffusion limited growth is possible in capillary tubes even on the ground. As predicted, compositionally uniform crystals were grown without initial transient region which is typical to directional solidification method. Although the TLZ method is a superior method, it has limitations and they were also studied theoretically and experimentally. Limitation by temperature gradient, limitation by zone length, limitation by constitutional supercooling, and limitation by convection in a melt were made clear. In microgravity, TLZ growth is free from limitation by convection in a melt. In this point, microgravity is beneficial for TLZ growth. We grew platy $In_xGa_{1-x}As$ (x: 0.08 – 0.13) crystals by the TLZ method on the ground for substrates of 1.3 μm wavelength laser diodes. In platy crystal growth, convection in a melt was suppressed by limiting thickness of platy crystals to 2 mm and large surface area was obtained by increasing width of platy crystals. Laser diodes on $In_xGa_{1-x}As$ (x: 0.08 – 0.13) substrates showed excellent temperature stability in output power as predicted by Ishikawa et al. (1993). Now 50×50 mm² platy crystals were reproducibly grown and mass production of laser diodes with emitting wavelength of 1.3 μm is expected.

6. Acknowledgments

We are grateful to Dr. Adachi for the numerical analysis of the TLZ growth, to Dr. Y. Kondo, Dr. M. Arai, Dr. Y. Kawaguchi and Dr. F. Kano for laser fabrication and characterization, to Dr. H. Aoki, Dr. T. Hosokawa, Dr. S. Yamamoto and Dr. M. Matsushima for wide platy crystal growth. This work was supported by the New Energy and Industrial Technology Development Organization (NEDO).

7. References

Adachi, S.; Ogata, Y.; Koshikawa, N.; Matsumoto, S.; Kinoshita, K.; Yoshizaki, I.; Takayanagi, M.; Yoda, S.; Kadowaki, A.; Tsuru, T.; Miyata, H. & Muramatsu, Y. (2004a). Numerical analysis of growth rates in the traveling liquidus-zone method. J. Crystal Growth, Vol.270, No.1-2 (September 2004), pp. 42-49, ISSN 0022-0248

Adachi, S.; Ogata, Y.; Koshikawa, N.; Matsumoto, S.; Kinoshita, K.; Yoshizaki, I.; Takayanagi, M. & Yoda, S. (2004b). Latent heat effect on growth rate in the traveling liquidus-zone method. J. Crystal Growth, Vol.271, No. 1-2 (October 2004), pp. 22-28, ISSN 0022-0248

Arai, M.; Watanabe, T.; Yuda, M.; Kinoshita, K.; Yoda, S. & Kondo, Y. (2007). High-Characteristic-Temperature 1.3μm-Band Laser on an InGaAs Ternary Substrate Grown by the Traveling Liquidus-Zone Method. IEEE J. Selected Topics in Quantum Electronics, Vol.13 (2007), pp. 1295-1300, ISSN 1077-260X

Arai, M.; Kobayashi, W.; Fujisawa, T.; Yuda, M.; Tadokoro, T.; Kinoshita, K.; Yoda, S. & Kondo, Y. (2009). 10-Gbps Direct Modulation Using a 1.31µm Ridge Waveguide Laser on an InGaAs Ternary Substrate. Appl. Phys. Express, Vol.2, No.1 (January 2009) pp. 022101-1-3, ISSN 1882-0778

Bublik, V.T. & Leikin, V.N. (1978). Calculation of pseudobinary alloy semiconductor phase diagrams. physica status solidi (a), Vol.46, No.1 (March 1978), pp. 365-372, ISSN 1862-6319

Ishikawa, H. (1993). Theoretical gain of strained quantum well on an InGaAs ternary substrate. Appl. Phys. Lett., Vol.63, No.6, (August 1993), pp. 712-714, ISSN 0003-6951

Ishikawa, H. & Suemune, I (1994). Analysis of Temperature Dependent Optical Gain of Strained Quantum Well Taking Account of Carriers in the SCH Layer. IEEE Photonics Technology Lett., Vol.6, No.3, (March 1994), pp. 344-347, ISSN 0003-6951

Kinoshita, K.; Kato, H.; Matsumoto, S.; Yoda, S.; Yu, J.; Natsuisaka, M.; Masaki, T.; Koshikawa, N.; Nakamura, Y.; Nakamura, T.; Ogiso, A.; Amano, S.; Goto, K.; Arai, Y.; Fukazawa, T.; Kaneko, M. & Itami, T. (2000). InAs-GaAs interdiffusion measurements. J. Jpn. Soc. Microgravity Appl., Vol.17, No.2 (April 2000), pp.57-63, ISSN 0915-3616

Kinoshita, K.; Kato, H.; Iwai, M.; Tsuru, T.; Muramatsu, Y. & Yoda, S. (2001). Homogeneous In0.3Ga0.7As crystal growth by the traveling liquidus zone method. J. Crystal Growth, Vol.225, No.1 (May 2001), pp.59-66, ISSN 0022-0248

Kinoshita, K.; Hanaue, Y.; Nakamura, H.; Yoda, S.; Iwai, M.; Tsuru, T. & Muramatsu, Y. (2002). Growth of homogeneous mixed crystals of In0.3Ga0.7As by the traveling liquidus-zone method, J. Crystal Growth, Vol.237-239, Part 3 (April 2002), pp.1859-1863, ISSN 0022-0248

Kinoshita, K.; Ogata, Y.; Adachi, S.; Koshikawa, N.; Yoda, S.; Miyata, H & Muramatsu, Y. (2004). A new crystal growth method for growing homogeneous mixed crystals of In0.3Ga0.7As: the traveling liquidus-zone (TLZ) method. Adv. in Astronautical Sciences, Vol.117 (2004), pp.865-872, ISSN 0065-3438

Kinoshita, K.; Ogata, Y.; Koshikawa, N.; Adachi, S.; Yoda, S.; Iwai, M.; Tsuru, T. & Muramatsu, Y. (2005). Improvement of compositional homogeneity in In1−xGaxAs bulk crystals grown by the travelling liquidus-zone method. Int. J. Materials and Product Technology, Vol.22, No.1-3 (January 2005) pp. 95-104, ISSN 0268-1900

Kinoshita, K.; Ogata, Y.; Adachi, S.; Yoda, S.; Tsuru, T.;, Miyata, H. & Muramatsu, Y. (2006). Convection effects on crystallinity in the growth of In0.3Ga0.7As crystals by the traveling liquidus-zone method, Ann. New York Acad. Sci., Vol.1077 (September 2006), pp. 161-171, ISSN 0077-8923

Kinoshita, K.; Ueda, T.; Yoda, S.; Arai, M.; Kawaguchi, Y.; Kondo, Y.; Aoki, H.; Hosokawa, T.; Yamamoto, S. & Matsushima, M. (2008). High Quality InxGa1-xAs (x: 0.1 – 0.13) Platy Crystal Growth for Substrates of 1.3 µm Laser Diodes. Proceedings of 20th International Conference on Indium Phosphide and Related Materials, TuB1.2, ISBN 1424422582,Versailles, France, May 25-29 , 2008,

Kinoshita, K.; Yoda, S.; Aoki, H.; Hosokawa, T.; Yamamoto, S.; Matsushima, M.; Arai, M.; Kawaguchi, Y.; Kondo, Y. &Kano, F. (2010). Growth of large platy InGaAs crystals and fabrication of semiconductor laser diodes. Proceedings of 22nd International Conference on Indium Phosphide and Related Materials, TuB3-3, ISBN 9781424459193, Takamatsu, Japan, May 31-June 4, 2010

Kinoshita, K. & Yoda, S. (2011). Growth of homogeneous semiconductor mixed crystals by traveling liquidus-zone method. J. Crystal Growth, Vol.318, No. 1(March 2011), pp. 1026-1029, ISSN 0022-0248

Nakajima, K.; Kusunoki, T. & Takenaka, C. (1991). Growth of ternary InxGa1-xAs bulk crystals with a uniform composition through supply of GaAs, J. Crystal Growth, Vol.113, No.3-4 (September 1991), pp. 485-490, ISSN 0022-0248

Nakamura, H.; Hanaue, Y.; Kato, H.; Kinoshita, K. & Yoda, S. (2003). A one-dimensional model to predict the growth conditions of InxGa1-xAs alloy crystals grown by the traveling liquidus-zone method. J. Crystal Growth, Vol.258, No.1-2 (October 2003), pp. 49 – 57, ISSN 0022-0248

Nishijima, Y.; Tezuka, H. & Nakajima, K. (2005). A modified zone growth method for an InGaAs single crystal. J. Crystal Growth, Vol.280, No.3-4 (July 2005), pp. 364-371, ISSN 0022-0248

Sell, H.J. (1991). Growth of GaInAs bulk mixed crystals as a substrate with a tailored lattice parameter. J. Crystal Growth, Vol.107, No.1-4 (January 1991), pp. 396-402, ISSN 0022-0248

Tiller, W.A.; Jackson, K.A.; Rutter, J.W. & Chalmers, B. (1953). The redistribution of solute atoms during the solidification of metals. Acta Metallurgica, Vol.1, No.7 (July 1953), pp. 428-437. ISSN 0001-6160

10

Development of 2″ AlN Substrates Using SiC Seeds

O.V. Avdeev et al.*

Nitride Crystals Ltd., Saint-Petersburg,, 194156
Russia

1. Introduction

The unique properties of the group III-nitrides (Edgar et al., 1999; Jain et al., 2000; Kasap & Capper, 2006) make them the best semiconductor material for

- optoelectronic devices emitting light in the visible and UV spectral ranges (Orton & Foxon, 1998), including sources for general illumination (Craford, 2005; Liu, 2009; Miyajima et al., 2001; Nakamura et al., 2000; Schubert & Kim, 2005; Schubert, 2006; Taguchi, 2003; Zukauskas et al., 2002),

- photodetectors for these spectral ranges, including solar-blind UV detectors,

- high power/high frequency electronic devices capable of operating at high temperatures and in harsh environment (Bennett et al., 2004; Morkoc, 1998; Pearton et al., 2000; Shur, 1998; Skierbiszewski, 2005; Xing et al., 2001).

To fully exploit the potential of the group III-nitrides in optoelectronics and communication technology, two problems are to be solved:

1. difficulties of doping group III-nitrides, especially attaining the high-level p-doping - in 1999 AlN was even called an "undopable" material (Fara et al., 1999) (p-doping is also a problem for other wide bandgap semiconductors - oxides such as ZnO and chalcogenides such as ZnSe;) and

2. the lack of large high crystalline quality native substrates with required electrical properties.

1.1 Substrates for III-nitride devices
The requirements for a good substrate are:

- lateral and vertical[1] lattice matching,

- thermal expansion coefficient (TEC) matching,

---

*T.Yu. Chemekova, E.N. Mokhov, S.S. Nagalyuk (Nitride Crystals Ltd., Saint-Petersburg,, 194156, Russia), H. Helava, M.G. Ramm (Helava Systems Inc., 181 E Industry Court, Site B, Deer Park, NY 11729, USA), A.S. Segal (STR Group - Soft-Impact Ltd., Saint-Petersburg, 194156, Russia), A.I. Zhmakin (A.F. Ioffe Physical Technical Institute, RAS, Saint-Petersburg, 194021, Russia) and Yu.N. Makarov (Nitride Crystals, Inc., Richmond VA 23238, USA)

[1] A vertical lattice mismatch generates additional crystalline defects by upsetting the epitaxial layers, including inversion domain boundaries and stacking faults (Kasap & Capper, 2006).

- chemical compatibility,
- large size ("a size agreeable to an industry" (Schujman & Schowalter, 2010) - at least 2″),
- affordable price.

Sometimes additional features are desirable, e.g., the cleavability for laser diodes or the high electrical resistivity for field effect transistors (FETs).

A substrate determines the crystal orientation, polarity, polytype, surface morphology; the difference in the chemical composition of the substrate and the epitaxial layer could lead to the contamination of the layer with substrate elements.

The lattice mismatch for the semiconductor layer with the lattice constant a is measured by the misfit parameter $f_m = (a - a_s)/a_s$, where a_s is the lattice constant of the substrate. It leads to misfit strain in the grown layer that should be accommodated if the layer thickness exceeds the critical one[2] (according to either the energy minimization theory of Frank-Van der Merwe (Frank & van der Merwe, 1949) or the force balance theory of Matthews-Blakeslee (Matthews & Blakeslee, 1974), see, for example, (Jain et al., 1997; Jain & Hayes, 1991; Zhmakin, 2011b)) via the introduction of misfit dislocations, the modulation of the free surface profile (waviness (Freund, 1995)), the composition modulation (Seol et al., 2003) or the interdiffusion between the layer and the substrate (Lim et al., 2000); formation of the V-grooves and the random stacking faults (Cho et al., 2000) and the mosaic structure of slightly misoriented subgrains in the epitaxial films (Srikant et al., 1997) are also observed. These mechanisms could be cooperative as well as competitive (Hull et al., 2002). In addition to the misfit strain, surface free energies and interface energies are the factors that determine the growth mode (Frank-van der Merve, Stranski-Kristanov, Volmer-Weber, polycrystalline, columnar) (Gilmer et al., 1998; Wadley et al., 2001). Nitride layers grown on the substrates with the large lattice mismatch can also include twin crystals (Matsuoka, 2005; Rojas et al., 1998).

Homoepitaxy provides a better control over the surface morphology and the defect density compared to the heteroepitaxial growth (Cao et al., 2007; Kirchner et al., 1999). The best substrates for GaN-based devices are GaN and AlN. AlN ones are superior for the structures with high Al content, e.g., UV devices (AlN is transparent down to about 200 nm), FETs and both vertical and lateral Schottky and p-i-n diodes (Luo et al., 2002) (unintentionally doped (UID) AlN crystals are insulators[3] in contrast to GaN ones that have a high background

---

[2] The critical thickness is the equilibrium entity, thus metastable metamorphic layers (Hull & Stach, 1996) of higher thickness could exist at low temperatures: e.g., recently an n-type AlGaN layer on the bulk AlN substrate with a thickness more than an order of magnitude greater than the critical value was reported (Grandusky et al., 2009).

[3] The concentration of the residual oxygen is high for the sublimation grown AlN crystals (Herro et al., 2006); however, most of oxygen is incorporated not as a substitutional shallow donor, but as impurity-forming complexes with point and extended defects located deep in the gap (Freitas, 2005; Schultz, 2010; Slack et al., 2002). Below a critical concentration, the presence of oxygen leads to the formation of Al vacancies; at higher concentrations a defect based on octahedrally coordinated Al atoms is formed (Kasap & Capper, 2006).
The effect of the oxygen contamination is more severe on the thermal conductivity of AlN due to the phonon scattering on oxygen defects whose nature does depend on the oxygen concentration (Kasap & Capper, 2006; Kazan et al. , 2006a). For example, the substitutional defect associated with aluminum vacancy serves as a very efficient center for the phonon scattering (Rojo et al., 2001).

concentration of free electrons (Freitas, 2005))[4]. AlN substrates are also the best choice for for nitride-based power devices due to the high thermal conductivity.

The high thermal conductivity of AlN is attributed to the simplicity of the atomic structure, it's low atomic mass, the existence of strong covalent bonding, and low anharmonicity within the lattice (Dinwiddie et al., 1989). Note that the thermal conductivity of AlGaN alloys is smaller than that of pure GaN, to say nothing of AlN, due to the high degree of the disorder in the system (Liu & Balandin, 2005).

The cooling of devices on AlN substrates could be effectively better than of that on SiC substrates that have higher thermal conductivity since the former do not require a buffer layer between the device structure and the substrate that is highly defective and thus presents considerable thermal resistance (Schowalter et al., 2004).

AlN crystals, possessing the highest surface and bulk acoustic wave velocity known for piezoelectric materials, a small temperature coefficient of delay, and a large piezoelectric coupling coefficient, are needed for the surface acoustic wave (SAW) devices[5]. Aluminum nitride for such applications has an evident advantage - AlN crystals could be used up to $1150\,°C$ (Fritze, 2011) (conventional piezoelectric materials are not suited for high-temperature applications, for example, α-quartz αSiO_2 undergoes a phase transformation at 573 $°C$ while lithium niobate $LiNbO_3$ and lithium tetraborate $Li_2B_4O_7$ decompose at 300 $°C$). AlN crystals are also considered as perspective material for the integration of semiconductor electronic and SAW devices (Cleland et al., 2001).

1.2 Foreign substrates

In the lack of native GaN or AlN substrates, a number of foreign ones are used in practice leading to low quality of the grown epitaxial layers in terms of large threading dislocation density, inversion domain boundaries, partial dislocations bounding stacking faults, bowing of the structure, cracks (Cherns, 2000; Speck, 2001; Speck & Rosner, 1999; Wu et al., 1996; 1998). The near-interfacial region of the film could contain a mixture of cubic and hexagonal GaN (Wu et al., 1996).

The island coalescence is considered as the main mechanism, at least during MBE growth, of the dislocation appearance (Waltereit et al., 2002). Similar conclusion is reached in the study of the early stages of HVPE epitaxy - all the edge and most of the mixed threading dislocation are originated from the island junctions of the high temperature (HT) buffer layer (Golan et al., 1998). An increase of the smoothness was also observed and attributed to the subsequent surface and bulk reconstruction. A strong correlation between the final threading dislocation

---

[4] For example, in GaN crystals grown by the hydride (halide) vapour-phase epitaxy (HVPE) an unintentional n-type conductivity originates from the background doping by silicon and oxygen from the quartz elements of the reactor or from the process gases (Paskova et al., 2010). Fe compensating doping allows one to achieve semi-insulating properties of the layer in this growth method (Vaudo et al., 2003) with the the lowest free carrier concentration reported so far $5 \cdot 10^{13}$ cm^{-3} (Paskova et al., 2009).

However, the performance parameters of AlGaN/GaN heterostructure field-effect transistor (HFET) structure formed by depositing a layer of AlGaN on a relatively thick semi-insulating GaN epitaxial layer are greatly improved by replacing a GaN epitaxial layer with a highly resistive AlN epitaxial layer in the device structure, in particularity, parasitic conduction in the GaN epilayer, leakage current through the GaN epilayer, and the channel electrons spillover into the GaN epilayer have been completely eliminated and the drain current collapse has been reduced (Fan et al., 2006).

[5] Even polycrystal AlN films obtained, for example, by low-cost sputtering are of great interest since the corresponding SAW devices could operate at frequencies above 1GHz (Epelbaum et al., 2004). However, the properties of elasic waves in polycrystal depend on the grain size and orientation.

density in the thick films and the initial island density in the high temperature buffer layer was registered in the MOCVD grown GaN layers on the sapphire substrate (Fini et al., 1998). Early stages of the film growth frequently occur via coherent (dislocation-free) island formation due to the existence of the energy barrier to the introduction of dislocations; dislocations arise at the island edges for large enough islands since those edges are characterized by the large stress (Eisenberg & Kandel, 2002).

A review of the study of the morphology of the heteroepitaxially grown GaN layers using the scanning tunnel microscopy could be found in Ref. (Bakhtizin et al., 2004).

Threading dislocation are electrically active with the charge density of approximately 2 elementary charges per nanometer dislocation length (Müller et al., 2006); surface depressions caused by the high strain-energy density near dislocations are observed on the surface of GaN films (Heying et al., 1999). Edge and mixed (screw/edge) dislocations in nitride materials act as nonradiative recombination centers and as conduits for charge transport resulting in leakage currents and breakdown (Amano et al., 2003; Davis et al., 2002).

While high threading dislocation density is acceptable for light emitting diodes (LEDs)[6] — usually their small effect on the performance is attributed to the potential fluctuations related to the indium composition fluctuations (resulting from the poor In incorporation in the epitaxial layer during the growth or from the phase separation due to the large miscibility gap (I-hsiu & Stringfellow, 1996; Karpov, 1998; Korcak et al., 2007)) observed in the active layers made from the ternary solid solution InGaN and in the corresponding QWs (Christen et al., 2003; Limb, 2007; Mukai et al., 2001) that provide the localization of carriers and reduce their in-plane diffusion to the non-radiative recombination centers — the high dislocation density, however, is fatal for laser diodes and power transistors; even for the LEDs including indium-free violet ones (Usikov et al., 2003) it leads to the drop in the efficiency and is the main factor of the device failure (Karpov, 2009; Roycroft et al., 2004).

The most important foreign substrates are sapphire Al_2O_3 and silicon carbide (6H-SiC or 4H-SiC). The list of other substrates used for group III nitride epitaxy includes GaAs, AlAs, GaP, ZnO, MgO, Mg(Zn) Fe_2O_4, $LiGaO_2$, $LiAlO_2$, SiO_2, $NdGaO_3$, $ScAlMgO_4$, ZrO_2, Mo, glass, quartz glass (Jain et al., 2000; Kukushkin et al., 2008; Miskys et al., 2003).

Sapphire is a good choice for the nitride layers from the crystallographic point of view: crystal orientations of sapphire and gallium nitride grown on c-plane (0001) are parallel, with the unit cell of GaN being rotated by 30° about c axis with respect to the unit cell of sapphire; the $(1\bar{1}00)$ axis of GaN is parallel to the $(1\bar{2}10)$ sapphire axis (Jain et al., 2000).

However, sapphire is inferior (due to the high TEC mismatch and the lattice mismatch being about 15% that corresponds to a critical thickness of less than a monolayer (Jain et al., 2000) compared to 3.5 % for 6H SiC) in the quality of the grown nitride layers leading to higher threading dislocation density.

The insulating nature of sapphire restricts the device architecture, excluding the vertical die design. Bipolar devices fabricated on the sapphire substrates should employ mesa structures with the lateral geometry of the anode and cathode electrodes (with both contacts placed in the same plane) and are especially prone to the current crowding effect - a nonhomogenous in-plane distribution of the current density, especially at the vicinity of the edge of the metal electrodes.

This effect is one of the limiting factors of the efficiency of light emitting diodes, but it is also of concern in other semiconductor devices, e.g., bipolar transistors and Schottky diodes (Chen

---

[6] Threading dislocations with vacancies at Ga sites (Hovakimian, 2009) acting as non-radiative recombination centers (Choi et al., 2004) present more serious problem for GaN/AlGaN system than for GaN/InGaN one (Akasaki & Amano, 2006).

et al., 2007; Paskova et al., 2010). The current crowding can lead to the localized overheating and the formation of thermal hot spots, lowering the internal quantum efficiency of LED and affecting the series resistance of the diode as well as can result in a premature device failure (Bogdanov et al., 2008; Bulashevich et al., 2007; Evstratov et al., 2006).

Another weak point of sapphire as a substrate for the light emitting diodes is a low refractive index of sapphire in comparison to III-nitrides that leads to waveguiding of light emitted in the active region of the LED and thus to the decrease of the light extraction efficiency (Karpov, 2009).

Nevertheless, at present time over 80 % of LEDs are fabricated on sapphire 2" substrates (Russel, 2006) due to the low cost and availability.

A relatively new approach to the effective use of the sapphire substrates that has been developed for the nitrides light-emitting diodes (LEDs) growth is the nano-patterning of the surface of the substrate: it is expected that the strain induced by the lattice misfits between the GaN epilayers and the sapphire substrates can be effectively accommodated via the nano-trenches (Yan et al., 2007) (the patterning also enhances light extraction from the device (Zhmakin, 2011a)).

A few years ago a possibility to directly obtain aluminum nitride layers on the sapphire by nitriding α-SiO_2 surface on large (2 inch) substrate by chemical reaction with N_2 - CO gas mixture under carbon-saturation conditions was demonstrated (Fukuyama et al., 2006).

The drawback of SiC substrates is the high cost and the surface roughness on the scale of a few atomic bilayers, leading to a stacking-mismatch defects at the interface between SiC and III-nitride layers in the form of a 2H-polytype (Karpov, 2009; Potin et al., 1998).

Reducing the high cost of SiC substrates while conserving the high thermal conductivity is possible by inserting a thin film of monocrystalline SiC onto polycrystalline SiC (Pecz et al., 2008) which has the thermal conductivity lower than that of single crystal but close to it (Franko & Shanafield, 2004). Polycrystalline SiC is produced by the low cost process of the SiC powder sintering (Medraj et al., 2005). Bonding the thin sapphire layer to polycrystalline AlN (P-AlN) retains the epitaxial template for the growth while improving the thermo-mechanical properties of the substrate (Pinnington et al., 2008). Another example of composite substrates is SiC semiconductor-on-insulator (SOI) structures that are the low cost 3C-SiC (111) substrates fabricated by the conversion of silicon SOI by carbonization of the surface —a reaction with propane and hydrogen at high temperatures (Steckl et al., 1997).

Epitaxy of InGaN/GaN layers on foreign substrates with the large lattice mismatch requires a preliminary growth of a buffer layer or multiple buffer layers (Miskys et al., 2003; Xi et al., 2006). The common example of the use of a double buffer layer is the growth of AlGaN with a starting low-temperature AlN layer on the GaN templates on sapphire which themselves contain the low-temperature GaN buffer layer (Kuwano et al., 2004) or an introduction of an insulating AlN sub-buffer layer on the semi-insulating SiC substrate under the GaN buffer layer (Shealy et al., 2002).

The buffer layer frequently has an amorphous-like structure with small crystallites providing the very smooth morphology (Matsuoka, 2005). AlN is frequently used as the buffer layer material for sapphire and SiC substrates; buffer layers from GaN, AlGaN (including graded AlGaN conducting buffer layers (Moran et al., 2000)), ZnO (Jain et al., 2000; Ougazzaden et al., 2008) or platinum nanocluster (Oh et al., 2009) are used for sapphire while HfN buffer layers for Mo substrates (Okamoto et al., 2009).

The threading dislocation density could be greatly reduced using the lateral epitaxial overgrowth (LEO, the term ELOG is also used) developed in 1990s (Beaumont et al., 1998; Davis et al., 2002; Nam et al., 1997; 1998). This method involves patterning of the substrate and the initial vertical growth in the "windows" etched in the mask with subsequent growth of material of higher quality laterally over the mask patches. The modification of LEO - the so called pendeo-epitaxy (Davis et al., 2001; Zheleva et al., 1999) - is free of the two major LEO drawbacks (cracking of thick layers and void formation on the top of mask stripes (Wang et al. , 2001)). The crystallographic tilt in the overgrown material is also significantly reduced and the diffusion of impurities from the mask is avoided (Davis et al., 2002).

The extremely low threading dislocation density could be obtained via two sussessive ELO steps with the mask of the second step positioned over the windows etched in the mask during the first step (Pearton et al., 2000).

Growth of GaN or AlN directly on Si surface usually results in the formation polycrystalline films, probably due to the prior formation of amorphous SiN_x layer (Davis et al., 2001a). The large lattice (17%) and TEC (33%) mismatch between Si and GaN cannot be fully accommodated by an AlN buffer layer (Lin et al., 2008), still, involving additionally nitridation (Yamabe et al., 2009) or carbonization of the surface or SiC coating (Kukushkin et al., 2008), some GaN device grown on Si substrates such as HEMTs show an acceptable performance.

In MBE the best results (an order of magnitude decrease of the threading dislocation density in GaN epilayers) are obtained when the growth is initiated by exposing the Si substrate to NH_3 first (Louarn et al., 2009). Usually Si (111) surface (Davis et al., 2001a) is used while an integration with Si microelectronics requires growth on Si (100). In the GaN layers grown on (111) Si substrate inclusions of the cubic nitride phase are frequently observed (Jain et al., 2000).

1.3 Templates & *pseudo bulk* substrates

Considerable efforts were directed to the development of *templates* (the foreign substrates with the deposited thin nitrides layers) (Gautier et al., 2008; Miskys et al., 2003) sometimes referred to as *MOVPE-derived GaN substrates* (Davis et al., 2002) and *pseudo bulk* or *freestanding* nitride substrates (Lee et al., 2004; Nikolaev et al., 2000; Weyers et al., 2008) obtained by separation of the thick nitride layers from the sacrificial substrate after the growth (by laser-assisted lift-off (LLO) (Lee et al., 2004; Paskova et al., 2004) or by etching, e.g, in aqua regia for GaAs substrates) or during the growth on sapphire substrates with patterned GaN seeds by spontaneous self-separation (Tomita et al., 2004).

The quality of the epitaxial layers grown on the templates in comparison to those grown on the basis substrate is under discussion. Recently Ashraf et al. (Ashraf et al., 2008) have used a number of characterization techniques (diffraction interference contrast optical microscopy, scanning electron microscopy, micro-Raman scattering, X-ray diffraction) to asses the quality of the thick GaN layers grown by HVPE directly on the sapphire substrate (with the optimal nucleation layer deposition at low temperatures and low pressures on the nitridated surface) and on the GaN/Al_2O_3 templates prepared by GTS-metalorganic chemical vapour deposition (MOCVD) process (Grzegorczyk et al., 2005) (GTS stands for the gallium treatment step which is a long exposition of the substrate surface to TMGa at high temperature). The authors found that the layers did not significantly differ in the surface morphology and the structural quality, but the layers grown on the MOCVD templates suffered from cracking in few cases while no cracking occurred in the HVPE layers directly grown on sapphire.

1.4 Epitaxial layers and devices on single-crystal native III-nitride substrates

Characterization of AlGaN epilayers with 40 and 50 % concentration of aluminum grown on the single-crystal AlN substrates by Migration Enhanced Metal Organic Chemical Vapour Deposition (MEMOCVD) using the observation of atomic steps in atomic force microscope scans of epilayers and the measurements of FWHM of X-ray diffraction curves demonstrated an excellent crystallographic quality of the epilayers, the dislocation density of AlGaN layers was estimated to be in mid $10^6 cm^{-2}$ range (Schowalter et al., 2006). The comparison of photoluminescence of the GaN layers deposited on the Al face and on the N face of the single-crystal AlN substrate showed that in the former case photoluminescence is consistent with that of the homoepitaxial Ga-face GaN while in the latter an existance of the tail localized states was found (Tamulatis et al., 2003).

The studies of deep-UV emission of AlGaN quantum wells (Gaska et al., 2002) as well as AlGaN UV (Nishida et al., 2004a; Ren et al., 2007; Xi et al. , 2006a) and InGaN MQW green (Cartwright et al., 2006) LEDs grown on the bulk AlN substrates and blue and UV LED on the bulk GaN substrates (Cao et al., 2004; Cao et al. , 2004a; Du, Lu, Chen, Xiu, Zhang & Zheng, 2010) as well as cyan and green LEDs grown on a-plane (Detchprohm et al., 2008) and m-plane (Detchprohm et al., 2010) GaN bulk substrates prove the superiority of the native substrates, e.g., the luminescence intensity of the quantum well grown on bulk AlN was higher that that of the quantum well grown on SiC by a factor of 28, the noticeable improvement over LEDs grown on sapphire in device impedance and thermal characteristics (Ren et al. , 2007a), the reduction in current-voltage differential resistance and in turn-on voltage (Paskova et al., 2010). The emission spectrum of AlGaN-based UV-LEDs on a bulk AlN substrate under the high current injection is much more stable than that of LEDs fabricated on the conventional substrate (Nishida et al., 2004a;b). Recently J. J. Grandusky et al. have demonstrated mid-UV LED fabricated from pseudomorphic layers on the bulk AlN substrates (Grandusky et al., 2010).

High-quality green (Miyoshi et al., 2009) and violet and near-UV (Perlin et al., 2005) laser diodes have been fabricated on the bulk GaN substrates, in the latter case substrate were grown by the HNPSG.

Studies of high-electron mobility transistors (HEMTs) with the ALGaN channel[7] grown on different substrates also demonstrate the superiority of the single crystal AlN substrates (Hu et al., 2003), in particularity, the use of AlN substrates improved the crystalline quality of the AlGaN layer and lowered the sheet resistance of the 2-dimensional electron gas (Hashimoto et al., 2010).

The substrates cut from the bulk crystals along the specific crystallographic plane can have different orientation (polar, semipolar or nonpolar) (Lu et al., 2008)[8], enhancing the freedom of the device design (Mymrin et al., 2005; Schowalter et al., 2006; Stutzmann et al., 2001): an engineer, using the spontaneous and piezoelectric polarization that is a nonlinear function of the strain and the composition of nitride materials, can tailor the surface and interface charges

---

[7] The channel layer substitution of a wider bandgap AlGaN for a conventional GaN in high electron mobility transistors is an effective method of drastically enhancing the breakdown voltage (Nanjo et al., 2008).

[8] For a long time attempts to grow the nitride epitaxial layers on the nonpolar planes were unsuccessful producing the low quality films (Karpov, 2009). However, recently I. Satoh et al. (Satoh et al., 2010) demonstrated the possibility to fabricate the non-polar AlN substrate by heteroepitaxial growth on m-plane SiC substrates.

Although the dislocation density of the epitaxial layers grown on m-plane was an order of magnitude higher than that on c-plane substrate, the emission intensity was 25 times greater, evidently due to the elimination of the spatial separation of the electron and holes wavefunctions in the quantum wells induced by the built-in polarization field that leads to the decrease of the radiation recombination rate - see, for example, (Chakraborty et al., 2005; McAleese et al., 2006; Ram-Mohana et al., 2006).

High efficiency non-polar m-plane InGaN light emitting diodes and laser diodes have been demonstrated also by M. S. Schmidt et. al. (Schmidt et al. , 2007;a).

to get the desired properties of the heterostructure, for example, to achieve two-dimensional electron gas without modulation doping (Ambacher et al., 2003).

Note that an "effective" cost of single crystal bulk AlN substrates could be rather low if repeated use (removal of the AlN substrate by laser-lift off and recycling[9]) is realized.

The most mature growth technology for bulk group III - nitride crystals is the sublimation growth. Currently, the reproducible growth of large AlN bulk single crystals has been achieved (Bondokov et al., 2008; Helava et al., 2007; Mokhov et al., 2005; Raghothamachar et al., 2006; 2003). Still, a number of problems such as improvement of the crystal seeding and stoichiometry, the reduction of impurities concentration, the maintenance of stable conditions during the long growth remains.

The present chapter reports advances in AlN sublimation growth and its modelling. Both numerical and experimental approaches to understanding and optimization of the sublimation growth of AlN bulk single crystals are reviewed. A developed two-stage technology for the growth of large AlN crystals on SiC seeds that allows to exploit the best features of crucibles made from different materials is described. The superiority of single crystal AlN substrates for the growth of the group-III nitride epitaxial layers and the device performance is demonstrated. The chapter is organized as follows. The next section contains a review of approaches to growth of bulk group III - nitride crystals from liquid - melt or solution - and vapour phases. The sublimation growth of bulk AlN crystals and modelling of this process are considered in the sections 3 & 4, respectively. The developed by the authors two-stage growth procedure based on using SiC seeds and two crucibles made from different materials is described in the section 5. Results of the assessment of the quality of the grown crystals and fabricated single-crystal 2in AlN substrates are presented in the section 6. The section 7 summarizes the results.

2. Growth of bulk group III nitride crystals from liquid and vapour phases

Although the first AlN crystals were synthesized by F. Briegler and A. Gúther using the reaction between molten aluminum and nitrogen about a century and a half ago (see, for example, (Dalmau, 2005; Dalmau & Sitar, 2005) and the small needles of GaN were synthesized by R. Juza and H. Hahn in 1938 by passing ammonia over hot gallium (Jain et al., 2000) (and, similar, the AlN needle crystals were obtained by flowing nitrogen over the compacted AlN powder), growth of bulk GaN, AlN and AlGaN crystals is difficult (Denis et al., 2006); InN bulk crystals have not been demonstrated so far - the thermal instability of the group III - nitride compound increases as one goes down the group III column of the Periodic Mendeleev system (Schowalter et al., 2004).

Bulk group-III nitride crystals could not be congruently grown from the stoichiometric melt under practically acceptable environment conditions (temperature and pressure) as most semiconductors not due to the high melting temperature itself, as sometimes stated, but due to the decomposition of the crystals occurring at much lower temperature resulting from the strong bonding of nitrogen molecule and the low free energy of the crystal (Krukowski,

---

[9] LLO is usually performed to separate group III-nitride structure from the sapphire substrate by a short pulse of UV laser — either the eximer KrF laser at 248 nm or the third harmonic (255 nm) of the Nd:YAG laser — that locally heats the nitride layer causing its decomposition into metal and nitrogen. In case of AlN substrate having a relatively low transparency at short wavelengths an additional operation could be needed — a preliminary thinning of the substrate by chemical etching (Schujman & Schowalter, 2010).

1997)[10]. The congruent melting of gallium nitride has been achieved recently under the severe experimental conditions (the pressure of 6 Gpa and temperature about 2200° C (Utsumi et al., 2003) (so far grown crystals are smaller than 100 μm).

Bulk single group III- nitride crystals could be grown either from solution or from vapour phase. The former is known in three variants: High Nitrogen Pressure Solution Growth (HNPSG), and two "low" pressure techniques — ammonothermal growth and flux growth. Two vapour phase growth methods are halide (hydride) vapour phase epitaxy (HVPE) and sublimation growth. These growth techniques are briefly considered in this section, except sublimation growth that is treated in the next section.

2.1 High Nitrogen Pressure Solution Growth

The solubility of nitrogen in the Ga melt is very low (Nord et al., 2003) and the formation of the N_2 bubbles is possible in the supersaturated GaN liquid (Krukowski, 1997). The nitrogen dissolution in metal melt could be increased by two orders of magnitude —as well known in iron- and steelmaking — by dissolving the nitrogen radicals instead of nitrogen itself, thus ammonia is preferable as an ambient gas (Kawahara et al., 2005).

The growth of the centimeter-size GaN crystals was achieved at high temperatures and ultra-high N_2 pressures that provide a sufficient concentration of nitrogen in the Ga melt (High Nitrogen Pressure Solution Growth - HNPSG) (Grzegory, 2001; Porowski & Grzegory, 1997). The GaN crystals grown by this method have a very low threading dislocation density of $10^2 cm^{-2}$. Similar temperature and pressure are used in the Pressure-Controlled Solution Growth - PCSG (Denis et al., 2006). The HNPSG without an intentional seeding produces the needle-like AlN crystals (Bockowski, 2001). Both needle-like and bulk form of AlN single crystals up to 1 cm and 1 mm, respectively, were grown at high nitrogen pressure of the order of 1 GPa and temperatures up to 2000 K (Bockowski et al., 2001).

Recently (Al,Ga)N bulk single crystals have been grown from the Al/Ga melt under high nitrogen pressure (up to 10 kbar) at high temperature (1425 - 1780 °C) with an aluminum content from 0.22 to 0.91 (Belousov, 2010; Belousov et al., 2010; 2009). The largest crystal was $0.8 \times 0.8 \times 0.8$ mm^3. The distinct feature of this study is the use of pre-reacted polycrystalline (Al,Ga)N or AlN pellets. The composition of the growing crystal was controlled by the proper choice of the pressure and temperature.

Note, however, that the high-pressure requirement limits the scalability of the HNPSG to the growth of small area crystals.

2.2 Ammonothermal growth and flux growth

The extreme parameters needed for the HNPSG and the PCSG are reduced in the ammonothermal (Purdy et al., 2003; Yoshikawa et al., 2004) and the alkali metal flux methods (Aoki et al., 2002; Onda et al., 2002; Song et al., 2003; Yano et al., 2000). The former is similar to the well-known hydrothermal growth of quartz crystals (Iwasaki & Iwasaki, 2002) (that closely reproduces the growth of amethyst in nature (Carter & Norton, 2007)) with supercritical ammonia instead of water. This method belongs to a wide class of methods called *solvothermal*, another member of this class of growth techniques is *glycothermal* growth from glycerinated solutions (Adekore et al., 2006; Callahan & Chen, 2010). Being a low-temperature

---

[10] For example, the melting point of GaN is 2300° C at the pressure of 6GPa; at lower pressure GaN dissociates into metallic gallium and nitrogen gas or a state where nitrogen is dissolved in liquid gallium (Ohtani et al., 2007); at atmospheric pressure GaN decomposes at 1150 K (Ehrentraut & Fukuda, 2008).

process, solvothermal growth minimize the incidence of the temperature-induced point defects.

The relatively low growth rates of the solvothermal methods are partially compensated by the ability to grow multiple crystals (for example, over a hundred in the case of ZnO and over a two thousands in the case of quartz) in a single run.

Hundreds of different compounds are grown by the hydrothermal method, some of them at ambient conditions such as aluminum potassium sulfate and potassium dihydrogen sulfate (KDP). Hydrothermal growth of the low-defect ZnO crystals requires a high oxygen overpressure (about 50 atm) (Nause & Nemeth, 2005), but thus far remains a unique example of the industrial growth of widebandgap semiconductors by the solvothermal method (Ehrentraut et al., 2006).

In solvothermal method a liquid polar solvent (water in hydrothermal and ammonia in ammonothermal growth) forms metastable intermediate products with the solute (nutrient). Mineralizers are needed to increase the solubility of the nutrient.

No growth of GaN crystals is observed in the pure Ga solution. It is necessary, similar to other hydrothermal-type processes, to add either lithium as a transporting agent or either acidic or basic *complexing agents - mineralizers* (Callahan & Chen, 2010) such NH_4X (where X= Cl, Br, I) (Purdy et al., 2003; Yoshikawa et al., 2004) or gallium triiodide with CuI or LiI (Purdy et al., 2003). In the first approach chemical reactions occur in the solution involving such compounds as $LiGa(NH_2)_2$ and $LiNH_2$. Acidic mineralizers effectively increase the reaction rate by increasing the amount of anions in the solution. A mixture of alkali metal amide and iodide has been successfully used in (Ketchum & Kolis, 2001) while neither of these mineralizers alone could provide GaN growth. The growth mechanism involves formation of the intermediate soluble Ga-imide complex. For the growth of AlN crystals a Ca_3N_2 or Na flux has been used (Yano et al., 2000).

In the sodium flux method Na acts as a catalyst that releases electrons easily. It is speculated that nitrogen in N_2 molecule absorbed onto the Ga-Na melt surface receives electrons from Na, that weakens the N_2 bonds and causes the dissociation of N_2 into two negatively charged radicals at much lower temperature and pressure (Aoki et al., 2002). The use of lithium instead of sodium is more promising since the ability of the former to fix nitrogen is higher and that allows one to achieve the growth of GaN single crystals under the pressure of 1-2 atm (Song et al., 2003).

The sodium flux growth is performed in either a closed or in an open tube (Aoki et al., 2002; Onda et al., 2002). In the former the only source of nitrogen is a solid precursor such as NaN_3 powder, thus the pressure decreases during the process as nitrogen is being consumed for the GaN growth. In the latter N_2 or its mixture with NH_3 serves as a nitrogen source and the pressure can be either kept constant or varied with the time by a prescribed law (Onda et al., 2002). NH_3 is superior over N_2, however, the size of GaN crystals grown in $NH_3 - N_2$ mixture is smaller then in pure N_2.

Cathodoluminescence spectra show that the GaN crystals grown in the open system are of higher quality (Aoki et al., 2000). Sometimes the formation of the intermetallic compound $Ga_{39}Na_{22}$ is observed (Aoki et al., 2000). Black color of grown crystals is explained either by the nitrogen deficiency or by the oxygen impurity (Aoki et al. , 2002a; Aoki et al., 2000).

Change of the crystal shape from prismatic to platelet with increasing the Na/(Na+Ga) ratio has been studied in (Aoki et al., 2000; Yamane et al., 1998). An agglomeration of small crystals at high values of this ratio is explained by the drastic increase of the supersaturation. It can lead to the growth instability due to the constitutional supercooling similar to processes in pure Ga melt HNPSG (Grzegory, 2001; Grzegory et al., 2002). Polycrystallization occurs in the seeded growth when the pressure exceeds the threshold value for the unseeded nucleation (Iwahashi et al., 2003). The growth rate of GaN crystals is anisotropic being higher in c direction. Usually N-polar face is smooth while Ga polar face is rough, corrugated with macrosteps (Frayssinet et al., 2001; Skromme et al., 2002). However, the reverse pattern could also be observed (Yamane et al., 1998) probably due to the impurity incorporation: in (Grzegory, 2001) the growth instability has been observed on the Ga-polar face without doping and on the N-polar face in the presence of Mg.

It is speculated (Grzegory et al., 2002) that the nucleation rates at the different faces can vary greatly due to the different geometry of the 2D nuclei (hexagonal or square, differing in the number of atoms in the nucleus as well as the number of the broken bonds). In the Li flux method (Song et al., 2003) liquid gallium infiltrates into porous Li_3N and reacts to produce $GaLi_3N_2$ and metal Li. In the case of the excess of Li_3N no growth of GaN has been observed and only $GaLi_3N_2$ has been formed. The authors consider two possible mechanisms of GaN growth - direct reaction of $GaLi_3N_2$ with gallium and dissolution of $GaLi_3N_2$ in Li-Ga melt to form ternary system - and conclude that the latter is the one that is most probable.

The growth rate of GaN in the ammonothermal technique is rather small (not greater than 2μ/hour (Fukuda & Ehrebtraut, 2007)). Frequently a columnar growth occurs yielding the crystals of poor quality (Waldrip, 2007; Wang et al. , 2001a). Still, "tremendous progress over the last decade" has been recently reviewed in (Avrutin et al., 2010; Ehrentraut & Fukuda, 2010). It is claimed that since the ammothermal growth occurs at near thermodynamic equilibrium with the extremely low supersaturation, the high crystalline quality can be expected (Ehrentraut & Kagamitanii, 2010). However, the growth of the large GaN crystals by the ammonothermal method using the HVPE-grown free-standing substrates gives disappointing results: the density of dislocations in the grown crystals are two order of magnitude larger than in HVPE seed (Callahan & Chen, 2010; Ohtani et al., 2007).

A new method called Electrochemical Solution Growth (ESG) based on the transport of the nitrogen ion N^{3-} in the molten chloride salt is being developed now (Waldrip, 2007); so far only millimeter-size GaN crystals were produced. A reaction between Ga and Li_3N under NH_3 atmosphere via the formation of $LiNH_2$ is used to grow GaN crystals by T. Hirano et al. (Hirano et al., 2009).

The ammonothermal growth of polycrystalline Aln at temperatures between 525° and 550° in alkaline conditions (using potassium azide KN_3 as the mineralizer) was reported by B. T. Adekore et al. (Adekore et al., 2006). The thickness of the layers grown on the GaN seed in 21 days varied from 100 to 1500 μm. For the growth of AlN crystals a Ca_3N_2 or Na flux has been used (Yano et al., 2000). The growth using AlN wires as a starting material shows the very high contamination of oxygen, probably due to the intrinsic oxidized Al surface. Precipitation of single crystalline AlN from Cu-Al-Ti solution was studied in (Yonemura et al., 2005). The larges pencil type crystal has 3mm in length and 0.2 mm in its diameter. An AlN platelet (1.5 mm diameter) was also obtained by the regrowth technique. Evidently, the solution growth of AlN cannot be developed in a near future to the production scale due to the difficult control of process and low growth rate.

The alkali metal flux growth has been used in the liquid phase epitaxy (LPE) (Kawahara et al., 2005). 3 μm-thick MOCVD-GaN layers with the threading dislocation density $1.3 \cdot 10^6$ were used to grow the 500 μm crystals that were almost dislocation free. PL intensity of the LPE-GaN with Na flux and Ca-Na mixed flux was 47 and 86 times, respectively, greater than that of the seed crystal. LPE was also used to grow the hexagonal or prismatic platelets at ambient pressure with NH_3 as a nitrogen source; the growth anisotropy was found to be comparable to that in bulk Na flux growth and much smaller than in HNPSG (Meissner et al., 2004).

2.3 Halide (hydride) vapour phase epitaxy - HVPE

A HVPE reactor consists of the two main zones: the source zone where chloride gas of group III metal is formed and the growth zone where it is mixed with NH_3 to grow the nitride crystal. This method, including its variant *iodine* vapour phase epitaxy (IVPE) (Cai et al., 2010), and corresponding mathematical models are well documented (see, in addition to the just cited chapter, for example (Dmitriev & Usikov, 2006; Hemmingsson et al., 2010; Segal et al., 2009) and the references therein), thus only a few comments are in order here.

The uniqueness of HVPE is the applicability of this method to both growth of thick substrates and epitaxial heterostructures due to an extremely wide range of growth rates (1 - 150 μm/hour), the low cost compared to the MOCVD, an ability to grow the heavily doped p-layers, an absence of the carbon contamination.

At present, however, the freestanding AlN films grown by HVPE are of inferior crystalline quality: the typical value of the x-ray rocking curve Full Width at Half Maximum (FWHM) is at least an order of magnitude larger than that of AlN substrate cut out from bulk AlN boule (Cai et al., 2010; Freitas, 2010); the self-separated thick (85 μm) AlN films grown recently by the three-step modification of HVPE that include the formation of numerous voids at the interface between an AlN layer and the sapphire substrate has the dislocation density on order of $10^9 cm^{-2}$ (Kumagai et al., 2008); the sublimation-grown bulk AlN crystals usually are transparent while the HVPE-grown ones are opaque (Cai et al., 2010), Tabl. 37.16.

The reverse breakdown voltage of the m-i-m structure on bulk AlN was an order of magnitude greater than that on free-standing GaN (Luo et al., 2002) proving high potential of Al-Ga-N system for high power rectifiers.

3. Sublimation growth of AlN crystals

Sublimation[11] (also *sublimation - recondensation*) growth (or physical vapour transport - PVT) of AlN is the most mature technology of the bulk nitride crystal growth (Dalmau & Sitar, 2005; 2010) (sublimation is also used to grow AlN fibers (Bao et al., 2009) and other nitride compounds, for example, the titanium nitride crystals (Du, Edgar, Kenik & Meyer, 2010); sublimation growth of GaN crystals is less successful (Kallinger et al., 2008)[12]). Probably the first application of PVT is the growth of the single cadmium sulfide crystals more than a half of a century ago.

The growth of other wide bandgap materials such as ZnO using physical vapour transport also was reported (Rojo et al., 2006). Note that the earlier attempts to grow ZnO by sublimation were performed using much lower temperatures and the sublimation activators such as H_2O,

---

[11] "Sublimation" refers to the direct formation of the vapour from the solid phase; however, usually it is implicitly assumed that solid and vapour are the same substance. Thus the use of this term for the process in question is not strictly correct: AlN does not *sublime* but rather *decomposes*.

[12] Evidently, the source of the problem is the nitrogen pressure over GaN surface that is six orders of magnitude higher than that over AlN (Freitas, 2010).

The gallium vapour is generated either from the molten gallium or by the thermal decomposition of the GaN powder (Waldrip, 2007). In order to suppress the dissociation of GaN, NH_3 gas is used in addition to nitrogen. It is possible to grow only pellets of GaN with the size up to several square millimeters due to the depletion of the source (Ohtani et al., 2007). There was a great interest in this method in 1960s and 1970s that has been lost due to the great progress in producing pseudo-bulk GaN substrates by HVPE.

$HgCl_2$ or ZnX_2 (where X = Cl, Br or I) gases so that both sublimation and growth involve reversible chemical reaction (Rojo et al., 2006).

The simplest case for the analysis is the so called congruent (diffusionless) vapor transport (Abernathy et al., 1979): sublimation at the source and condensation at the growing crystal surface are congruent (i.e. there is no change in the composition), no foreign gases are involved and the vapour stoichiometry is preserved across the growth facility. Thus there is no relative motion of the vapour components - the diffusion plays no role, the transport is provided by the "Stefan wind" ("drift transport" (Karpov et al., 1999)) and the growth rate is maximal (Brinkman & Carles, 1998). An experimental path to the "Dryburgh" regime is the decreasing the pressure in the reactor (Wolfson & Mokhov, 2010). The growth rate under the "vacuum" conditions (the growth cell was placed in a special container with a background pressure maintained at the level about 10^{-4} Torr) corresponds to the growth rate in nitrogen atmosphere at temperatures about 350–400 K higher (Karpov et al., 1999).

In the other extreme case where inert gas is the predominant component in the vapour, the growth rate is directly proportional to the partial pressure difference at the source and at the crystal. Polycrystalline AlN frequently is grown by the sublimation method with the grain size increasing and the number of grain per unit area decreasing in the first few mm of growth (Noveski et al., 2004b).

The overall reaction of AlN sublimation growth can be written as

$$(AlN)_{\text{solid}} \xrightarrow[T_{growth}+\Delta T]{Sublimation} Al_{vapour} + \frac{1}{2}N_2 \xrightarrow[T_{growth}]{Deposition} (AlN)_{\text{solid}}$$

This method developed by G.A. Slack and T.F. McNelly in 1976 (Slack & McNelly, 1977) (whose largest crystal was 10 mm long by 3mm diameter) now provides the growth rates up to 1 mm/hr (Rojo et al., 2002)) and the high crystal quality (the threading dislocation density is lower than 1000 cm² and FWHM of the rocking curve is less than 10 arcsec in the best samples (Raghothamachar et al., 2003)).

Either a spontaneous nucleation (a self-seeding growth) without any attempt to control the crystal orientation or an intentional seeding (homoepitaxial (Hartmann et al., 2008) or heteroepitaxial (Lu, 2006; Miyanaga et al., 2006)) can be exploited. The SiC substrates are often used (Lu et al., 2008; Mokhov et al., 2002), other substrates —sapphire, tantalum carbide (TaC) and niobium carbide (NbC) —also have been tried (Lee, 2007).

The decomposition of SiC at high temperature affects the growth morphology and could provoke the growth of polycrystalline AlN (Noveski et al., 2004a). Fig. 1 clearly shows the graphitization of the SiC substrate (silicon evaporation) propagating from its lower side.

Both the crucible and the source usually are cylindrical, however, the conical crucibles are used sometimes (Slack & McNelly, 1977) as well a central hole in the powder source to increase the source surface area (Wang et al., 2006).

In contrast to the bulk SiC sublimation growth, there is no evidence of the polytypism in the bulk AlN wurtzite 2H polytype structure that has the lowest formation energy (Bondokov et al., 2007).

The high growth temperature and the highly reactive Al vapour create a problem in selection of crucible material that should have melting point well above 2300 C, a reasonable degree of chemical compatibility with AlN, relatively low vapor pressures, and the relatively small thermal expansion coefficient (Slack et al., 2004).

Different crucibles have been tried including ones made from refractory transition metals (W, Ta, Nb, Zr) and graphite coated with SiC, NbC or TaC (Dalmau & Sitar, 2005; Lu, 2006) revealing their weak points: e.g., the boron nitride growth environment results in the highly

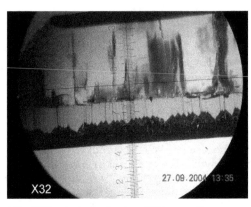

Fig. 1. Graphitization of the SiC substrate during AlN growth.

anisotropic growth (Dalmau & Sitar, 2005), the coated graphite crucibles deteriorate at high temperatures and carbon detrimentally influence the growth morphology as in the pure C crucibles; the crucibles made from nitrides or carbides suffer severe cracking. Among the most successful ones are the crucibles made from transition metals and their carbides (Lu, 2006).

The nature of other components of the growth facility such as heating elements along with the crucible material defines an environment that is responsible for the crystal contamination with different impurities (compatibility of the reactor materials has been considered, among others, by B. Epelbaum et al. (Epelbaum et al., 2002) and C.M. Balkas et al. (Balkas et al., 1997)).

The duration of growth process is limited by the degradation of the crucible (de Almeida & Rojo, 2002; Wang et al., 2006) and of the graphite insulation (Cai et al., 2006) as well as by continuous operation of the source that requires the temperature gradients within the source to be as low as possible (Bogdanov et al., 2003; 2004).

The growth temperature influence the size of AlN nuclei (Yazdi et al., 2006) and thus the crystal morphology (Sitar et al., 2004). The dislocations in the crystal can arise both during growth or after the growth in the course of the thermomechanical stress relaxation (Bogdanov et al., 2003; Klapper, 2010; Kochuguev et al., 2001; Zhmakin et al., 2000). The effect of substrate misorientation and buffer layers on growth modes and defects in AlN sublimed onto 6H-SiC substrates were studied in Refs. (Shi et al., 2001), (Yakimova et al., 2005); different growth modes were related to the low mobility of AlN adatoms on the crystal surface.

The grown bulk AlN crystals (the typical growth rate is about 100 μ/hour) usually have the rough side surface while the top surface could be facetted. The crystals are transparent with colour from yellow or amber to glass-clear (Fig. 2) having the Bragg FWHM 60 - 150 arc sec (Helava et al., 2007).

The reddish samples turned out to contain Fe^{2+} impurity (Ilyin et al., 2010). The below band-gap absorption bands limiting UV transparency are attributed to the point defects (Bickermann et al., 2010), for example, the band-to-impurity absorption manifesting itself as yellow coloration is thought to be related to either the doubly negative charged state V_{Al}^{2-}, the isolated aluminum vacancies $(V_{Al})^{3-/2-}$ (Hung et al., 2004; Sedhain et al., 2009) or the Al vacancy-impurity complexes (Lu et al., 2008).

(a)   (b)

(c)

Fig. 2. Examples of grown bulk AlN crystals.

4. Modelling of bulk AlN growth

Since the AlN sublimation growth is implemented in a tightly closed crucible under high-temperature, it is difficult to study and control the growth process *in situ*, thus the importance of the mathematical modelling is evident (Bogdanov et al., 2003; 2004; Chen et al., 2008; de Almeida & Rojo, 2002; Ern & Guermond, 2004; Wellmann et al., 2006; Wu et al., 2005; Zhmakin, 2004).

The aims of numerical simulation are to explain and to predict the growth behaviour. Numerical simulation is not a substitute for experiment, but a complement to it. Numerical models can provide detailed information on the flow, temperature and concentration fields, strain in the crystals etc. which can be measured experimentally only partly or not at all. On the other hand, numerical models depend on experimental data (materials properties, boundary conditions and so on). Moreover, numerical predictions are unreliable unless models are validated using experimental data in the widest possible range of macroscopic parameters.

"Modelling" and "Simulation" are frequently used as synonyms. In Computational Fluid Dynamics (CFD) community, however, the former usually refers to the development or modification of a model while the latter is reserved for the application of the model (AIAA, 1998).

A model should relate the process specification (equipment geometry, materials properties and transport coefficients, technological parameters such as the heating power and heater position, external electromagnetic fields, orientation of the growth facility etc.) to its outputs: crystals yield, crystal quality, process duration and production costs.

If one needs a single criteria (ideally, quantifiable) to estimate the practical usefulness of the simulation, the best choice is probably the reliability of a computer prediction (Oden, 2002). It should be stressed, however, that this parameter characterizes not the model itself, but the simulation, being depending on the adequacy of the model and the accuracy of the computations as well as on the particular aim of the simulations. Evidently, the same results could be considered successful if one is interested in unveiling some trend - and unsatisfactory if the goal is to find, for example, the exact position of the inductor coils in the growth furnace. The straightforward use of the model is referred to as a direct problem (see Table 1).

Process Specification		Process Output
Equipment geometry	$\Longrightarrow$ Direct (Insight)	Growth rate
Materials properties		Crystal quality
Process parameters		Process duration
	$\Longleftarrow$ Inverse (Optimization)	Production costs

Table 1. Simulation scheme

From a practical point of view the reversed formulation is more useful: how one should change the equipment design or the process parameters to improve the crystal quality or to reduce production costs, for example. The simplest way is a "try-and-error" approach: to use one's intuition to introduce changes into the process specification or into the growth facility design/size, perform simulation and evaluate results. Sometimes, especially at the early development stages, an even simpler "blind search" (Luft et al., 1999) approach is exploited which essentially is a screening of a range of parameters.

A more systematic way is to state an inverse problem by indicating

1. which geometry characteristics of the reactor or operating conditions (control parameters) could be varied and

2. what criteria should be used to measure the success of the optimization.

The *inverse* problems are, unfortunately, ill-conditioned (*not-well-posed*) and their solution requires some kind of a regularization (Tihonov & Arsenin, 1977) that frequently is just the restriction on the space of possible solutions.

Mathematical models, as well as numerical methods, used for simulation of crystal growth are essentially the same as in other Computational Continuum Mechanics (CCM) applications (heat transfer, fluid dynamics, electromagnetics, elasticity). Both the block-structured and the unstructured grids are used to solve practical problems.

The advantage of the unstructured grids is the relative ease with which the complex geometry can be treated. This approach needs the minimum input description of the domain to be discretized and is not tied closely to its topology in contrast to the block-structured grid. The required CPU time to attain the prescribed accuracy may be less than for the block-structured approach due to the much lesser total number of the grid cells as a direct sequence of the second advantage of the unstructured grids — the easiness of an adaptive mesh refinement, which allows one to place the cells exactly where needed.

The main difference between simulation of the thin film growth and the bulk crystal growth is that in the former case the computational domain can be considered fixed due to the small thickness of the epitaxial layer. Numerical study of the bulk crystal growth requires the use of either the moving grids or a regeneration of the grid. The latter approach is attractive when one can exploit a quasi-stationary approximation of the growth processes (the characteristic time of the crystal shape changes is large compared to the hydrodynamic/thermal time).

A simulation of the crystal growth requires solution of the conjugated multidisciplinary problem. The key sub-problem is the computation of the fluid flow coupled to the global heat transfer in the growth facility. Frequently, the global solution is used to the specify boundary conditions for a smaller imbedded computational domain where a more elaborate physical model is considered.

4.1 Low-Mach number (hyposonic) equations

The low-Mach number Navier–Stokes equations seem to be the most adequate model for gas flows with essentially subsonic velocities and large temperature variations (Makarov & Zhmakin, 1989). These equations provide the results identical to the full compressible Navier–Stokes computations while reducing greatly CPU time. Often (when gas mixtures used are not diluted) a CFD problem can not be decoupled from the mass transfer one (Egorov & Zhmakin, 1998).

The hyposonic flow equations follow from the full compressible Navier–Stokes equations under the following assumptions (Makarov & Zhmakin, 1989):

1) the Mach number is small $M^2 \ll 1$;
2) the hydrostatic compressibility parameter $\varepsilon = gL/R_g T_0$ is small $e \ll 1$;
3) the characteristic time τ is large compared to an acoustic time scale $\tau \gg L/a$

and for N_s-species mixture flows may be written in the following vector form:

$$\nabla \cdot \rho \mathbf{V} = 0$$
$$\nabla \cdot \left(\rho \mathbf{V}\mathbf{V} + p\hat{I} - \hat{\tau} \right) - (\rho - \rho_0)\, \mathbf{g} = 0$$
$$\nabla \cdot (\rho \mathbf{V}h + \mathbf{q}) = 0$$
$$\nabla \cdot (\rho \mathbf{V}c_s + \mathbf{J}_s) = W_s, \ s = 1, 2, \ldots, N_s$$

where ρ is the mixture density, $\mathbf{V}$ is the mixture mass-averaged velocity, h is the mixture specific enthalpy, c_s is the mass-fraction of s-th species, p is the dynamic pressure, τ is the viscous stress tensor, $\mathbf{q}$ is the heat flux, $\mathbf{J_s}$ is the mass diffusion flux of the s-th species.

The constitutive relations required to close the system are the state equations for the perfect gas mixture with the variable specific heat:

$$\rho \frac{R}{m} T = p_0 = const, \ 1/m = \sum_{s=1}^{N_s} (c_s/m_s)$$

$$h = \sum_{s=1}^{N_s} c_s h_s(T), \ h_s(T) = h_s^0 + \int_{T^0}^{T} Cp_s(T) dT$$

and the relations for the molecular transfer fluxes: the viscous stress tensor

$$\hat{\tau} = -\frac{2}{3}\mu \left(\nabla \cdot \mathbf{V} \right) \hat{I} + \mu \left(\nabla \mathbf{V} + \mathbf{V}\nabla \right)$$

the heat flux

$$\mathbf{q} = -\lambda \nabla T + \sum_{s=1}^{N_s} h_s \mathbf{J}_s + p_0 \sum_{s=1}^{N_s} k_s^T \mathbf{J}_s / (\rho c_s)$$

and the diffusion flux of sth component

$$\mathbf{J}_s = -\rho D_s \left(\nabla c_s + \frac{m_s}{m} k_s^T \frac{\nabla T}{T} \right)$$

Generally both homogeneous and heterogeneous chemical reactions are to be taken into account, the latter resulting in the highly nonlinear boundary conditions.

4.2 Conjugate heat transfer

The steady-state temperature distribution inside the solid block without heat sources satisfies the scalar equation of thermal conductivity:

$$\nabla \cdot (-\lambda_{solid} \nabla T) = 0$$

The heat conduction is the simplest heat transfer mechanism. Still, two aspects of heat conduction in crystal growth problems should be mentioned.

Firstly, one needs to account for the anisotropic thermal conductivity for certain crystals and for the solid blocks manufactured from the pyrolytic graphite. In the first case the degree of the anisotropy, being determined by the crystal composition and the crystallographic symmetry, is usually not large. The pyrolytic graphite is obtained by the pyrolysis of hydrocarbon gas at high temperature in the vacuum furnace and has a layered structure with the highly ordered hexagonally arranged carbon atoms in planes and the randomly oriented atoms in the perpendicular direction. The ratio of the values of the thermal conductivity in the different directions is 100-400, depending on the material quality.

Whether anisotropy forces one to consider a three-dimensional problem for the geometrically two-dimensional configuration depends on the crystal symmetry and the orientation of its principal axes. For example, if the symmetry axis coincides with the [0001] axis of a hexagonal crystal such as SiC, the solution should be isotropic with respect to rotations around the axis and the two-dimensional formulation is valid.

Secondly, some parts of the facility could be modelled as the porous medium. Powder source is used in a number of techniques such as the metal flux method for growth of bulk GaN crystals from Li-Ga-N liquid phase (Song et al., 2004), the ammonothermal method for GaN (Yoshikawa et al., 2004), the sublimation growth of single crystals of wide bandgap semiconductors (SiC, AlN) (Bogdanov et al., 2003; Dhanaraj et al., 2003). The granular or fibrous medium is often used for the insulation. The usual approach in the computation of the global heat transfer in the facility is to treat the porous media using the effective thermal conductivity. For a given porous structure this quantity is a function of the pressure and the temperature that determine the relative contribution of the solid matrix conduction, heat conduction through the medium (gas) filling the pores and radiation to the total heat transfer. Experimental data being rather scarce, especially at high temperatures, the main problem is to formulate a model that could adequately extrapolate the effective thermal conductivity beyond the measured range of the pressures and temperatures (Daryabeigi, 1999; Kitanin et al., 1998). The effective thermal conductivity could be two orders of magnitude smaller than that of bulk material that is evidently favorable for the use of the porous media as insulation. However, it has a detrimental effect on the optimal heating of the SiC powder source in the sublimation method (Kitanin et al., 1998). The composition, the porosity and the thermal conductivity of the SiC powder vary during the growth process (Karpov et al., 2001a).

Often radiative heat exchange through a non-participating fluid between solid surfaces can be accounted for under the assumption of the gray-diffusive surface radiation. All solid blocks are assumed to be opaque, while the external boundaries of the gaseous domain may be semi-transparent. Computation of the total radiative flux incoming to the given small surface element requires knowledge of the configuration factors (view factors). Calculation of these view factors via an integration over the complex geometry of the emitting area with account for the shadowing effect is described in details in (Dupre et al., 1990). If the view factors are known, the total radiative flux incoming to the surface element $i(i = \overline{1, N_e}$, where N_e is the total number of surface elements on the boundary) can be calculated as a sum

$$q_i^{in} = \sum_{j=1}^{N_e} q_j^{out} F_{ij},$$

where q^{in} and q^{out} are the radiation fluxes to the wall and from the wall, F_{ij} are the view factors. For the semitransparent external boundary the radiative flux out from the objects inside the region could be calculated by the Stefan-Boltzmann law and definitions of emissivity (E), reflectivity (R) and transmissivity (T) as

$$q_I^{out} = \sigma E_i T_I^4 + R_i q_I^{in} + T_I \sigma T_a^4,$$

where σ is the Stefan-Boltzmann constant and T_a is the ambient temperature.

4.3 Boundary conditions

Boundary conditions at the surfaces where heterogeneous reactions occur are formulated under assumptions that growth is limited by mass transport to the surface, the properties of the adsobtion layer are identical to those of the solid phase and are described by the following equations:

total zero flux of the inert gas

$$\rho u c_0 + J_0 = 0$$

equations relating the total species fluxes and the rates of the heterogeneous reactions

$$\rho u c_i + J_i = M_i \sum_{r=1}^{N_r} v_{ir} \dot{w}_r , i = 1, \ldots, N_k^r$$

the mass action law

$$\prod_{i=1}^{N_k^r} X_i^{v_{ir}} = K_r , r = 1, \ldots, N_r$$

the normalization condition

$$\sum_{i=0}^{N_k^r} X_i = 1,$$

where N_k^r is the number of the gas phase species participating in the surface reactions, subscript "0" refers to the inert ('carrier') gas, N_E is the total number of elements, N_S is the number of solid state phases, $N_r = N_k^r - N_E + N_S$ is the number of reactions, v_{ir} is the stoichiometric coefficient of ith component in rth reaction, $\dot{w}_r$ is the rate of rth heterogeneous reaction, u is Stephan velocity, c_i and M_i are mass concentration and molar mass of ith component, J_i is the normal diffusion flux of ith component, K_r is the equilibrium constant of rth heterogeneous chemical reaction.

When experimental data on the reaction constants are absent, their values could be estimated using thermodynamic properties of individual materials as

$$\ln K_r = \frac{1}{RT} \sum_{i=1}^{N_k^r + N_s} \nu_{ir} G_i - \ln \left(\frac{p_s}{p} \right) \sum_{i=1}^{N_k^r} \nu_{ir} ,$$

where G_i is the Gibb's potential of the Ith component at normal pressure.

Theoretical analysis of AlN PVT growth was probably first performed in (Dryburgh, 1992). In this paper, the surface decomposition of low-reactive nitrogen was noticed as the rate-limiting stage of the AlN evaporation/ growth. kinetic mechanism. An one-dimensional model of the process was developed, no transport effects being taken into account. In contrast, most of the later studies assumed the AlN growth rate to be limited by the species transport (Noveski et al., 2004b; Wu et al., 2004) and the AlN growth rate was found using the Hertz-Knudsen equation for the interface Al fluxes at the AlN surfaces. Under the additional assumption of a low Al content in the vapor, an approximate explicit relationship for the AlN growth rate was derived.

These studies, unfortunately, neither clarified the boundaries of the kinetically- and transport-limited approximations nor accounted for the mass exchange between the crucible and the ambient through the small gaps and/or through the porous crucible walls that may essentially affect the process. Besides, they do not consider the evolution of the AlN crystal and of the source and the corresponding gradual change of the growth conditions. The evolution effects are also important as they determine the crystallization front shape that, in turn, affects distributions of dislocations and other defects in the crystal (a slightly convex crystallization front is preferable). A model of AlN sublimation growth that does not rely on the kinetically or transport limited approximations and describes all the above effects within a single approach was developed in (Karpov et al., 2001; Karpov et al., 1999; Segal et al., 2000) where an one-dimensional stationary model was considered.

The developed evolutionary model for AlN growth relies on the following assumptions:

- there are only Al and N_2 species in the gas phase (volatile impurities are negligible);

- the growth rate of the AlN crystal is determined by the local vapor composition and temperature but independent of the surface orientation (the isotropic growth);

- the evaporation of the AlN source occurs from the surface only (dense polycrystalline sources is used rather than porous sources);

- the evolution of the AlN source and the crystal occurs much slower than the transfer processes (the quasi-stationary transfer).

The model of AlN sublimation growth is based on the conventional description of heat and radiation transfer, gas flow dynamics, and species diffusion in the growth system coupled with the reduced quasi-thermodynamic description of the surface kinetics at the AlN surfaces. The latter was earlier applied to the modelling of other growth techniques (see, for example, (Segal et al., 2004) and references therein). As applied to AlN sublimation growth, it utilizes the extended Hertz-Knudsen relationships (Segal et al., 1999) for two reactive gaseous species, Al and N_2

$$J_i = \alpha_i(T)\beta_i(T)(P_i^w - P_i^e)$$

Here, J_i are the interface molar fluxes, $\alpha_i(T)$ are the temperature-dependent sticking probabilities,

$$\beta_i(T) = (2\pi\mu_i RT)^{-1/2}$$

are the Hertz-Knudsen collision factors, μ_i are the molar masses, R is the gas constant, P_i^w are the species partial pressures at the interface, P_i^e are the quasi-equilibrium (thermodynamic) species pressures, and subscript i indicates a particular species (Al or N_2).

The Al sticking probability is assumed unity due to its high reactivity. In contrast, the N_2 sticking probability is very low. In (Karpov et al., 2001), it was fitted as a function of temperature using data of (Dreger et al., 1962) on the AlN evaporation in vacuum (more recent data of (Fan & Newman, 2001) confirmed the derived approximation)

$$\alpha_{N_2}(T) = \frac{3.5 \exp -30000/T}{1 + 8 \cdot 10^{15} \exp 55000/T}.$$

The pressures P_i^e satisfy the mass-action law equation

$$(P_{Al}^e)^2 \, P_{N_2}^e = K(T),$$

where $K(T)$ is the equilibrium constant for the surface reaction

$$2Al + N_2 \Leftrightarrow 2AlN(s)$$

The model was validated in (Segal et al., 2000) by the comparison with the experimental data on the AlN growth rate as a function of the temperature and the pressure; it is implemented as software package Virtual Reactor^{TM} (Bogdanov et al., 2001; STR-soft, 2000). Virtual Reactor^{TM} provides an accurate simulation of all major physical-chemical phenomena relevant to this method such as resistive or RF heating ; conductive, convective and radiative heat transfer; mass transfer in gas and porous media; heterogeneous chemical reactions at the catalytic walls and on the surface of powder granules; deposits formation; formation of elastic strain and dislocations in the growing crystal; the evolution of the crystal and of the deposit shape, including partial facetting of the crystal surface. The problem is solved using a quasi-stationary formulation.

Temperature distributions in the tungsten and in the graphite furnaces for PVT growth of 2 inch diameter AlN boules are shown on Figs. 3, 4, respectively.

Computations revealed, in particularity, that the growth characteristics are extremely sensitive to the temperature distribution in the crucible, for which reason an accurate prediction of this distribution is of primary importance for successful modeling. Figure 5 illustrates the high accuracy of temperature prediction.

Good agreement with experimental data proves the adequacy of the model. The small deviation of the points and the curve is probably due to some uncertainty in the thermal and optical properties of the materials involved at high temperatures.

The distribution of the Al vapor molar fraction in the gaps between the AlN source (bottom), the seed (top), and the walls of the carbonized tantalum crucible in the graphite furnace is shown on Fig. 6.

If the growth occurs in a hermetically closed and chemically inert crucible, the inside static pressure is spontaneously established to provide the conservation of the initial difference between the total numbers of aluminum and nitrogen atoms in the crucible. This quantity is constant since the moment of the sealing of the crucible because the vapour-solid mass

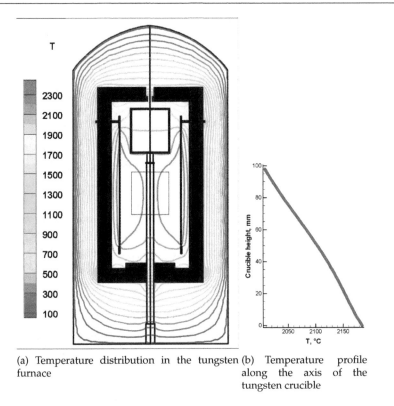

(a) Temperature distribution in the tungsten furnace

(b) Temperature profile along the axis of the tungsten crucible

Fig. 3. Temperature distribution in the tungsten furnace.

exchange occurs stoichiometrically (at the AlN surfaces) or does not occur at all (at the inert crucible' walls). l
Due to the existence of the small gap between the crucible body and lid, the ambient pressure strongly affects the growth process, which is largely related to the notion of critical pressure. In equilibrium

$$J_i = 0$$
$$P_i^e = P_i^w$$

and the partial pressures of the two species can be found from the system of two equations:

$$(P_{Al}^w)^2 P_{N_2}^w = K(T)$$
$$P_{Al}^w + P_{N_2}^w = P,$$

where P is the total pressure in the crucible. Analysis shows that if

$$P > P^\star(T) = 3/2[2K(T)]^{1/3}$$

with $P^\star$ denoting the critical pressure, then the system has two solutions corresponding to the Al-rich and N-rich vapor. Since the vapor composition in the crucible is established due to

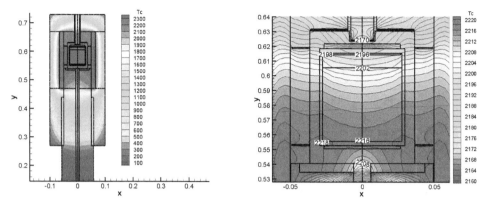

(a) Temperature distribution in the graphite furnace (b) Temperature distribution in the carbonized tantalum crucible

Fig. 4. Temperature distribution in the graphite furnace.

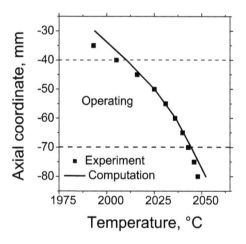

Fig. 5. Temperature at the center of the crucible lid vs. crucible coordinate at the vertical axis, computed (solid line) and pyrometrically measured (points).

mass exchange with the nitrogen ambient, the N-rich branch should be selected. If $P = P^{\star}(T)$, then there is the unique solution corresponding to the stoichiometric vapor, and if $P < P^{\star}(T)$, there is no solution. The latter means that in a hermetically closed crucible the equilibrium pressure is always higher than the critical pressure.

In a non-hermetic crucible, the equilibrium is impossible if the ambient pressure is so low that the related pressure inside the crucible is lower than the critical pressure. In this case, both the AlN source and seed evaporate, with the Al/N_2 vapor coming out from the crucible

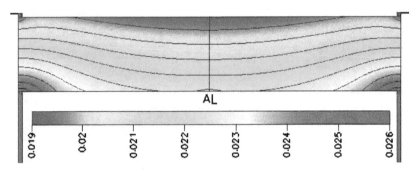

Fig. 6. Al vapor molar fraction.

to the ambient (see Fig. 7 where the pressure inside the crucible and AlN crystal growth rate vs. ambient pressure are shown for different gaps between the side crucible wall and the lid). Mass exchange between the crucible and the ambient occurs through a narrow

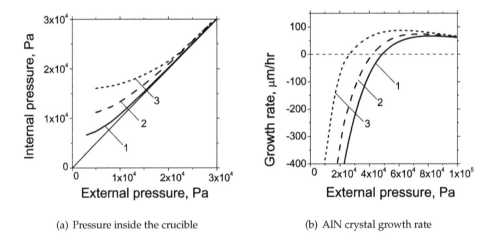

(a) Pressure inside the crucible   (b) AlN crystal growth rate

Fig. 7. Pressure inside the crucible (a) and AlN crystal growth rate (b) vs. ambient pressure at different gaps between the side crucible wall and the lid: 1 - gap length λ is 1 mm, gap thickness δ is 100 μm, 2 - λ = 1 mm, δ = 50 μm, 3 - λ = 10 mm, δ = 100 μm.

ring gap between the crucible side wall and the lid. The internal and external pressures are close to each other at a sufficiently high external pressure but considerably deviate as it decreases, depending on the gap hydraulic resistance ζ that is proportional to the gap length and inversely proportional to the third power of the gap thickness. Al and N_2 evaporate from the AlN source and then either deposit on the AlN crystal or escape from the crucible. The ratio of the deposited and escaped material depends on the external pressure and on ζ. Figure 7 shows the computed dependencies of the AlN growth rate on the ambient pressure at the crystal center.

As the external pressure decreases, a higher fraction of the material escapes from the crucible and the growth rate decreases. At a sufficiently low ambient pressure P_a^0, the crystal begins evaporating, with all vapors coming out from the crucible through the gap (negative growth rates). The value P_a^0 decreases with the gap hydraulic resistance (at a sufficiently high ζ, the crystal does not evaporate at an arbitrary small ambient pressure). The local AlN evaporation/growth rate is determined by the local supersaturation, both for the source and seed. The local interface flux on an AlN surface can be approximated as

$$J \approx \frac{\left(P_{Al}^w\right)^2 P_{N_2}^w / K(T) - 1}{4/\left[3\beta_{Al}(T)P_{Al}^w\right] + 1/\left[3\alpha_{N_2}\beta_{N_2}(T)P_{N_2}^w\right]}$$

Here, the quantity in the numerator is the local supersaturation that represents the driving force for the local AlN evaporation/growth while the denominator corresponds to the local kinetic resistance. The distribution of the supersaturation over the crystal surface determines its evolution (Fig. 8). Black lines with arrows are the streamlines, gray lines are the

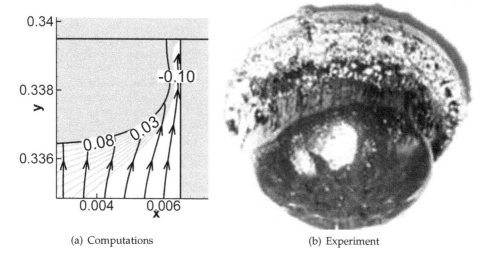

(a) Computations   (b) Experiment

Fig. 8. Shape of AlN boule after 20 hrs of growth.

supersaturation isolines, numbers at the crystal surface are the local supersaturation values. Note that the absolute values of supersaturation are rather small due to the smallness of the relative temperature difference between the source and the seed.

The higher the local supersaturation is, the faster the crystal grows there. At the very periphery, supersaturation is negative due to the species exchange between the crucible and the ambient through the ring gap (here, the vapor is enriched by N_2 and depleted in Al) and the crystal evaporates, taking a mushroom shape; the vapor escapes from the crucible through the gap (the last right streamline is directed to the gap). Computational results are in good agreement with the observed crystal shapes. Using this model, we have optimized the growth conditions and crucible design, which eventually favored the growth of 2″ diameter and up to 10 mm long AlN boules with a slightly convex shape providing a low defect content in the crystal.

5. Experimental set up for AlN bulk growth

Before AlN seeds become available, SiC (4H and 6H of both (0001)C and (0001)Si orientations (Mokhov et al., 2002)) seeds were used. SiC has a small a-lattice mismatch with AlN - 0.96% for 6H-SiC and 1.2% for 4H-SiC. However, SiC substrates are known to degrade at high temperatures required for the AlN growth.

A two-stage technology to exploit the best features of different crucibles by avoiding interaction of W with Si and C that form easily melted eutectics and by limiting the incorporation of Si and C in C-rich environment was applied (the AlN crystals grown on SiC seeds in C-containing ambient contain a lot of Si and C impurities that, in particularity, determine the color of the crystal):

1. seeding and initial growth of the 2-3 mm long AlN crystals on the SiC seeds in the TaC crucibles in graphite equipment and

2. growth of bulk AlN crystals on the AlN seeds in tungsten crucibles and equipment.

High-quality AlN seeds of large diameter are currently unavailable while use of AlN seeds of a smaller diameter requires long multi-time lateral overgrowth of the crystals to reach the desired diameters, as the lateral overgrowth angle was found not to exceed 10-15 degrees — the diameter enlargement of AlN boules is often associated with defect generation (Bondokov et al., 2006) or the crack formation (Schujman & Schowalter, 2010).

Another kind of the two-stage procedure was reported by M. Strassburg et al. (Strassburg et al., 2004) where the temperature was gradually ramped between the two stages and by Z. Sitar et al. (Sitar et al., 2004) who used vaporization of Al (use of the metal vapour source is called *direct synthesis method* by K. Nishino et al. (Nishino et al., 2002))[13] at the first and the AlN powder source at higher temperature at the second stage; two-stage growth was also used by R. Dalmau et al. (Dalmau et al., 2005) with stages differing in the growth temperature and, thus, growth rate.

5.1 Pre-growth processing

Preparation of W crucibles includes annealing of W ones to remove the adsorbed impurities, while for Ta crucibles a pre-carbonization is necessary (Fig. 9). These crucibles are remarkably thermally and chemically stable and can endure over 3000 hours of the cumulative AlN growth in graphite (tungsten) equipment.

The presence of oxygen is detrimental in AlN growth due to the formation of oxynitrides and enhanced formation of stacking faults (de Almeida & Rojo, 2002; Majewski & Vogl, 1998) that can induce shallow electronic states (Northrup, 2005) and decrease the thermal conductivity. Oxygen has a negligible effect on the growth rate itself, but it can, at low temperature (e.g., during the heating of the system), provoke generation of Al_2O_3 inclusions (Karpov et al., 2003) that, in turn, causes surface roughness (Kazan et al., 2006). The addition of the hydrogen to the nitrogen during growth is beneficial (Karpov et al., 2001); alternatively, the sources could be processed to reduce the oxygen content. The source of oxygen in the sublimation AlN growth is the hydroxides and oxides on the surface of AlN particles (Edgar et al., 2008).

High-purity AlN sources were prepared from the commercially available AlN powders either by the annealing in the N_2 atmosphere or by the sublimation-recrystallization (Epelbaum

---

[13] Such method provides a high growth rate (5 mm/hour) but the long-term growth is impossible due to the formation of a nitride layer over the metallic Al source and resulting drop of the Al vapour pressure (Lu, 2006).

Fig. 9. Ta (dark gray) crucible (diameter 63 mm) completely covered with a layer of TaC-Ta$_2$C ceramics (gold) that is converted into TaCN during the growth to prevent the formation of easily melted eutectics (Ta-Si and Ta-Al) (Vodakov et al., 2003).

et al., 2004; Helava et al., 2007), producing, respectively, dense porous or polycrystalline AlN source[14].

The content of impurities in the AlN samples was studied using glow discharge mass spectrometry (GDMS). Sublimation-recrystallization was found to be superior and accepted as a standard technique.

SiC seeds were cut from 6H SiC bulk crystals as 0.5 mm thick round plates of different diameters (15-50 mm). The SiC seeds are mechanically lapped and mounted onto the crucible lid using the C-based glue. The technique of the lateral overgrowth of AlN crystals from small-diameter AlN seeds can also be attempted starting from 0.5 mm thick and 15-18 mm diameter AlN seeds from previously grown AlN bulk single crystals. These seeds are similarly mechanically processed and mounted onto the crucible lid with the AlN-based glue.

---

[14] In a two-step procedure developed by L. Du & J.H. Edgar (Du & Edgar, 2010) the low temperature (< 1000 ° C) annealing was followed by a high temperature (> 1900 ° C) sintering aimed at the reduction of the specific surface area of the AlN powder through the particle agglomeration. Sintering could be applied not only to raw AlN powder, but also to the flakes obtained by pressing the AlN powder (Han et al., 2008).

Evidently, the initial contamination of the AlN powder depends on the process of its synthesis. AlN powder is usually obtained either by the carbothermal reduction of Al$_2$O$_3$ or by direct nitridation of metals, both methods having very long reaction times (hours or even days) - see (Radwan & Miyamoto, 2006) and references therein. The recently proposed microwave-assisted synthesis with addition of ammonium chloride to produce HCl as an intermediate product requires much shorter time - tens of minutes (Angappan et al., 2010). Another fast method is the combustion of ultrafine aluminum powder in air; in this case additives could increase the yield of AlN (Gromov et al., 2005).

5.2 Seeding and initial growth

Use of SiC seeds in pre-carbonized Ta crucibles in graphite RF-heated furnaces is necessarily accompanied by the diffusion of Si and, to a lesser degree, of C into AlN. This process along with the lattice and TEC mismatch results in generation of defects at the SiC-AlN interface, such as micropipes inherited from SiC, dislocations, cracks, and others. Thus in-house sublimation grown *thick* low dislocation density (dislocation etch pits density $1\text{-}4\cdot10^{3}cm^{-2}$ after 30 min etching in molten KOH) micropipe-free SiC substrates (60 mm in diameter) have been used (Fig. 10). X-ray diffractometry and topography of the grown AlN layers show that

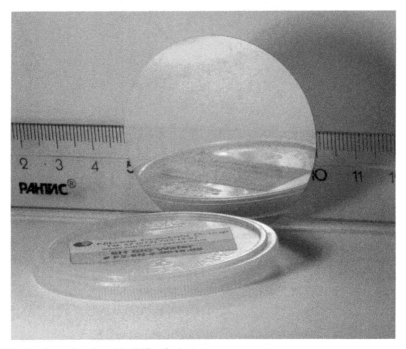

Fig. 10. Thick micropipe-free 2in SiC substrates.

FWHMs of the rocking curves in ω-scan to lie in the range of 60-120 arcsec. At the same time, the c-parameter of the crystal lattice is found to vary as 4.984-4.988 A, which suggests that the AlN layers contain much impurity (the reference value of the c-parameter for pure AlN is 4.982). X-ray microanalysis with SEM shows 5-6% wt of Si and 1% wt of C in the AlN layers. Concentration of other impurities is less than 100 ppm.

It is known that the typical temperature of AlN growth on SiC seeds is lower by 200-300 °C than that on AlN seeds for the comparable growth rates (Epelbaum et al., 2001). Analysis of the three-phase thermodynamic equilibrium in the system Al/N/Si/C(vapor)-AlN(solid)-SiC(solid) allows finding gaseous species that may be responsible for the more intensive AlN growth on SiC. Calculation of the equilibrium partial pressures of the most volatile species in this system (more than 30 gaseous species were considered) showed that there are two "cross" volatile species (AlNC and Si_2N) that may intensify AlN growth on SiC.

5.3 Growth of bulk AlN crystals starting from AlN seeds

The 2-3 mm thick AlN layers (Fig. 11) are separated from the SiC substrate and mounted on the tungsten crucible lid with the AlN-based glue. AlN bulk crystals 10-20 mm long are

Fig. 11. 2in AlN seed grown on 60 mm micropipe-free SiC substrate.

grown in the W crucibles in the W resistively heated furnace (Mokhov et al., 2005), the growth temperature was varied in the range of 2050-2250 °C, the N_2 pressure was varied in the range of 0.5-1 atm, and the AlN growth rate was of 50-150 μm/hr. Long AlN crystals grow for several dozens of hours and often in several runs.

The evolution of the crystal quality via the improvement of the growth regime from the 2-inch diameter 10 mm long bulk AlN crystal having a single-crystal core of about 40 mm diameter and a polycrystalline rim (Avdeev et al., 2010; 2011) to the "good enough" single crystal (and, hence, the substrate) is illustrated by Fig. 12.

6. Properties of sublimation-grown AlN

Currently the technology provides stable reproducible growth of up to 2" diameter and 10-15 mm long AlN single crystals. Post-growth processing includes calibration, slicing into wafers, mechanical lapping-polishing, finishing chemical-mechanical polishing (CMP) to remove the subsurface damage due to the mechanical polishing that extends up to 4000 Å below the surface (Chen et al., 2008; Freitas, 2005), and characterization.

Impurities such as oxygen, silicon, carbon, boron contribute to the absorption and emission bands below the bandgap (Senawiratne et al., 2005) and thus reduce the AlN transparency in the UV spectral range. The content of impurities in the AlN seeds grown on SiC substrates

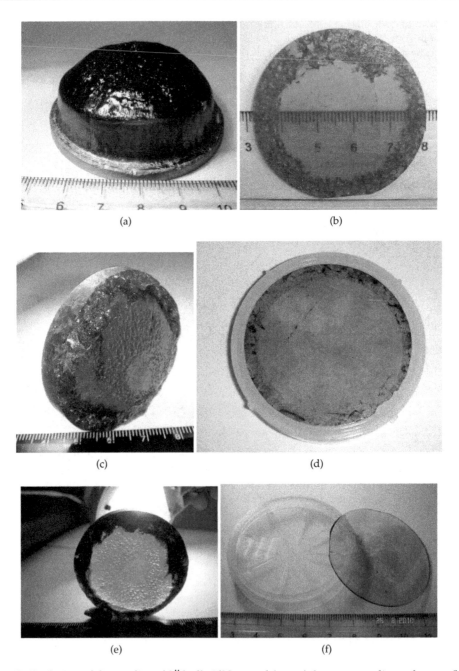

Fig. 12. Evolution of the quality of 2″ bulk AlN crystal (a, c, e) & corresponding substrate (b, d, f).

are rather high: about 5% Si and about 0.6 % C. However, the impurity concentration rapidly drops during the subsequent growth of the bulk AlN as could be seen from Fig. 13 where the variation of the concentration of aluminum, nitrogen, carbon, silicon and oxygen along the crystal thickness is shown.

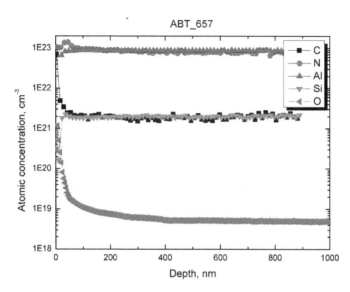

Fig. 13. Concentration of impurities along the crystal (Courtesy of V.V. Ber, Ioffe Physical Technical Institute, Russian Academy of Sciences).

X-ray diffractometry of the AlN substrates gives FWHMs of the rocking curves in ω-scan of about 2-5 arcmin. No impurities in concentrations higher than 100 ppm were found in the substrates. Typical data of X-ray analysis are presented in (Mokhov et al., 2005).
The resistivity of Si-doped AlN is generally lower than 10^5 Ohm·cm. These values are much lower than the reference resistivity of undoped semi-insulating AlN (10^{11}-10^{13} Ohm·cm). This result is attributed to the effect of the residual impurities (primarily Si) that still remain in the AlN bulk crystals grown on the SiC seeds. Repeated use of the initial AlN layer separated from the "primary" crystal (Chemekova et al., 2008) results in high purity material with the resistivity in the range of $3 \cdot 10^9$ - $3 \cdot 10^{11}$ Ohm · cm, which is rather close to the reference values.
Measurements of the transmittance spectrum of the substrates with thickness of 400 μm in the UV range have shown that most of the crystals have the average transmittance of 50-60% and demonstrate the sharp cut-off between 250nm and 320nm.
Selective etching in KOH/NaOH eutectic solution at 450°C reveals the presence of large grains with dislocation grain boundaries and individual dislocations with a low density (Fig. 14). The grain dimensions and the local dislocation densities can vastly differ.
The high crystallographic quality of the single crystal AlN substrate is confirmed by the Laue photo (Fig. 15, a) and the photo of the substrate in polarized light (Fig. 15, b).
The typical X-ray rocking curves of the 2in substrate at two different points are presented in Fig. 16.

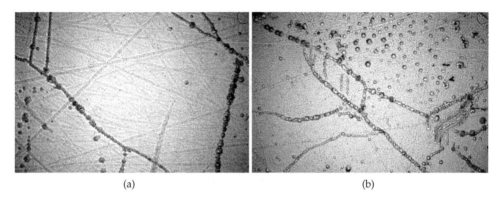

(a)   (b)

Fig. 14. Etched AlN surface (Courtesy of A. Polyakov, GIREDMET, Moscow).

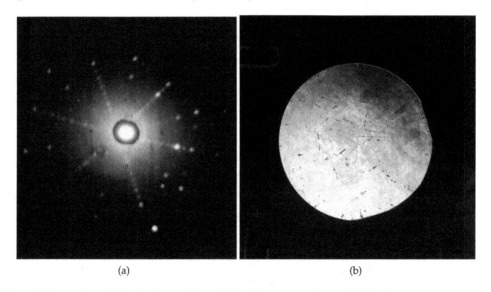

(a)   (b)

Fig. 15. Laue photo (a) and photo of 2in AlN substrate in polarized light (b).

6.1 Epitaxial layers

The best characteristic of the substrate is the quality of the grown epitaxial layers and the performance of the fabricated devices. The epitaxial structures deposited at bulk AlN have been studied using transmittance and reflectance optical microscopy, high-resolution XRD diffraction, cross-polarization and cathodoluminescence. The CL spectrum of the MOCVD grown 0.3 μm AlGaN layer (Fig. 17) has FWHM of about 10 nm that indicates layer-by-layer growth of the epitaxial film.

The morphology of 0.5μm AlN epitaxial layer grown on the 2in wafer could be assessed by the AFM images shown on Fig. 18. One could see large step bunches associated with the miscut of the wafer (1° for this wafer) off the (0001) axis. On the terraces between step bunches there are nice atomic steps.

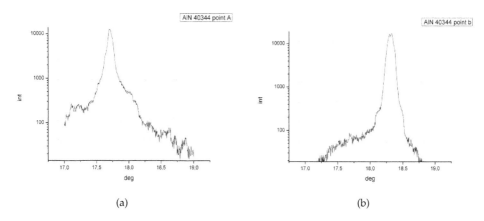

Fig. 16. X-ray rocking curves at different points of 2in substrate.

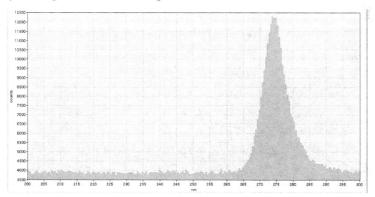

Fig. 17. CL spectrum of LED structure AlGaN/AlN (Courtesy of Prof. Asif Khan , University of South Carolina).

6.2 Light-emitting diodes

The emission of 5QW UV LED grown on bulk AlN substrate has a peak about 352 nm with FWHM 8 nm. The emission intensity was 4 times greater than that of the identical structure grown on sapphire (Fig. 19).

It should be stressed that the MOCVD growth procedure on the single crystal AlN substrate was not specially developed - the one optimized for the growth group III nitride epitaxial layers on sapphire was used.

UV LED emitting at 360 nm was grown by chloride VPE is shown in Fig. 20. This LED has a rather long lifetime: only a slight degradation is observed during long time operation (Fig. 21).

6.3 SAW applications

The surface acoustic wave velocity of the grown bulk AlN crystals measured at different frequencies and extrapolated to the zero frequency yields a value of about 5750 m/s.

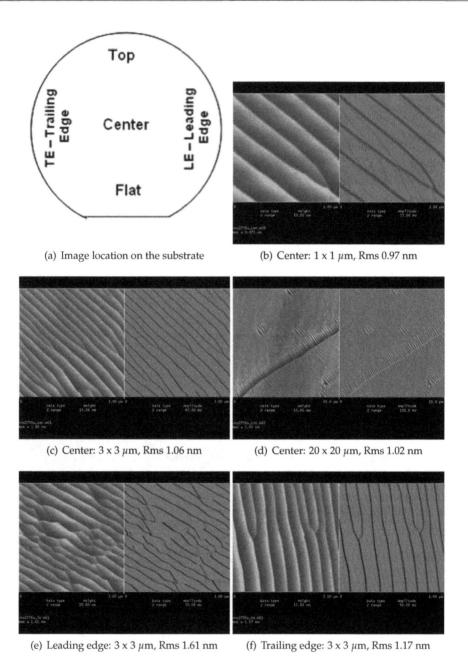

(a) Image location on the substrate   (b) Center: 1 x 1 μm, Rms 0.97 nm

(c) Center: 3 x 3 μm, Rms 1.06 nm   (d) Center: 20 x 20 μm, Rms 1.02 nm

(e) Leading edge: 3 x 3 μm, Rms 1.61 nm   (f) Trailing edge: 3 x 3 μm, Rms 1.17 nm

Fig. 18. AFM of AlN epitaxial layer on 2 inch bulk AlN substrate (Courtesy of A. Allerman, Sandia National Laboratory)

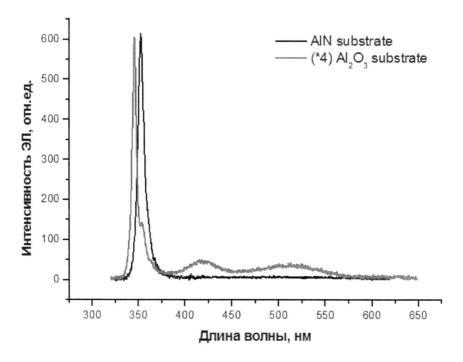

Fig. 19. EL intensity, that of UV LED on the sapphire substrate is increased fourfold (Courtesy of V.V. Lundin, Ioffe Physical Technical Institute, Russian Academy of Sciences).

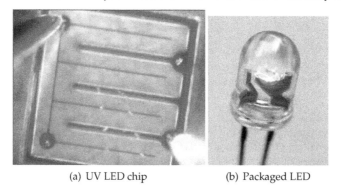

(a) UV LED chip   (b) Packaged LED

Fig. 20. 360 nm UV LED (Courtesy of The Fox Group, Inc.).

A simple regular electrode structure for SAW devices has been proposed in ref. (Biryukov et al., 2007). The structure consists of an interdigital transducer in the form of a ring placed on the Z cut of a hexagonal piezoelectric crystal (Fig. 22). Finite thickness electrodes produce the known slowing effect for a SAW in comparison with this SAW on a free surface. The closed

Radiometric Power Maintenance Chart (T$_{ambient}$ = 21°C)

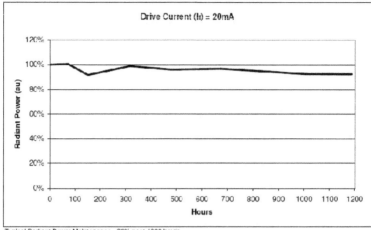

Typical Radiant Power Maintenance >90% past 1200 hours

Fig. 21. Radiant power degradation (Courtesy of The Fox Group, Inc.).

slow electrode region with the *fast* surrounding region forms an open waveguide resonator structure with the acoustic field concentrated in the electrode region. If the radius of the structure - ring waveguide resonator (RWR) - is large enough for a given wavelength, an acceptable level of radiation losses can be reached. The electrical admittance of such resonator does not have sidelobes. Such device has been manufactored using a transparent pale brown

Fig. 22. Ring waveguide resonator on surface acoustic waves.

colored AlN single crystal wafer of 21 mm diameter and a thickness of 850 μm (Biryukov et al., 2009). Excitation of radial modes has been investigated. Experiments demonstrated the excellent device performance (high selectivity and a large Q-factor, estimated to be about 2700). The electrical admittance frequency dependence did not have sidelobes.

7. Conclusions

The technology of sublimation growth of AlN bulk crystals on the SiC seeds based on the successive use of crucibles made from different materials is described. Currently the technology provides stable growth of crystals up to 2" diameter and producing of single crystal 2in AlN substrates.   The superiority of such "good enough" AlN substrates for crystalline quality of the grown epitaxial layers and the device performance is demonstrated.

8. Acknowledgments

The authors thank J.H. Edgar (Kansas State University), T. Bogart (Penn State University), D. Yoo (Georgia Institute of Technology), A. Allerman (Sandia National Laboratory), Asif Khan (University of South Carolina), V.V. Ber and V.V. Lundin (Ioffe Physical Technical Institute, Russian Academy of Sciences), A. Polyakov (GIREDMET, Moscow), The Fox Group, Inc. (Warrenton, VA) for experimental data.
The authors are grateful to M.V. Bogdanov, S.Yu.   Karpov, A.V. Kulik, and M.S. Ramm (Soft-Impact, Ltd.) for useful discussions.

9. References

Abernathy, J. R. et al. (1979). Congruent (diffusionless) vapour transport, *J. Crystal Growth* 47: 145–154.

Adekore, B. T. et al. (2006). Ammonothermal synthesis of aluminum nitride crystals on group III-nitride templates, *J. Electron. Mater.* 35: 1104–1111.

AIAA (1998). Guide for the verification and validation of computational fluid dynamics simulations, AIAA G-077-1998.

Akasaki, I. & Amano, H. (2006).   Breakthroughs in improving crystal quality of GaN and invention of the p-n junction blue-light-emitting diode, *Jap. J. Appl. Phys.* 454: 9001–9010.

Amano, H. et al. (2003). Group III nitride-based UV light emitting devices, *phys. stat. sol. (a)* 195: 491–495.

Ambacher, O. et al. (2003).   Electronics and sensors based on pyroelectric AlGaN/GaN heterostructures. Part A: Polarization and pyroelectronics, *phys. stat. sol. (c)* 0: 1878–1907.

Angappan, S. et al. (2010).   Aluminum nitride by microwave-assisted synthesis: Effect of added ammonia chloride, *Int. J. Self-Propag. High-Temp. Synthesis* 19: 214–220.

Aoki, M. et al. (2002). CrN single-crystal growth using Cr-Ga-Na ternary melt, *J. Crystal Growth* 246: 133–138.

Aoki, M. et al. (2002a). GaN single crystal growth using high-purity Na as a flux, *J. Crystal Growth* 242: 70–76.

Aoki, M. et al. (2000). Growth of GaN single crystals from a Na-Ga melt at $750°$ C and 5 MPa of N_2, *J. Crystal Growth* 218: 7–12.

Ashraf, H. et al. (2008). Thick GaN layers grown by HVPE: Influence of the templates, *J. Crystal Growth* 310: 3957–3963.

Avdeev, O. V. et al. (2010). Manufacturing of bulk AlN substrates, *in* P. Capper & P. Rudolph (eds), *Crystal Growth Technology: Semiconductors and Dielectrics*, Wiley-VCH Verlag GmbH & Co. KGaA, ISBN: 978-3-527-32593-1, pp. 121–136.

Avdeev, O. V. et al. (2011).   Growth of bulk AlN crystals, *in* P. K. Bhattacharya et al. (eds), *Comprehensive Semiconductor Science and Technology*, Vol. 3, Elsevier Science Ltd, ISBN-10: 0444531432, ISBN-13: 9780444531438, pp. 282–301.

Avrutin, V. et al. (2010). Growth of bulk GaN and AlN: Progress and challenges, *Proceedings of the IEEE* 98: 1302 – 1315.

Bakhtizin, R. Z. et al. (2004). Scanning tunneling microscopy studies of III-nitride thin film heteroepitaxial growth, *Phys. Usp.* 47: 371–424.

Balkas, C. M. et al. (1997). Sublimation growth and characterization of bulk aluminum nitride single crystals, *J. Crystal. Growth* 179: 363–370.

Bao, H. Q. et al. (2009). The sublimation growth of AlN fibers: transformations in morphology & fiber direction, *Appl. Phys. A* 94: 173–177.

Beaumont, B. et al. (1998). Mg-enhanced lateral overgrowth of GaN on patterned GaN/sapphire substrate by selective metal organic vapor phase epitaxy, *MRS Internet J. Nitride Semicond. Res.* 3: 20.

Belousov, A. (2010). High pressure crystal growth, thermodynamics and physical properties of $Al_xGa_{1-x}N$ semiconductors, Dr. Sci. thesis, Eidgenössische Technische Hochschule, Zürich.

Belousov, A. et al. (2010). $Al_xGa_{1-x}N$ bulk crystal growth: Crystallographic properties and p-T phase diagram, *J. Crystal Growth* 312: 2585–2592.

Belousov, A. et al. (2009). Bulk single-crystal growth of ternary $Al_xGa_{1-x}N$ from solution in gallium under high pressure, *J. Crystal Growth* 311: 3971–3974.

Bennett, H. S. et al. (2004). Radio-frequency and analog/mixed-signal circuits and devices for wireless communications, *IEEE Circuits & Devices Magazine* pp. 38–51.

Bickermann, M. et al. (2010). UV transparent single-crystalline bulk AlN substrates, *phys. stat. sol. (c)* 7: 21–24.

Biryukov, S. V. et al. (2007). Ring waveguide resonator on surface acoustic waves, *Appl. Phys. Lett.* 90: 173503.

Biryukov, S. V. et al. (2009). Ring waveguide resonator on surface acoustic waves: First experiments, *J. Appl. Phys.* 106: 126103.

Bockowski, M. (2001). Growth and doping of GaN and AlN single crystals under high nitrogen pressure, *Cryst. Res. Technol.* 36: 771–787.

Bockowski, M. et al. (2001). Crystal growth of aluminum nitride under high pressure of nitrogen, *Mater. Sci. Semicond. Processing* 4: 543–548.

Bogdanov, M. V. et al. (2008). Coupled modeling of current spreading, thermal effects and light extraction in III-nitride light-emitting diodes, *Semicond. Sci. Technol* 23: 125023.

Bogdanov, M. V. et al. (2003). Advances in modeling of wide-bandgap bulk crystal growth, *Cryst. Res. Techn.* 38: 237–249.

Bogdanov, M. V. et al. (2001). Virtual reactor: a new tool for SiC bulk crystal growth study and optimization, *Mat. Sci. Forum* 353-356: 57–61.

Bogdanov, M. V. et al. (2004). Industrial challenges for numerical simulation of crystal growth, *Centr. Europ. J. Phys.* 2: 183.

Bondokov, R. T. et al. (2006). Defect content evaluation in single-crystal AlN wafers, *Mater. Res. Soc. Symp. Proc.* 892: FF30.

Bondokov, R. T. et al. (2007). Fabrication and characterization of 2-inch diameter AlN single-crystal wafers cut from bulk crystals, *Mater. Res. Soc. Symp. Proc.* 955: I03.

Bondokov, R. T. et al. (2008). Large-area AlN substrates for electronic applications: An industrial perspective, *J. Crystal Growth* 310: 4020–4026.

Brinkman, A. W. & Carles, J. (1998). The growth of crystals from the vapour, *Progr. Crystal Growth Character. Mater.* pp. 169–209.

Bulashevich, K. A. et al. (2007). Current spreading and thermal effects in blue LED dice, *Phys. Stat Solidi (c)* 4: 45–48.

Cai, D. et al. (2006). Thermal environment evolution and its impact on vapor deposition in large diameter AlN bulk growth. AIAA 2006-3825.

Cai, D. et al. (2010). Vapor growth of III nitrides, In: Springer Handbook of Crystal Growth, ISBN 978-3-540-74182-4, Eds. Dhanaraj et al., part F, 1243–1280.

Callahan, M. J. & Chen, Q. S. (2010). Hydrothermal and ammonothermal growth of ZnO and GaN, Ibid., part C, 655–689.

Cao, X. A. et al. (2004). Growth and characterization of blue and near-ultraviolet light-emitting diodes on bulk GaN, *Proc. SPIE* 5530: 48–53.

Cao, X. A. et al. (2007). Homoepitaxial growth and electrical characterization of GaN-based Schottky and light-emitting diodes, *J. Crystal Growth* 300: 382–386.

Cao, X. A. et al. (2004a). Electrical characteristics of InGaN/GaN light-emitting diodes grown on GaN and sapphire substrates, *Appl. Phys. Lett.* 85: 7–9.

Carter, C. B. & Norton, M. G. (2007). *Ceramic Materials, ISBN 978-0-387-46271-4*, Springer New York.

Cartwright, A. N. et al. (2006). Ultrafast carrier dynamics and recombination in green emitting InGaN MQW LED, *Mater. Res. Soc. Symp. Proc.* 916: DD04.

Chakraborty, A. et al. (2005). Demonstration of nonpolar m-plane InGaN/GaN light-emitting diodes on free-standing m-plane GaN substrates, *Jap. J. Appl. Phys.* 44: L173–L175.

Chemekova, T. Y. et al. (2008). Sublimation growth of 2 inch diameter bulk AlN crystals, *phys. stat. sol. (c)* 5: 1612–1614.

Chen, J. R. et al. (2007). Theoretical analysis of current crowding effect in Metal/AlGaN/GaN Schottky diodes and iIts reduction by using polysilicon in anode, *Chin. Phys. Lett.* 24: 2112–2114.

Chen, X. F. et al. (2008). Surface preparation of AlN substrates, *Cryst. Res. Technol.* 43: 651–655.

Cherns, D. (2000). The structure and optoelectronic properties of dislocations in GaN, *J. Phys.: Condens. Matter* 12: 10205–10212.

Cho, H. K. et al. (2000). Phase separation and stacking faults of $In_xGa_{1-x}N$ layers grown on thick GaN and sapphire substrate by metalorganic chemical vapor deposition, *J. Crystal Growth* 220: 197–203.

Choi, Y. S. et al. (2004). Effect of dislocations on the luminescence of GaN/InGaN multi-quantum-well light-emitting-diode layers, *Mater. Lett.* 58: 2614–2617.

Christen, J. et al. (2003). Optical micro-characterisation of group-III-nitrides: correleation of structural, electronic and optical properties, *phys. stat. sol. (c)* 0: 1795–1815.

Cleland, A. N. et al. (2001). Single-crystal aluminum nitride nanomechanical resonators, *Appl. Phys. Lett.* 79: 2070–2072.

Craford, M. G. (2005). LEDs for solid state lighting and other emerging applications: status, trends and challenges, *Proc. of SPIE* 5941: 594101.

Dalmau, R. F. (2005). Aluminum nitride biulk crystal growth in a resistively heated reactor, Ph. D. Thesis, North Carolina State University.

Dalmau, R. et al. (2005). AlN bulk crystals grown on SiC seeds, *J. Crystal Growth* 281: 68–74.

Dalmau, R. & Sitar, Z. (2005). Sublimation growth of AlN crystals, In: Encyclopedia of Materials: Science and Technology, Eds. K. H. J. Buschow et. al., pp. 1-9, Elsevier.

Dalmau, R. & Sitar, Z. (2010). AlN bulk crystal growth by physical vapor transport, In: Springer Handbook of Crystal Growth, ISBN 978-3-540-74182-4, Eds. Dhanaraj, G. et al., part D, 821–843.

Daryabeigi, K. (1999). Analysis and testing of high temperature fibrous insulation for reusable launch vehicles, AIAA 99-1044.

Davis, R. F. et al. (2001). Pendeo-epitaxial growth of thin films of gallium nitride and related materials and their characterization, *J. Crystal Growth* 225: 134–140.

Davis, R. F. et al. (2001a). Conventional and pendeo-epitaxial growth of GaN (0001) thin films on Si (111) substrates, *J. Crystal Growth* 231: 335–341.

Davis, R. F. et al. (2002). Gallium nitride materials - progress, status, and potential roadblocks, *Proc. IEEE* 90: 993–1005.

de Almeida, V. F. & Rojo, J. C. (2002). Simulation of transport phenomena in aluminum nitride single-crystal growth, Oak Ridge Nat. Lab., Tech. Rep. ORNL/TM-2002/64, 1–32.

Denis, A. et al. (2006). Gallium nitride bulk crystal growth processes; a review, *Mater. Sci. Eng.* R50: 167–194.

Detchprohm, T. et al. (2008). Green light emitting diodes on a-plane GaN bulk substrates, *Appl. Phys. Lett.* 92: 241109.

Detchprohm, T. et al. (2010). Wavelength-stable cyan and green light emitting diodes on nonpolar m-plane GaN bulk substrates, *Appl. Phys. Lett.* 96: 051101.

Dhanaraj, G. et al. (2003). Silicon carbide crystals - part I: Growth and characterization, *in* K. Byrappa & T. Ohashi (eds), *Crystal Growth Technology*, William Andrew Publishing, pp. 181–232.

Dinwiddie, R. B. et al. (1989). Thermal conductivity, heat capacity, and thermal diffusivity of selected commercial AlN substrates, *Int. J. Thermophys.* 10: 1075.

Dmitriev, V. & Usikov, A. (2006). Hydride vapor phase epitaxy of group III nitride materials, *in* Z. C. Feng (ed.), *III-nitride semiconductor materials*, ISBN-10: 1860946364, ISBN-13: 978-1860946363, World Scientific Publishing Company, pp. 1–40.

Dreger, L. H. et al. (1962). Sublimation and decomposition studies on boron nitride and aluminum nitride, *J. Phys. Chem.* 66: 1556–1559.

Dryburgh, P. M. (1992). The estimation of maximum growth rate for aluminium nitride crystals grown by direct sublimation, *J. Crystal Growth* 125: 65–68.

Du, L. & Edgar, J. H. (2010). Thermodynamic analysis and purification for the source materials in sublimation crystal growth of aluminum nitride, *Mater. Res. Soc. Symp.* 1202: I05–08.

Du, L. et al. (2010). Sublimation growth of titanium nitride crystals, *J. Mater. Sci.: Mater. Electron.* 21: 78–87.

Du, X. Z. et al. (2010). UV light-emitting diodes at 340 nm fabricated on a bulk GaN substrate, *Chin. Phys. Lett.* 27: 088105.

Dupret, F. et al. (1990). Global modeling of heat transfer in crystal growth furnaces, *Int. J. Heat Mass Transfer* 33: 1849–1871.

Edgar, J. H. et al. (2008). Native oxides and hydroxides and their implications for bulk AlN crystal growth, *J. Crystal Growth* 310: 4002–4006.

Edgar, J. H. et al. (eds) (1999). *Properties, processing and applications of GaN and related semiconductors*, INSPEC, the IEE, London.

Egorov, Y. & Zhmakin, A. (1998). Numerical simulation of low-Mach number gas mixture flows with heat and mass transfer using unstructured grid, *Comput. Mat. Sci.* 11: 204–220.

Ehrentraut, D. & Fukuda, T. (2008). Bulk zinc oxide and gallium nitride crystals by solvothermal techniques, *in* Y. Fujikawa et al. (eds), *Frontiers in Materials Research*, Springer, pp. 111–120.

Ehrentraut, D. & Fukuda, T. (2010). The ammonothermal crystal growth of gallium nitride - a technique on the up rise, *Proceedings of the IEEE* 98: 1316 – 1323.

Ehrentraut, D. & Kagamitanii, Y. (2010). Acidic ammonothermal growth technology for GaN, *in* Hull, R. et al. (eds), *Technology of Gallium Nitride Crystal Growth*, , ISBN 978-3-642-04830-2, Vol. 133 of *Springer Series in Materials Science*, Springer Berlin Heidelberg, pp. 183–203.

Ehrentraut, D. et al. (2006). Solvothermal growth of ZnO, *Prog. Cryst. Growth Charact. Mater.* 52: 280–335.

Eisenberg, H. R. & Kandel, D. (2002). Origin and properties of the wetting layer and early evolution of epitaxially strained thin films, *Phys. Rev. B* 66: 155429.

Epelbaum, B. et al. (2002). Sublimation growth of bulk AlN crystals: materials compatibility and crystal quality, *Mater. Sci. Forum* 433–436: 983–986.

Epelbaum, B. et al. (2001). Seeded PVT Growth of Aluminum Nitride on Silicon Carbide, *Mat. Sci. Form* 389-393: 1445–1448.

Epelbaum, B. M. et al. (2004). Growth of bulk AlN crystals for SAW devices, Proc. Second Int. Symp.on Acoustic Wave Devices for Future Mobile Comm. Systems, Chiba, March 2004, 157–162.

Ern, A. & Guermond, J. L. (2004). Accurate numerical simulation of radiative heat transfer with application to crystal growth, *Int. J. Numer. Meth. Engineer.* 61: 559–583.

Evstratov, I. Y. et al. (2006). Current crowding effects on blue LED operation, *Phys. stat. sol.(c)* 3: 1645 – 1648.

Fan, Z. Y. et al. (2006). AlGaN/GaN/AlN quantum-well field-effect transistors with highly resistive AlN epilayers, *Appl. Phys. Lett.* 88: 073513.

Fan, Z. Y. & Newman, N. (2001). Experimental determination of the rates of decomposition and cation desorption from AlN surfaces, *Mat. Sci. Eng. B* 87: 244–248.

Fara, A. et al. (1999). Theoretical evidence for the semi-insulating character of AlN, *Appl. Phys. Lett.* 85: 2001–2003.

Fini, P. et al. (1998). The effect of growth environment on the morphological and extended defect evolution in GaN grown by metalorganic chemical vapor deposition, *Jap. J. Appl. Phys.* 37: 4460–4466.

Frank, F. C. & van der Merwe, J. (1949). One dimensional dislocations. II. Misfit monolayers and oriented overgrowth, *Proc. Roy. Soc.(London) A* 198: 216–225.

Franko, A. & Shanafield, D. J. (2004). The thermal conductivity of polycrystalline aluminum nitride (AlN) ceramics, *Ceramica* 50: 247–253.

Frayssinet, E. et al. (2001). Evidence of free carrier concentration gradient along the c-axis for undoped GaN single crystals, *J. Crystal Growth* 230: 442–447.

Freitas, J. A. (2005). Bulk and homoepitaxial films of III-V nitride semiconductors: Optical studies, *J. Ceramic Process. Res.* 6: 209–217.

Freitas, J. A. (2010). Properties of the state of the art of bulk III-V nitride substrates and homoepitaxial layers, *J. Phys. D: Appl. Phys.* 43: 073301.

Freund, L. (1995). Evolution of waviness of the surface of a strained elastic solid due to stress-driven diffusion, *Int. J. Solid Structures* 32: 911–923.

Fritze, H. (2011). High-temperature piezoelectric crystals and devices, *J. Electroceram.* 26: 122–161.

Fukuda, T. & Ehrebtraut, D. (2007). Prospects for the ammonothermal growth of large GaN crystals, *J. Crystal Growth* 305: 304–310.

Fukuyama, H. et al. (2006). Single crystalline aluminum nitride films fabricated by nitriding α-SiO$_2$, *Appl. Phys. Lett.* 100: 024905.

Gaska, R. et al. (2002). Deep-ultraviolet emission of AlGaN/AlN quantum wells on bulk AlN, *Appl. Phys. Lett.* 81: 4658–4660.

Gautier, S. et al. (2008). AlGaN/AlN multiple quantum wells grown by MOVPE on AlN templates using nitrogen as a carrier gas, *J. Crystal Growth* 310: 4927–4931.

Gilmer, G. H. et al. (1998). Thin film deposition: fundamentals and modeling, *Comput. Mater. Sci.* 12: 354–380.

Golan, Y. et al. (1998). Morphology and microstructural evolution in the early stages of hydride vapor phase epitaxy of GaN on sapphire, *Appl. Phys. Lett.* 73: 3090–3092.

Grandusky, J. J. et al. (2010). Properties of mid-ultraviolet light emitting diodes fabricated from pseudomorphic layers on bulk aluminum nitride substrates, *Appl. Phys. Express* 3: 072103.

Grandusky, J. J. et al. (2009). Pseudomorphic growth of thick n-type $Al_xGa_{1x}N$ layers on low-defect-density bulk AlN substrates for UV LED applications, *J. Crystal Growth* 311: 2864–2866.

Gromov, A. A. et al. (2005). Nitride formation during combustion of ultrafine aluminum powder in air. I. Effect of additives, *Combustion, Explosion and Shock Waves* 41: 303–314.

Grzegorczyk, A. P. et al. (2005). Influence of sapphire annealing in trimethylgallium atmosphere on GaN epitaxy by MOCVD, *J. Crystal Growth* 283: 72–80.

Grzegory, I. (2001). High pressure growth of bulk GaN from solutions in gallium, *J. Phys.: Condens Matter* 13: 6875–6892.

Grzegory, I. et al. (2002). Mechanisms of crystallization of bulk GaN from the solution under high N_2 pressure, *J. Crystal Growth* 246: 177–186.

Han, Q. et al., Q. H. (2008). Polarity analysis of self-seeded aluminum nitride crystals grown by sublimation, *J. Electron. Mater.* 37: 1058–1063.

Hartmann, C. et al. (2008). Homoepitaxial seeding and growth of bulk AlN by sublimation, *J. Crystal Growth* 310: 930–934.

Hashimoto, S. et al. (2010). Epitaxial layers of AlGaN channel HEMTs on AlN substrates, *SEI Techn. Rev.* 71: 83–87.

Helava, H. et al. (2007). Growth of bulk aluminum nitride crystals, *phys. stat. sol. (c)* 4: 2281–2284.

Hemmingsson, C. et al. (2010). Growth of III-nitrides with halide vapor phase epitaxy (HVPE), In: Springer Handbook of Crystal Growth, ISBN 978-3-540-74182-4, Eds. Dhanaraj, G. et al., part D, 869–896.

Herro, Z. D. et al. (2006). Growth of large AlN single crystals along the [0001] direction, *Mater. Res. Soc. Symp. Proc.* 892: FF21.

Heying, B. et al. (1999). Dislocation mediated surface morphology of GaN, *J. Appl. Phys.* 85: 6470–6476.

Hirano, T. et al. (2009). Synthesis of GaN crystal by the reaction of Ga with Li_3N in NH_3 atmosphere, *J. Crystal growth* 311: 3040-3043.

Hovakimian, L. B. (2009). Strong-coupling theory of mobility collapse in GaN layers, *Appl. Phys. A* 96: 255–257.

Hu, X. et al. (2003). AlGaN/GaN heterostructure field-effect transistors on single-crystal bulk AlN, *Appl. Phys. Lett.* 82: 1299–1301.

Hull, R. et al. (2002). Interaction between surface morphology and misfit dislocations as strain relaxation modes in lattice-mismatched heteroepitaxy, *J. Phys.: Condens. Matter* 14: 12829–12841.

Hull, R. & Stach, E. A. (1996). Equilibrium and metastable strained layer semiconductor structures, *Current Opinion in Solid State & Mater. Sci.* 1: 21–28.

Hung, A. et al. (2004). An *ab initio* study of structural properties and single vacancy defects in wurtzite AlN, *J. Chem. Phys.* 120: 4890–4896.

I-hsiu, H. & Stringfellow, G. B. (1996). Solid phase immiscibility in GaInN, *Appl. Phys. Lett.* 69: 2701–2703.

Ilyin, I. V. et al. (2010). Deep-level defects in AlN single crystals: EPR studies, *Mater. Sci. Forum* 645–648: 1195–1198.

Iwahashi, T. et al. (2003). Effects of ammonia gas on threshold pressure and seed growth for bulk GaN single crystals by Na flux method, *J. Crystal Growth* 253: 1–5.

Iwasaki, F. & Iwasaki, H. (2002). Historical review of quartz crystal growth, *J. Crystal Growth* 237-239: 820.

Jain, S. C. et al. (1997). Misfit strain and misfit dislocations in lattice mismatched epitaxial layers and other systems, *Phil. Mag. A* 75: 1461–1515.

Jain, S. C. & Hayes, W. (1991). Structure, properties and applications of Ge_xSi_{1x} strained layers and superlattices, *Semicond. Sci. Technol.* 6: 547–576.

Jain, S. C. et al. (2000). III-nitrides: growth, characterization, and properties, *J. Appl. Phys.* 87: 965–1006.

Kallinger, B. et al. (2008). Vapor growth of GaN using GaN powder sources and thermographic investigations of the evepoarting behaviour of the source material, *Cryst. Res. Techn.* 43: 14–21.

Karpov, D. et al. (2001a). Mass transport and powder source evolution in sublimation growth of SiC bulk crystals, *Mat. Sci. Forum.* 353-356: 37–40.

Karpov, S. Y. (1998). Suppression of phase separation in InGaN due to elastic strain, *MRS Internet J. Nitride Semicond. Res.* 3: 16.

Karpov, S. Y. (2009). Bandgap engineering of III-nitride devices on low-defect substrates, in Z. C. Feng (ed.), *III-Nitride Devices and Nanoengineering*, Imperial College Press, London, pp. 367–398.

Karpov, S. Y. et al. (2001). Effect of reactive ambient on AlN sublimation growth, *phys. stat. sol. (a)* 188: 763–767.

Karpov, S. Y. et al. (2003). Role of oxygen in AlN sublimation growth, *phys. stat. sol. (c)* 0: 1989–1992.

Karpov, S. Y. et al. (1999). Sublimation growth of AlN in vacuum and in a gas atmosphere, *phys. stat. sol. (a)* 176: 435–438.

Kasap, S. & Capper, P. (eds) (2006). *Springer Handbook of Electronic and Photonic Materials, ISBN 978-0-387-29185-7*, Springer.

Kawahara, M. et al. (2005). A systematic study on the growth of GaN single crystals using Na-based flux method, *J. Ceramic Process. Res.* 6: 146–152.

Kazan, M. et al. (2006). What causes rough surface in AlN crystals?, *J. Crystal Growth* 290: 44–49.

Kazan, M. et al. (2006a). Phonon dynamics in AlN lattice contaminated by oxygen, *Diamond Rel. Mater.* 15: 1525–1534.

Ketchum, D. R. & Kolis, J. W. (2001). Crystal growth of gallium nitride in supercritical ammonia, *J. Crystal Growth* 222: 431–434.

Kirchner, C. et al. (1999). Homoepitaxial growth of GaN by metal-organic vapor phase epitaxy: A benchmark for GaN technology, *Appl. Phys. Lett.* 75: 1098–1100.

Kitanin, E. et al. (1998). Heat transfer through source powder in sublimation growth of SiC crystal, *Mater. Sci. Eng.* B55: 174–183.

Klapper, H. (2010). Generation and propagation of defects during crystal growth, in G. Dhanaraj et al.(eds), *Springer Handbook of Crystal Growth, ISBN 978-3-540-74182-4*: 93–116.

Kochuguev, S. K. et al. (2001). Solution of crystal growth problems using adaptive unstrucured grids, in M. Baines (ed.), *Numerical Methods for Fluid Dynamics VII*, ISBN-10: 095249292X, ISBN-13: 978-0952492924, Oxford University Press, pp. 363–369.

Korcak, S. et al. (2007). Structural and optical properties of an $In_xGa_{1-x}N$/GaN nanostructure, *Surf. Sci.* 601: 3892–3897.

Krukowski, S. (1997). Thermodynamics and high-pressure growth of (Al, Ga, In)N single crystals, *Diamond and Related Materials* 6: 1515–1523.

Kukushkin, S. A. et al. (2008). Substrates for epitaxy of gallium nitride; new materials and techniques, *Rev. Adv. Mater. Sci* 17: 1–32.

Kumagai, Y. et al. (2008). Self-separation of a thick AlN layer from a sapphire substrate via interfacial voids formed by the decomposition of sapphire, *Appl. Phys. Express* 1: 045003.

Kuwano, N. et al. (2004). Formation and annihilation of threading dislocations associated with stress in hetero-structure of GaN and AlGaN, *IPAP Conf. Series* 4: 21–24.

Lee, H. et al. (2004). Preparation of freestanding GaN and GaN template by hybride vapor phase epitaxy, *IPAP Conf. Series* 4: 25–27.

Lee, R. G. (2007). Transmission electron microscopy and thermal residual stress analysis of AlN crystal. MS thesis, Texas Tech UniverSity.

Lim, Y. S. et al. (2000). Strain-induced diffusion in a strained $Si_{1-x}Ge_x$/Si heterostructure, *Appl. Phys. Lett.* 77: 4157–4159.

Limb, J. B. (2007). Design, fabrication and characterization of III-nitride pn junction devices, PhD thesis, Georgia Institute of Technology.

Lin, G. Q. et al. (2008). Influence of AlN buffer thickness on GaN grown on Si(111) by gas source molecular beam epitaxy with ammonia, *Chin. Phys. Lett.* 25: 4097–4100.

Liu, W. & Balandin, A. A. (2005). Thermal conduction in $Al_xGa_{1-x}N$ alloys and thin films, *J. Appl. Phys.* 97: 073710.

Liu, Y. S. (2009). An overview of the development of major light sources; from light bulbs to solid state lighting, *in* Z. C. Feng (ed.), *III-Nitride Devices and Nanoengineering*, Imperial College Press, London, pp. 1–20.

Louarn, A. L. et al., Vézian, S., Semond, F. & Massies, J. (2009). AlN buffer layer growth for GaN epitaxy on (111) Si: Al or N first?, *J. Crystal Growth* 311: 3278–3284.

Lu, P. (2006). Sublimation growth of AlN bulk crystals and high-speed CVD growth of SiC epilayers, and their characterization, Ph. D. thesis, Kansas State Univ.

Lu, P. et al. (2008). Seeded growth of AlN on SiC substrates and defect characterization, *J. Crystal Growth* 310: 2464–2470.

Luft, J. et al. (1999). Microbatch macromolecular crystallization on a thermal gradient, *J. Crystal Growth* 196: 447–449.

Luo, B. et al. (2002). High breakdown M-I-M structures on bulk AlN, *Solid-State Electron.* 46: 573–576.

Majewski, J. A. & Vogl, P. (1998). Polarization and band offsets of stacking faults in AlN and GaN, *MRS Internet J. Nitride Semicond. Res.* 3: 21.

Makarov, Y. & Zhmakin, A. (1989). On flow regimes in VPE reactors, *J. Crystal Growth* 94: 537–551.

Matsuoka, T. (2005). Progress in nitride semiconductors from GaN to InN - MOVPE growth and characteristics, *Superlatt. and Micristruct.* 37: 19–32.

Matthews, J. W. & Blakeslee, A. E. (1974). Defects in epitaxial multilayers.I. Misfit dislocations, *J. Crystal Growth* 27: 118–125.

McAleese, C. et al. (2006). Electric fields in AlGaN/GaN quantum well structures, *phys. stat. sol. (b)* 243: 1551–1559.

Medraj, M. et al. (2005). Understanding of AlN sintering through computational thermodynamics combined with experimental investigation, *J. Mat. Process. Technol.* 161: 415–422.

Meissner, E. et al. (2004). Growth of GaN crystals and epilayers from solutions at ambient pressure, *IPAP Conf. Series* 4: 46–49.

Miskys, C. R. et al. (2003). Freestanding GaN-substrates and devices, *phys. stat. sol. (c)* 0: 1627–1650.

Miyajima, T. et al. (2001). GaN-based blue laser diodes, *J. Phys.: Condens. Matter* 13: 7099–7114.

Miyanaga, M. et al. (2006). Single crystal growth of AlN by sublimation method, *SEI Techn. Rev.* 63: 22–25.

Miyoshi, T. et al. (2009). 510–515 nm InGaN-based green laser diodes on *c*-plane GaN substrates, *Appl. Phys. Express.* 2: 062201.

Mokhov, E. N. et al. (2005). Sublimation growth of AlN bulk crystals in Ta crucibles, *J. Crystal Growth* 281: 93–100.

Mokhov, E. N. et al. (2002). Growth of AlN bulk crystals by sublimation sandwich method, *Mat. Sci. Forum* 433-436: 979–982.

Moran, B. et al. (2000). Growth and characterization of graded AlGaN conducting buffer layers on n^+ SiC substrates, *J. Crystal Growth* pp. 301–304.

Morkoc, H. (1998). *Wide Band Gap Nitrides and Devices*, Springer, Berlin.

Mukai, T. et al. (2001). Nitride light-emitting diodes, *J. Phys.: Condens. Matter* 13: 7089–7098.

Müller, E. et al. (2006). Electrical activity of dislocations in epitaxial ZnO- and GaN-layers analyzed by holography in a transmission electron microscope, *Mat. Sci. in Semicond. Processing* 9: 127–131.

Mymrin, V. F. et al. (2005). Bandgap engineering of electronic and optoelectronic devices on native AlN and GaN substrates: a modelling insight, *J. Crystal Growth* 281: 115–124.

Nakamura, S. et al. (2000). *The Blue Laser Diode (2nd ed.)*, ISBN-10: 3642085792, ISBN-13: 978-3642085796, Springer, N. Y.

Nam, O. H. et aL. (1997). Lateral epitaxy of low defect density GaN layers via organometallic vapor phase epitaxy, *Appl. Phys. Lett.* 71: 2638–2640.

Nam, O. H. et al. (1998). Lateral epitaxial overgrowth of GaN films on SiO_2 areas via metal-organic vapor phase epitaxy, *J. Electron. Mater.* 27: 233–237.

Nanjo, T. et al. (2008). Remarkable breakdown voltage enhancement in AlGaN channel high electron mobility transistors, *Appl. Phys. Lett.* 92: 263502.

Nause, J. & Nemeth, B. (2005). Pressurized melt growth of ZnO boules, *Semicond. Sci. Technol.* 20: S45–S48.

Nikolaev, A. et al. (2000). AlN wafers fabricated by hydride vapor phase epitaxy, *MRS Internet J. Nitride Semicond. Res.* 5S1: W6.5.

Nishida, T. et al. (2004a). AlGaN-based ultraviolet light-emitting diodes grown on bulk AlN substrates, *Appl. Phys. Lett.* 84: 1002–1003.

Nishida, T. et al. (2004b). High current injection to a UV-LED grown on bulk AlN substrates, *MRS Proc.* 798: Y1.3.

Nishino, K. et al. (2002). Bulk GaN by direct synthesis method, *J. Crystal Growth* 237-239: 922–925.

Nord, J. et al. (2003). Modelling of compound semiconductors: analytical bond-order potential for gallium, nitrogen and gallium nitride, *J. Phys.: Condens Matter* 15: 5649.

Northrup, J. E. (2005). Shallow electronic states induced by prismatic stacking faults in AlN and GaN, *Appl. Phys. Lett.* 86: 071901.

Noveski, V. et al. (2004a). Growth of AlN crystals on AlN/SiC seeds by AlN powder sublimation in nitrogen atmosphere, *MRS Internet J. Nitride Semicond. Res.* 9: 2.

Noveski, V. et al. (2004b). Mass transfer in AlN crystal growth at high temperatures, *J. Crystal Growth* 266: 369–378.

Oden, J. T. (2002). The promise of Computational Engineering and Science: will it be kept?, *IACM Express* (12): 12–15.

Oh, T. S. et al. (2009). Epitaxial growth of improved GaN epilayer on sapphire substrate with platinum nanocluster, *J. Crystal Growth* 311: 2655–2658.

Ohtani, N. et al. (2007). Crystal growth, *in* K. Nakahashi et al., *Wide Bandgap Semiconductors. Fundamental Properties and Modern Photonic and Electronic Devices.* ISBN 10-3-540-47234-7, Springer Berlin Heidelberg New York, pp. 329–446.

Okamoto, K. et al. (2009). Epitaxial growth of GaN on single-crystal Mo substrates using HfN buffer layers, *J. Crystal Growth* 311: 1311–1315.

Onda, M. et al. & Sasaki, T. (2002). Influence of pressure control on the growth of bulk GaN single crystal using a Na flux, *J. Crystal Growth* 237-239: 2112–2115.

Orton, J. W. & Foxon, C. T. (1998). Group III nitride semiconductors for short wavelength light-emitting devices, *Rep. Prog. Phys.* 61: 1–75.

Ougazzaden, A. et al. (2008). Growth of GaN by metalorganic vapor phase epitaxy on ZnO-buffered c-sapphire substrates, *J. Crystal Growth* 310: 944–947.

Paskova, T. et al. (2010). GaN substrates for III-nitride devices, *Proc. IEEE* 98: 1324–1338.

Paskova, T. et al. (2009). Polar and non-polar HVPE GaN substrates: Impact of doping on the structural, electrical and optical characteristics, *phys. stat. sol.* pp. S344–S347.

Paskova, T. et al. (2004). Growth, separation and properties of HVPE grown GaN films using different nucleation schemes, *IPAP Conf. Series* 4: 14–20.

Pearton, S. J. et al. (2000). Fabrication and performance of GaN electronic devices, *Mater. Sci. Eng.* 30: 55–212.

Pecz, B. et al. (2008). Composite substrates for GaN growth, *in* A. G. Cullis & P. A. Midgley (eds), *Microscopy of Semiconducting Materials*, ISBN 978-1-4020-8614-4, 2007, Springer Netherlands, pp. 53–56.

Perlin, P. et al. (2005). Properties of InGaN blue laser diodes grown on bulk GaN substrates, *J. Crystal Growth* 281: 107–114.

Pinnington, T. et al. (2008). InGaN/GaN multi-quantum well and LED growth on wafer-bonded sapphire-on-polycrystalline AlN substrates by metalloorganic chemical vapor deposition, *J. Crystal Growth* 310: 2514–2519.

Porowski, S. & Grzegory, I. (1997). Thermodynamical properties of III-V nitrides and crystal growth of GaN at high N_2 pressure, *J. Crystal Growth* 178: 174–188.

Potin, V. et al. (1998). Extended defects in nitride semiconductors, *J. Electron. Mater.* 27: 266–275.

Purdy, A. P. et al. (2003). Synthesis of GaN by high-pressure ammonolysis of gallium triiodide, *J. Crystal Growth* 252: 136–143.

Radwan, M. & Miyamoto, Y. (2006). Self-propagating high-temperature synthesis of AlN nanostructures and their sintering properties, *Trans. of JWRI* 35: 43–46.

Raghothamachar, B. et al. (2006). Characterization of bulk grown GaN and AlN single crystal materials, *J. Crystal Growth* 287: 349–353.

Raghothamachar, B. et al. (2003). X-ray characterization of bulk AlN single crystals grown by the sublimation technique, *J. Crystal Growth* 250: 244–250.

Ram-Mohana, L. R. et al. (2006). Wavefunction engineering of layered wurtzite semiconductors grown along arbitrary crystallographic directions, *Superlattices and Microstructures* 39: 455–477.

Ren, Z. et al. (2007). Heteroepitaxy of AlGaN on bulk AlN substrates for deep ultraviolet light emitting diodes, *Appl. Phys. Lett.* 91: 051116.

Ren, Z. et al. (2007a). AlGaN deep ultraviolet LEDs on bulk AlN substrates, *phys. stat. sol. (c)* 4: 2482–2485.

Rojas, T. C. et al. & García, R. (1998). Relaxation mechanism of InGaAs single and graded layers grown on (111)B GaAs, *Thin Solid Films* 317: 270–273.

Rojo, C. J. et al. (2006). Physical vapor transport crystal growth of ZnO, *Proc. of SPIE* 6122: 61220Q1.

Rojo, J. C. et al. (2001). Single-crystal aluminum nitride substrate preparation from bulk crystals, *Mat. Res. Soc. Symp. Proc.* 680: E2.1.

Rojo, J. C. et al. (2002). Progress in the preparation of aluminum nitride substrates from bulk crystals *Mat. Res. Soc. Symp. Proc.* 722: K1.1.1.

Roycroft, B. et al. (2004). Origin of power fluctuations in GaN resonant-cavity light-emitting diodes, *Opt. Express* 12: 736–741.

Russel, P. (2006). SiC, sapphire and GaN materials status in Opto and RF business, Proc. CS MANTECH Conf., Vancouver, 231-232.

Satoh, I. et al. (2010). Development of aluminum nitride single-crystal substrates, *SEI Techn. Rev.* (71): 78–82.

Schmidt, M. C. et al. (2007). Demonstration of *m*-plane InGaN/GaN laser diodes, *Jap. J. Appl. Phys.* 46: L190–L191.

Schmidt, M. C. et al. (2007a). High power and high external efficiency *m*-plane InGaN light emitting diodes, *Jap. J. Appl. Phys.* 46: L126–L128.

Schowalter, L. J. et al. (2006). Development of native, single crystal AlN substrates for device applications, *phys. stat. sol. (a)* 203: 1667âĂŞ1671.

Schowalter, L. J. et al. (2004). Fabrication of native, single-crystal AlN substrates, *IPAP Conf. Series* 4: 38–40.

Schubert, E. F. & Kim, J. K. (2005). Solid-state light sources getting smart, *Science* 308: 1274–1278.

Schubert, E. F. (2006). *Light-Emitting Diodes (2nd ed.)*, Cambridge University Press.

Schujman, B. S. & Schowalter, L. J. (2010). GaN-ready aluminum nitride substrates for cost-effective, very low dislocation density III-nitride LEDs, Crystal IS, Inc., Final Scientific/Technical Report DE-FC26-08-NT01578.

Schultz, T. (2010). Defect analysis of aluminum nitride, Dr. rer. nat. thesis, Technishen Universität, Berlin.

Sedhain, A. et al. (2009). The origin of 2.78 eV emission and yellow coloration in bulk AlN substrates, *Appl. Phys. Lett.* 95: 262104.

Segal, A. S. et al. (2009). Modeling analysis of AlN and AlGaN HVPE, *phys. stat. sol. (c)* 6: S329–332.

Segal, A. S. et al. (2000). On mechanisms of sublimation growth of AlN bulk crystals, *J. Crystal Growth* 211: 68.

Segal, A. S. et al. (2004). Surface chemistry and transport effects in GaN hydride vapor phase epitaxy, *J. Crystal Growth* 270: 384–395.

Segal, A. S. et al. 1999). Transport phenomena in sublimation growth of SiC bulk crystals, *Mater. Sci. Engineer.* B61–62: 40–43.

Senawiratne, J. et al. (2005). Raman, photoluminescence and absorption studies on high quality AlN single crystals, *phys. stat. sol. (c)* 2: 2774–2778.

Seol, D. J. et al. (2003). Computer simulation of spinodal decomposition in constrained films, *Acta Mater.* 51: 5173–5185.

Shealy, J. R. et al. (2002). An AlGaN/GaN high-electron mobility transistor with an AlN sub-bufer layer, *J. Phys.: Condens. Matter* 14: 3499–3509.

Shi, Y. et al. & Kuball, M. (2001). New technique for sublimation growth of AlN single crystals, *MRS Internet J. Nitride semicond. Res.* 6: 7.

Shur, M. S. (1998). GaN based transistors for high power applications, *Solid-State Electron.* 42: 2131–2138.

Sitar, Z. et al. (2004). Growth of AlN crystals by vaporization of Al and sublimation of AlN powder, *IPAP Conf. Series* 4: 41–45.

Skierbiszewski, C. (2005). From high electron mobility GaN/AlGaN heterostructures to blue-violet InGaN laser diodes. perspectives of MBE for nitride optoelectronics, *Acta Physica Polonia* 108: 635–651.

Skromme, B. et al. (2002). Optical characterization of bulk GaN grown by a Na-Ga melt technique, *J. Crystal Growth* 246: 299–306.

Slack, G. A. & McNelly, T. F. (1977). Growth of high purity AlN crystals, *J. Crystal Growth* 34: 263–279.

Slack, G. A. et al. (2002). Some effects of oxygen impurities on AlN and GaN, *J. Crystal Growth* 246: 287–298.

Slack, G. A. et al. (2004). Properties of crucible materials for bulk growth of AlN, *Mat. Res. Soc. Symp. Proc.* 798: Y10.74.

Song, Y. et al. (2004). Preparation and characterization of bulk GaN crystals, *J. Crystal Growth* 260: 327–330.

Song, Y. et al. (2003). Bulk GaN single crystals: growth conditions by flux method, *J. Crystal Growth* 247: 275–278.

Speck, J. S. (2001). The role of threading dislocations in the physical properties of GaN and its alloys, *Mater. Sci. Forum* 353-356: 769–778.

Speck, J. S. & Rosner, S. J. (1999). The role of threading dislocations in the physical properties of GaN and its alloys, *Physica B: Condensed Matter* 273-274: 24–32.

Srikant, V. et al. (1997). Mosaic structure in epitaxial thin films having large lattice mismatch, *J. Appl. Phys.* 82: 4286–4295.

Steckl, A. J. et al. (1997). Growth and characterization of GaN thin films on SiC SOI substrates, *J. Electron. Mater.* 26: 217 – 223.

STR-soft (2000). http://www.str-soft.com/products/Virtual_Reactor/.

Strassburg, M. et al. (2004). The growth and optical properties of large, high-quality AlN single crystals, *J. Appl. Phys.* 96: 5870–5876.

Stutzmann, M. et al. (2001). Playing with polarity, *phys. stat. sol. (b)* 228: 505–512.

Taguchi, T. (2003). Present status of white LED lighting technology in Japan, *J. Light & Vis. Env.* 27: 131–138.

Tamulatis, G. et al. (2003). Photoluminescence of GaN deposited on single-crystal bulk AlN with different polarities, *Appl. Phys. Lett.* 83: 3507–3509.

Tihonov, A. & Arsenin, V. (1977). *Solution of Ill-posed problems*, ISBN-10: 0470991240, ISBN-13: 978-0470991244, Wiley, New York.

Tomita, K. et al. (2004). Self-separation of freestanding GaN from sapphire substrates with stripe-shaped GaN seeds by HVPE, *IPAP Conf. Series* 4: 28–31.

Usikov, A. S. et al. (2003). Indium-free violet LEDs grown by HVPE, *phys. stat. solidi (c)* 0: 2265–2269.

Utsumi, W. et al. (2003). Congruent melting of gallium nitride at 6 GPa and its application to single crystal growth, *Nat. Mater.* 2: 735–738.

Vaudo, R. P. et al. (2003). Characteristics of semi-insulating Fe doped GaN substrates, *phys. stat. sol. (a)* 200: 18–21.

Vodakov, Y. A. et al. (2003). Tantalum crucible fabrication and treatment, US Patent 6,547,87.

Wadley, H. G. et al. (2001). Mechanisms, model and methods of vapor deposition, *Progr. Mater. Sci.* 46: 329–377.

Waldrip, K. E. (2007). Initial Exploration of Growth of InN by Electrochemical Solution Growth, SAND2010-0952, 20 pp.

Waldrip, K. E. (2007). Molten salt-based growth of bulk GaN and InN for substrates, SAND2007-5210, 27 pp.

Waltereit, P. et al. (2002). Heterogeneous growth of GaN on 6H-SiC (0001) by plasma-assisted molecular beam epitaxy, *phys. stat. sol. (a)* 194: 524–527.

Wang, C. et al. (2001). Influence of growth parameters on crack density in thick epitaxial lateral overgrown GaN layers by hydride vapor phase epitaxy, *J. Crysytal Growth* 230: 377–380.

Wang, T. C. et al. (2001a). Dislocation evolution in epitaxial multilayers and graded composition buffers, *Acta. Mater.* 49: 1599–1605.

Wang, X. et al. (2006). Powder sublimation and porosity evolution in sublimation crystal growth, AIAA 2006-3270.

Wellmann, P. et al. (2006). Modeling and experimental verification of SiC M-PVT bulk crystal growth, *Mater. Sci. Forum* 527–529: 75–78.

Weyers, M. et al. (2008). GaN substrates by HVPE, *Proc. of SPIE* 6910: 691001.

Wolfson, A. A. & Mokhov, E. N. (2010). Dependence of the growth rate of an AlN layer on nitrogen pressure in a reactor for sublimation growth of AlN crystals, *Semicond.* 44: 1383–1385.

Wu, B. et al. (2005). Design of an RF-heated bulk AlN growth reactor: Induction heating and heat transfer modeling, *Crystal Growth & Design* 5: 1491–1495.

Wu, B. et al. (2004). Application of flow-kinetics model to the PVT growth of SiC crystals, *J. Crystal Growth* 266: 303–312.

Wu, X. H. et al. (1996). Defect structure of metal-organic chemical vapor deposition-grown epitaxial (0001) GaN/Al$_2$O$_3$, *J. Appl. Phys.* 80: 3228 - 3237.

Wu, X. H. et al. (1998). Dislocation generation in GaN heteroepitaxy, *J. Crystal Growth* pp. 231–243.

Xi, Y. A. et al. (2006). Very high quality AlN grown on (0001) sapphire by metal-organic vapor phase epitaxy, *Appl. Phys. Lett* 89: 103106.

Xi, Y. A. et al. (2006a). AlGaN UV light-emitting diodes emitting at 340 nm grown on AlN bulk substrates, Proc. L. Eastman Conf. High Performance Devices, Cornell Univ., 54–55.

Xing, H. et al. (2001). Gallium nitride based transistors, *J. Phys.: Condens. Matter* 13: 7139–7157.

Yakimova, R. et al. (2005). Sublimation growth of AlN crystals: Growth mode and structure evolution, *J. Crystal Growth.* 281: 81–86.

Yamabe, N. et al. (2009). Nitridation of Si(1 1 1) for growth of 2H-AlN(0 0 0 1)/Si$_3$N$_4$ /Si(1 1 1) structure, *J. Crystal Growth*: 3049-3053.

Yamane, H. et al. (1998). Morphology and characterization of GaN single crystals grown in a Na flux, *J. Crystal Growth* 186: 8–12.

Yan, F. et al. (2007). High-efficiency GaN-based blue LEDs grown on nano-patterned sapphire substrates for solid-state lighting, *Proc. of SPIE* 6841: 684103.

Yano, M. et al. (2000). Growth of nitride crystals, BN, AlN and GaN by using a Na flux, *Diamond and Related Materials* 9: 512–515.

Yazdi, G. R. et al. (2006). Growth and morphology of AlN crystals, *Phys. Scr.* T126: 127–130.

Yonemura et al. (2005). Precipitation of single crystalline AlN from Cu-Al-Ti solution under nitrogen atmosphere, *J. Mater. Sci.: Mater. In Electron.* 16: 197–201.

Yoshikawa, A. et al. (2004). Crystal growth of GaN by ammonothermal method, *J. Crystal Growth* 260: 67–72.

Zheleva, T. et al. (1999). Pendeo-epitaxy - a new approach for lateral growth of gallium nitride films, *J. Electron. Mater.* 28: L5–L8.

Zhmakin, A. I. (2004). Heat transfer problems in crystal growth, Keynote lecture. 3rd Int. Symp. Adv. in Comp. Heat Transfer, 2004, Norway. CD Proc., Begell House, 24 pp.

Zhmakin, A. I. (2011a). Enhancement of light extraction from light-emitting diodes, *Phys. Reports* 498: 189–241.

Zhmakin, A. I. (2011b). Strain relaxation models, arXiv:1102.5000v1 [cond-mat.mtrl-sci].

Zhmakin, I. A. et al. (2000). Evolution of thermoelastic strain and dislocation density during sublimation growth of silicon carbide, *Diamond and Related Materials* 9: 446–451.

Zukauskas, A. et al. (2002). *Introduction to Solid State Lighting*, ISBN-10: 0471215740, ISBN-13: 978-0471215745, Wiley, New York.

11

Crystal Growth and Stoichiometry of Strongly Correlated Intermetallic Cerium Compounds

Andrey Prokofiev and Silke Paschen

Institute of Solid State Physics, Vienna University of Technology,
Wiedner Hauptstr. 8-10, 1040 Vienna
Austria

1. Introduction

Strongly correlated electron systems are among the most active research topics in modern condensed matter physics. In strongly correlated materials the electron interaction energies dominate the electron kinetic energy which leads to unconventional properties. Heavy fermion compounds form one of the classes of such materials. In heavy fermion compounds the interaction of itinerant electrons with local magnetic moments generates quasiparticles with masses up to several 1000 electron masses. This may be accompanied by exciting properties, such as unconventional superconductivity in a magnetic environment, non-Fermi liquid behavior and quantum criticality. Strong electronic correlations are responsible for physical phenomena on a low energy scale. Consequently, these phenomena have to be studied at low temperatures. This, in turn, requires ultimate quality of single crystals to avoid that the low temperature intrinsic properties are covered by extrinsic effects due to off-stoichiometry, impurities or other crystal imperfections.

The overwhelming majority of heavy fermion systems are cerium and ytterbium intermetallic compounds. In the present paper we discuss the crystal growth of three cerium compounds, $Ce_3Pd_{20}Si_6$, $CeRu_4Sn_6$ and $CeAuGe$. $Ce_3Pd_{20}Si_6$ undergoes an antiferromagnetic phase transition at low temperatures and shows a magnetic field induced quantum critical point [Takeda et al (1995), Strydom et al (2006)]. $CeRu_4Sn_6$ [Das & Sampathkumaran (1992)] appears to be a Kondo insulator [Aepli & Fisk (1992)] with a highly anisotropic energy gap. $CeAuGe$ is one of a the few cerium compounds showing a ferromagnetic phase transition at low temperatures [Pöttgen et al (1998), Mhlungu & Strydom (2008)]. This is of special interest in the context of quantum criticality, since the occurrence of quantum criticality on the verge of a ferromagnetic ground state is a subject of current debate.

Much attention in this paper is paid to the problem of stoichiometry. Single crystals of intermetallic compounds are grown at high temperatures, which facilitates the formation of thermal defects realized often as deviation from the stoichiometric composition. Thermal instabilities of some intermetallic phases require the use of flux techniques, i.e., growth from off-stoichiometric melts, which is another source of non-stoichiometry.

Sizeable non-stoichiometries can be detected by measuring the composition by chemical and physical analytical techniques, e.g. energy dispersive X-ray spectroscopy analysis (EDX). Tiny deviations from the stoichiometry, on the other hand, can be found only from an analysis of the physical properties of single crystals. Thus physical property measurements are not only the final purpose of a crystal growth but also a valuable diagnostic tool for further improvement of their quality. Therefore, in the paper the consideration of crystal growth is accompanied by the discussion of their physical properties.

Physical property measurements on $Ce_3Pd_{20}Si_6$ single crystals grown by the floating zone technique have been reported earlier [Goto et al (2009), Prokofiev et al (2009), Mitamura et al (2010)]. In Ref. Prokofiev et al (2009) a systematic study of the relationship between the growth technique, stoichiometry and physical properties of single crystals has been done. The crystal growth and stoichiometry of $CeRu_4Sn_6$ and $CeAuGe$ are reported for the first time. The physical properties of the $CeRu_4Sn_6$ single crystals were published partially in an author's earlier paper [Paschen et al (2010)].

2. Ce$_3$Pd$_{20}$Si$_6$

The recent observation of magnetic-field induced quantum criticality [Paschen et al (2007); Strydom et al (2006)] in the cubic heavy fermion compound $Ce_3Pd_{20}Si_6$ [Takeda et al (1995)] has attracted considerable attention. $Ce_3Pd_{20}Si_6$ crystallizes in a cubic $Cr_{23}C_6$-type structure with the space group $Fm\bar{3}m$ [Gribanov et al (1994)]. The cubic cell with $a=$ 12.161 Å [Gribanov et al (1994)]; 12.280 Å [Takeda et al (1995)] contains 116 atoms. The Ce atoms occupy two distinct sites of cubic point symmetry. At the octahedral $4a$ (tetrahedral $8c$) site Ce atoms are situated in cages formed by 12 Pd and 6 Si atoms (16 Pd atoms). These high coordination numbers allow to classify $Ce_3Pd_{20}Si_6$ as a cage compound. Strongly correlated cage compounds are of much interest as potential candidates for thermoelectric applications [Paschen (2006)].

Similar to the isostructural germanide compound $Ce_3Pd_{20}Ge_6$, two phase transitions - a presumably antiferromagnetic one at T_L of 0.15 K [Takeda et al (1995)], 0.17 K [Goto et al (2009)], or 0.31 K [Strydom et al (2006)] and a possibly quadrupolar one at T_U of 0.5 K [Strydom et al (2006)] - have been found in the silicide compound $Ce_3Pd_{20}Si_6$. Similar to the effect of magnetic field applied along [100] or [110] in $Ce_3Pd_{20}Ge_6$ [Kitagawa (1998)], a magnetic field shifts the two transitions of polycrystalline $Ce_3Pd_{20}Si_6$ in opposite directions: At the critical field of about 1 T the transition at T_L is suppressed to zero whereas the transition at T_U shifts to 0.67 K [Strydom et al (2006)]. The non-Fermi liquid behavior of the electrical resistivity observed at the critical field in the temperature range 0.1-0.6 K is an indication for the existence of a field-induced quantum critical point [Paschen et al (2007)].

Neutron scattering experiments on polycrystalline $Ce_3Pd_{20}Si_6$ have to date failed to detect any kind of magnetic order [Paschen et al (2008)]. Thus, to clarify the nature of the phases below T_L and T_U large single crystals of high quality are needed. In fact, in the first neutron scattering study on $Ce_3Pd_{20}Si_6$ single crystals [Paschen et al (2008)] the absence of signatures of magnetic order was attributed to a too low T_L value of the investigated specimen.

Since both phase transitions take place at rather low temperatures, disorder may influence them significantly. The discrepancy in the reported ordering temperatures (e.g. [Goto et al (2009); Strydom et al (2006); Takeda et al (1995)]) demonstrates this delicate dependence. Also the non-negligible difference in the reported unit cell constants [Gribanov et al (1994), Takeda et al (1995)] needs a clarification. This has motivated us to undertake a systematic investigation of the relation between crystal growth techniques/conditions, sample quality, and the resulting physical properties down to dilution refrigerator temperatures.

We show here that both phase transitions are extremely sensitive to small stoichiometry variations that result from different growth procedures.

To elucidate the discrepancies in the low-temperature data reported on the quantum critical heavy fermion compound $Ce_3Pd_{20}Si_6$ and reveal the compound's intrinsic properties, single crystals of varying stoichiometry were grown by various techniques – from the stoichiometric and slightly off-stoichiometric melts as well as from high temperature solutions using fluxes of various compositions. The results of this work on $Ce_3Pd_{20}Si_6$ have been partially reported earlier [Prokofiev et al (2009)]. Here a more detailed analysis including also information on new crystal growth experiments as well as the physical property measurements on new single crystals are reported.

2.1 Growth from the stoichiometric melt

To investigate the melting character of $Ce_3Pd_{20}Si_6$ a differential thermal analysis (DTA) run up to 1350°C was carried out (Fig. 1). There is only a single peak both on the heating and on the cooling curve, indicating congruent melting as previously reported [Gribanov et al (1994)]. A closer inspection of the shape of the melting peak (Fig. 1, inset) might, however, suggest merely a quasi-congruent melting character. The onset of melting occurs at $T_M \approx 1250$°C. Because of undercooling the crystallization begins about 100°C lower than the melting.

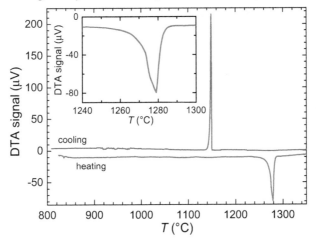

Fig. 1. DTA heating and cooling curves of $Ce_3Pd_{20}Si_6$. The inset shows a magnification of the melting peak. From Ref. Prokofiev et al (2009).

The floating zone technique with optical heating was used for crystal growth from the melt, with a pulling rate of 10 mm per hour and an upper rod rotation speed of 5 rpm. Due to the high density and the low surface tension of the $Ce_3Pd_{20}Si_6$ melt the floating zone was rather unstable, and the melt dropped down repeatedly (Fig. 2, right). However, one growth could be kept running long enough to grow an ingot of 25 mm length (Fig. 2, left).

Over the growth run the originally shiny and clear surface of the melt became more and more opaque, and a crust on the surface could be seen after some time. This is a sign of incongruent melting supposedly due to a peritectic reaction. The crust is the higher temperature phase, therefore it forms on the optically heated surface which is the hottest place of the melt. This observation can be explained by close proximity of the peritectic point to the temperature of the complete melting of the system with the total composition $Ce_3Pd_{20}Si_6$, as indicated already by the peculiar shape of the DTA curve (Fig. 1 inset). Therefore the formation of

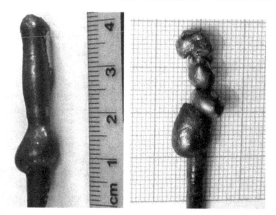

Fig. 2. Samples resulting from crystal growth experiments from the stoichiometric melt with upper and lower rod rotation. Left - Successfully grown single crystal ($sc1$); Right - The crystallized ingot experienced repeated dropping down of the melt because of the unfavorable combination of the high density and of the low surface tension of the melt.

a small amount of foreign phase on the hot surface results in a slight shift of the melt composition, which returns the crystallization process in the melt deep into the primary crystallization field of the $Ce_3Pd_{20}Si_6$ phase.

Due to the thin foreign phase film the surface of the recrystallized (lower) rod was yellowish. The EDX measurement detected Ce ($\sim$ 70 at.%), Pd ($\sim$ 22 at.%), and Si ($\sim$ 8 at.%) in this film. However no inclusions of foreign phases were found by scanning electron microscopy (SEM) inside the crystal. A single crystal grown in this way is specified here as $sc1$ (see Table 1 for the nomenclature of all crystals). In order to trace segregation effects we differentiate additionally between the part of this crystal grown at the beginning of the growth process (bottom part of the ingot, $sc1b$) and that grown at the end (top part, $sc1t$). For the growth of $sc1$ the lower purity starting materials (Ce 99.99 at.%, Pd 99.95 at.%) were used.

Further growth runs with rotation of the upper rod led to a permanent breaking of the surface crust. The melt leaked out through the cracks of the crust. Hence, the crust may serve as a quasi-crucible if it remains intact during the whole growth time. Based on this observation a growth run without rotation of the upper rod was carried out with the same growth parameters. The melt zone was quite stable in this experiment. A crystal grown by this procedure will be specified as $sc2$. For the growth of this crystal higher purity starting materials (Ce 99.99 at.%, Pd 99.998 at.%) were used.

Laue investigations show very good crystallographic perfection of both types of crystals grown from the melt. Before annealing an SEM/EDX investigation of both crystals was carried out. The polished cross-section of the $sc1$ rod is more homogeneous in composition than that of $sc2$. The surface of the latter had a well distinguishable 300 μm thick outer shell where the concentration of Ce was about 5% higher and that of Si somewhat lower than in the core region (Fig. 3, left). This shell may result from a partial dissolution and diffusion of the crust into the core of the rod. In the core region of $sc2$ diffuse 10 μm inclusions of supposedly the same shell phase, however with lower Ce concentration, were found (Fig. 3, right). These inclusions were not detected by X-ray powder diffraction and they disappeared after annealing, according to SEM analysis.

Fig. 3. Microstructure of $sc2$ before annealing.
Left - Surface shell (darker top part) with a 5% higher Ce concentration; Right - An inclusion with a lower Ce concentration in the core (shown by arrow).

After annealing of both crystals for 3 weeks at 900°C a second EDX investigation was carried out. At first we did the analysis *without* any reference sample. This yielded a stoichiometric Ce content (10.3 at.%) but an over-stoichiometric Si content (22.0 instead of 20.7 at.%) and an under-stoichiometric Pd content (67.7 instead of 69.0 at.%) for both $sc1$ and $sc2$, corresponding to a partial substitution of Pd by Si on their sites. These results motivated our efforts to grow crystals using the flux technique, where the stoichiometry of the crystals can be tuned through the variation of the flux composition. The results of further EDX investigations *with* a reference sample will be discussed in Sect. 2.4.

Sample name	Growth technique details
$sc1t$	from melt with rod rotation, top part
$sc1b$	from melt with rod rotation, bottom part
$sc2$	from melt without rod rotation
$sc3$	from Pd_5Si flux, molar $Ce_3Pd_{20}Si_6$ to Pd_5Si ratio 2:1
$sc4$	from Pd_5Si flux, molar $Ce_3Pd_{20}Si_6$ to Pd_5Si ratio 1:2
$sc5$	from slightly off-stoichiometric melt, with a 0.3 at.% Ce-excess and a 0.2 at.% Pd-deficiency
Sn flux crystals	using Sn or Sn-Pd fluxes

Table 1. Nomenclature of the $Ce_3Pd_{20}Si_6$ crystals grown by various techniques.

2.2 Flux growth

2.2.1 Tin flux

We tried at first a crystal growth with standard fluxes. The related compound $Ce_2Pd_3Si_5$ can be grown with Sn as flux at below 1100°C [Dung et al (2007)]. To check for the compatibility of Sn flux with $Ce_3Pd_{20}Si_6$, a mixture of $Ce_3Pd_{20}Si_6$ and Sn was heated up to 1100°C in a boron nitride crucible and then cooled slowly (1°C/h) down to 700°C. The solute-to-solvent ratio was 2:1. After crystallization the ingot was cut, polished, and investigated by SEM/EDX and X-ray powder diffraction (XRD).

The crystallization yielded relatively large single crystals of the non-stoichiometric phase $CePd_{2-x}Si_{2+x}$ with $x \approx 0.25$, incorporated in a matrix of Sn_4Pd. In addition, small inclusions of other phases were found. The experiment thus shows the inapplicability of Sn as a flux because its affinity to Pd leads to a destruction of the $Ce_3Pd_{20}Si_6$ phase.

However, other Sn-based flux compositions with a lower affinity to Pd can be found in the Sn-Pd binary phase diagram [Chandrasekharaiah (1990)]. A series of Sn-Pd compounds - Sn_4Pd, Sn_3Pd, Sn_2Pd - with low melting points (below 600°C) exists.

Similar experiments as the one with pure Sn flux were carried out using the above Sn-Pd compositions. In all cases the primarily crystallized phase was $CePd_{2-x}Si_{2+x}$ according to SEM/EDX. The x-value diminished with increasing Pd content in the flux, reaching about 0.05 for Sn_2Pd flux. The stoichiometric $CePd_2Si_2$ phase has the same Ce/Si ratio as the $Ce_3Pd_{20}Si_6$ phase but a strongly reduced Pd content. However, further increasing of the Pd concentration in the flux for tuning of Pd content in crystals was impossible: with the higher melting compound SnPd, only partial melting of the crucible content was observed at 1100°C. Thus, foreign flux growth was not successful.

2.2.2 Self-flux Pd_5Si

The reason for using the flux method was, on one hand, to obtain single crystals with exact stoichiometric composition and, on the other hand, the expectation that the off-stoichiometric melt would have a higher surface tension and hence the floating zone would be more stable than without flux. To avoid a contamination by foreign atoms we first opted for self flux growth. Since the concentration of thermal defects (e.g., Si - Pd substitutions) in the crystal is expected to decrease with decreasing growth temperature, we searched in the Ce-Pd-Si phase diagram (Fig. 4a) for low-melting (at first binary) compositions with an over-stoichiometric Pd content.

The phase Pd_5Si [Seropegin (2001)] which, according to a later study [Gribanov et al (2006)], appears to consists or two scarcely distinguished phases $Pd_{14}Si_3$ and $Pd_{84}Si_{16}$, fulfills all requirements: it melts at a relatively low temperature of 835°C [Chandrasekharaiah (1990)], has an over-stoichiometric ($>$ 20:6) Pd/Si ratio, and there are no stable Ce-containing intermediate phases between Pd_5Si and $Ce_3Pd_{20}Si_6$ in the Ce-Pd-Si ternary phase diagram (cross-section at 600°C [Gribanov et al (2006); Seropegin (2001)]).

As above, the floating zone technique with optical heating was used. The feed and the seed rods had the stoichiometric starting composition $Ce_3Pd_{20}Si_6$, and the zone was a molten mixture of $Ce_3Pd_{20}Si_6$ and Pd_5Si. Contrary to the melt growth, the floating zone was very stable, and its surface remained clear during the entire growth run. The latter means that the growth occurred within the $Ce_3Pd_{20}Si_6$ primary crystallization field. The stability of the melt zone allowed the rotation of the upper rod. The pulling rate was 0.6 mm/h. Two growth runs with different flux compositions (molar ratios $Ce_3Pd_{20}Si_6$ to Pd_5Si of 2:1 and 1:2) were carried out (Fig. 4b). The crystals were annealed for 3 weeks at 900°C. The corresponding samples are referred to as $sc3$ and $sc4$, respectively.

2.3 Growth from slightly off-stoichiometric melt

As the analysis of the composition and the properties of the crystals $sc1$ - $sc4$ grown from the melt and from flux has shown (see Sections 2.4, 2.5) the stoichiometry was strongly sensitive to the starting composition of the melt. For a fine correction of non-stoichiometry a growth from a slightly off-stoichiometric melt was carried out, too ($sc5$). For reasons discussed later, the feed rod for $sc5$ contained 0.3 at.% excess of Ce and 0.2 at.% deficiency of Pd. To avoid the floating zone instability the growth was carried out without rod rotation, as in case of $sc2$.

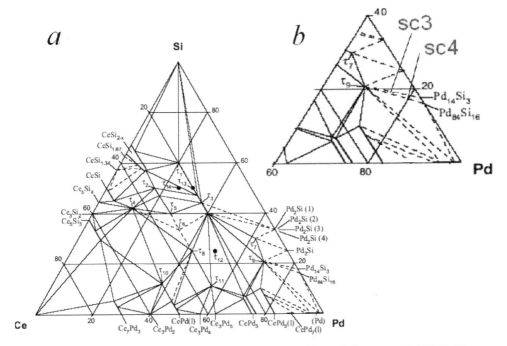

Fig. 4. a) Phase diagram of the Ce-Pd-Sn ternary system (from Gribanov et al (2006)). Here $Ce_3Pd_{20}Si_6$ is denoted as τ_9-phase. b) Magnified part of the phase diagram with the compositions of the solutions from that $sc3$ and $sc4$ were grown.

The pulling rate was 4 mm/h. Similar to $sc1$ we differentiate additionally between the bottom and top parts of the ingot, $sc5b$ and $sc5t$.

2.4 Composition of the grown crystals

The usual EDX technique without standards cannot determine absolute atomic concentrations with sufficient accuracy. This problem can, in principle, be solved by using a standard of exactly known composition. However, as such a sample is not readily available, we used, instead, the polycrystalline sample (pc) which will be shown below to be of best quality, according to the physical property measurements. Irrespective of whether its composition may be identified with the *exact* stoichiometry 3:20:6 or not it served as a practical guideline to establish the crystal composition–property relationship.

With additional improvements of our EDX setup (improved measurement statistics, counting time, beam current control) we can measure, with a high accuracy, *deviations* from the reference sample stoichiometry. The results are summarized in Table 2 where the measured lattice parameters are given, too. Because crystals grown from the Sn-containing fluxes are not the title phase and their compositions vary sizably they are simply represented by $CePd_{2-x}Si_{2+x}$ in Table 2.

Even though, in the absence of a real standard, there remains some uncertainty about the *absolute* values measured by EDX (which even for the reference sample pc differ from the calculated stoichiometry 3:20:6), *trends* in the composition of the investigated series of crystals

Sample name	Composition (at.%)			a (Å)
	Ce	Pd	Si	
sc1t	10.5	67.6	21.9	12.276
sc1b	9.9	67.7	22.4	12.272
sc2	9.9	68.0	22.1	12.272
sc3	9.2	68.7	22.1	12.233
sc4	7.4	70.1	22.5	12.180
sc5	10.7	68.0	21.3	12.277
pc	10.1	67.7	22.2	12.280
Exact 3:20:6 stoichiometry	10.3	69.0	20.7	
Sn flux crystals	$CePd_{2-x}Si_{2+x}$			
Standard deviation	0.24	0.20	0.20	
Systematic error				<0.002

Table 2. Composition and lattice parameter a of the investigated $Ce_3Pd_{20}Si_6$ samples.

can be discussed. While the Si content varies only weakly a stronger variation of the Ce and Pd content is observed, the Ce concentration nearly anticorrelating with the one of Pd (Fig. 5).

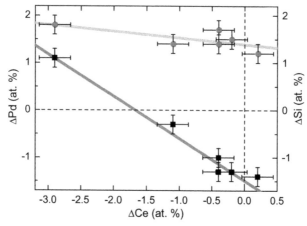

Fig. 5. Deviation of the Pd and Si content, ΔPd and ΔSi, from the exact 3:20:6 stoichiometry as function of the Ce non-stoichiometry ΔCe. The dashed lines indicate the exact stoichiometry. The meaning of the broad grey lines is discussed in the text. Error bars indicate the standard deviations given in Table 2. From Ref. Prokofiev et al (2009).

Ce *over*-stoichiometry can be realized by a substitution of Pd or Si atoms on their crystallographic sites by excess Ce-atoms or, alternatively, by vacancies on the Pd and/or Si sites. Since the lattice parameter of $sc1t$ and $sc5$ is smaller than that of the pc, the latter option is more probable, but a combination of both options cannot be excluded. If only Pd/Si vacancies were present, the Ce sublattice would remain occupied and fully ordered. Ce *under*-stoichiometry, on the other hand, can be associated either with Ce vacancies or, which is more probable taking into account the approximate anti-correlation between the Ce and Pd contents, with a partial Pd substitution on Ce sites. The broad dark grey line in Fig. 5 represents the Ce-Pd concentration relation for a crystallochemical model which, as an example, assumes a half-filling of the Ce vacancies V by excess Pd atoms:

$[Ce_{1-x}V_{0.5x}Pd_{0.5x}]_3Pd_{20}Si_6$. The same model was used to describe the ΔSi vs ΔCe relation (light grey line). The agreement with the data is excellent.

2.5 Influence of the growth technique on the physical properties

The resulting stoichiometry of the crystals as well as their physical properties show sizable dependence on the different growth techniques.

Neutron diffraction experiments were carried out on an oriented sample cut from sc1b. The crystal was confirmed to have excellent crystallinity but, contrary to our expectations from the investigations on polycrystalline samples, showed no phase transition down to the lowest accessed temperature of about 0.15 K [Paschen et al (2008)]. The present investigation shows that, for this very single crystal, this temperature was still too high.

The specific heat C_p was measured for sc1 - sc3 and sc5. Figure 6a shows the temperature dependence of C_p/T of these crystals, together with published data of a polycrystalline sample [Strydom et al (2006)]. $C_p/T(T)$ of the polycrystalline sample pc has a sharp peak with a maximum at $T_L = 0.31$ K and a shoulder at $T_U \sim 0.5$ K [Strydom et al (2006)]. These features get successively broadened and suppressed to lower temperatures in the crystals sc1t, sc5b, sc1b, and sc2, respectively. In sc3 no clear signature of T_L can be identified, suggesting that it has shifted further to lower temperatures and was further broadened or, alternatively, has vanished altogether. Due to the suppression of this lower temperature feature the anomaly at T_U, identified in all other samples as shoulder, now appears as a maximum.

From the specific heat measurements the best single crystals thus appear to be sc1t and sc5. With their rather sharp peaks at 0.22 K and 0.20 K, respectively, and a shoulder at about 0.4 K they demonstrate all features associated with the intrinsic behavior of $Ce_3Pd_{20}Si_6$.

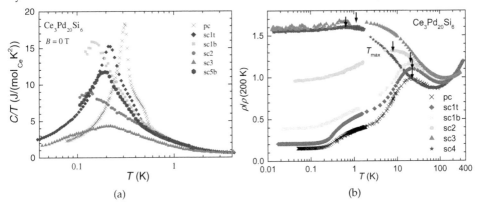

(a)   (b)

Fig. 6. a) Specific heat divided by temperature C_p/T plotted for all single crystals prepared here, and for a polycrystal [Strydom et al (2006)] for comparison, as function of temperature T on a logarithmic scale. The lower transition temperatures T_L are taken here as position of the maxima. The maximum in the data for sc3 can be identified with T_U (see text). b) Electrical resistivity of the $Ce_3Pd_{20}Si_6$ single crystals grown here normalized to the respective values $\rho_{200 K}$ vs temperature T. The arrows indicate the positions T_{max} of the (local) maxima. Data of a polycrystal [Strydom et al (2006)] are shown for comparison. From Ref. Prokofiev et al (2009).

The electrical resistivities of all single crystalline samples and of one polycrystalline sample are shown in Figs. 6b and 7 as function of temperature. While the resistivity of the polycrystalline sample drops to very low values at low temperatures, the residual resistivities of the single crystals are considerably higher. They increase in the sequence $pc \rightarrow sc5t \rightarrow sc1t \rightarrow sc5b \rightarrow sc1b \rightarrow sc2 \rightarrow sc3$. This is about the same sequence in which the temperature and the sharpness of the low-temperature phase transition of the C_p/T curves decreases (Fig. 6a). Thus, it is natural to assume that in the sequence $pc \rightarrow (sc1t$ and $sc5) \rightarrow sc1b \rightarrow sc2 \rightarrow sc3$ the lattice disorder increases. Since the starting material purity of $sc1t$ and $sc1b$ was lower than that of $sc2$, we conclude that the main reason of the disorder is a deviation of the sample composition from the exact stoichiometry 3:20:6 and not the concentration of foreign atoms. For single crystal $sc4$ the high temperature minimum occurs at lower temperatures than expected, leading to a slightly lower residual resistivity than for $sc3$.

In $sc1$ a remarkable increasing of the quality from the bottom to the top part of the crystal is observed. Only a small (top) part of the total crystal has an excellent quality and can be used for physical property investigations. $sc5$ is indeed ranked after $sc1t$ according to the C_p data but it is more homogeneous throughout the whole volume of the batch. A scenario which can explain the difference between the bottom ($sc1b$) and the top ($sc1t$) parts of the crystal $sc1$ is as follows. At first (bottom part of the ingot) the crystal phase captures less Ce and more Pd from the stoichiometric melt (Table 1). While the crystal grows the melt gets enriched by Ce and depleted by Pd. At the end of the crystallisation (top part) this change in the melt composition results in a shift of the crystal composition to a more stoichiometric one, in accordance with the law of mass action. It was this observation that motivated us to perform the off-stoichiometric growth with a little Ce excess and Pd deficiency that resulted in $sc5$. As Fig. 7 shows the low temperature (down to 2 K) relative resistivity of $sc5$ is comparable with the best sample $sc1t$ and the spatial (top-bottom) difference in the resistivity is much smaller for $sc5$ than for $sc1$.

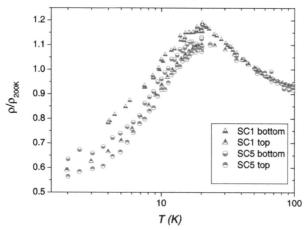

Fig. 7. Relative resistivities of the bottom and top parts of $sc1$ and $sc5$.

The highest residual resistance ratio (RRR, defined here as $\rho(200$ K$)/\rho(50$ mK$)$) and the sharpest and most pronounced phase transition features in C_p/T are found for the polycrystalline sample (pc) which therefore appears to be the most stoichiometric one. This can be easily understood by the specifics of preparation. The accuracy of the stoichiometric *total* composition of a polycrystalline sample is limited only by the accuracy of the weighing process of starting materials and by their purity. A possible high temperature non-stoichiometry of the main phase of an as-cast polycrystal is compensated by the presence of minor impurity phases, the phase separation being heterogeneous on a microscopic scale. This heterogeneity can be lifted by annealing at lower temperatures due to the short diffusion path. During single crystal growth, however, a macroscopic phase separation can occur, making annealing very inefficient. Actually the resistivity curves of $sc1b$ before and after annealing were practically identical.

The Ce content ΔCe varies by more than 3 at.% among all grown single crystals. We have revealed a systematic dependence of the residual resistance ratio, the lattice parameter, the (lower) phase transition temperature T_L, and the maximum in the temperature dependent electrical resistivity T_{max} with ΔCe. This clarifies the sizable variation in the values of T_L reported in the literature. We discuss the physical origin of the observed composition–property relationship in terms of a Kondo lattice picture. We predict that a modest pressure can suppress T_L to zero and thus induce a quantum critical point.

While no clear correlation between the physical properties and the Si content can be demonstrated, a pronounced dependence on the Ce content (or anti-correlated Pd content) is observed. Figure 8a shows the dependencies of the lattice parameter a and the residual resistance ratio RRR as function of the deviation from stoichiometry ΔCe. The largest deviations from the "intrinsic behavior" (which is most closely met by the polycrystal) are seen for $sc4$. It has the lowest Ce content of 7.4 at.% (ΔCe $= 2.9$ at.%). The polycrystalline sample (pc) which demonstrates the most pronounced phase transition features and $sc1t$ with the second-sharpest features have the highest Ce contents, – and the lowest deviation from the exact Ce stoichiometry. The deviations ΔCe of $sc1t$ and $sc5$ lie on the other side of the stoichiometry line. One may argue that only the absolute value of ΔCe is relevant with respect to the composition–disorder–property relationship since *any* non-stoichiometry is usually associated with lattice imperfection. To test this conjecture we plot, in Fig. 8a, the data point for $sc1t$ also mirrored through the ΔCe $= 0$ axis (open symbol). Indeed this point fits nicely into the overall $RRR(\Delta$Ce$)$ behavior.

As it was pointed out in Sec. 2.4, the Ce-understoichiometry may be realized by two ways: either by Ce-vacancies or by Ce-vacancies partially (to 50%) filled by Pd atoms. Both cases of Ce under-stoichiometry correspond, in a Kondo lattice description, to Kondo holes. In the Kondo coherent state at low temperatures Kondo holes act as strong scattering centers, decreasing the RRR. This effect is seen in Fig. 8a. Since also $sc1t$ has a reduced RRR, over-stoichiometry appears to be indeed realized by the combination of excess Ce and Pd/Si vacancies. Excess Ce, just as Ce holes, creates Kondo disorder. In addition to reducing RRR, Kondo disorder is expected to successively suppress the temperature T_{max} where the high-temperature incoherent Kondo scattering with an approximate $\rho \propto -\ln T$ behavior crosses over to coherent Kondo scattering at low temperatures. In samples $sc3$ and $sc4$ Kondo disorder is so strong that $\rho \propto -\ln T$ is observed in a wide temperature range, and $\rho(T)$ shows only a tiny drop below 2 K. Figures 6b and 7 show that a sizable suppression of T_{max} occurs in our sample series $pc - (sc1t$ and $sc5) - sc1b - sc2 - sc3 - sc4$ only from $sc2$ on. Thus, in addition to Kondo disorder, there appears to be a second effect influencing T_{max} in the opposite direction. This can be identified as a volume effect: the Ce vacancies or (smaller) Pd atoms on Ce sites in Ce under-stoichiometric samples as well as the Pd or Si vacancies in the over-stoichiometric sample $sc1t$ lead to a shrinkage of the crystal lattice, which is seen in Table 2 and in Fig. 8a. The decrease in volume with decreasing Ce content corresponds to positive (chemical) pressure. Using the bulk modulus of the closely related compound $Ce_3Pd_{20}Ge_6$ [~ 137.5 GPa at 150 K; Nemoto et al (2003)] we can convert the lattice shrinkage in our off-stoichiometric single crystals into a hypothetical external pressure p.

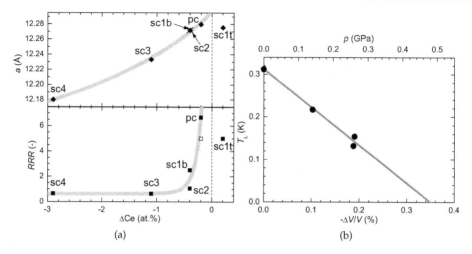

(a)   (b)

Fig. 8. a) Dependence of the lattice parameter a and the residual resistance ratio RRR on the deviation ΔCe from the stoichiometric Ce content. The open symbol in the lower panel represents the data point of $sc1t$, mirrored through the $\Delta Ce = 0$ line. The thick grey lines are guides to the eye. b) Lower transition temperature T_L vs relative volume shrinkage $-\Delta V/V$ of the different $Ce_3Pd_{20}Si_6$ single crystals with respect to the volume of the polycrystal pc. On the upper axis the corresponding pressure as estimated via the bulk modulus of $Ce_3Pd_{20}Ge_6$ [Nemoto et al (2003)] is given. The line represents a linear fit to the data and its extrapolation to $T_L = 0$. From Ref. Prokofiev et al (2009).

Figure 9 shows T_{max} (full dots) vs the relative volume shrinkage $-\Delta V/V$ of our single crystals with respect to our polycrystal (pc) (bottom axis) and vs a hypothetical external pressure calculated via the bulk modulus [Nemoto et al (2003)] (top axis). The step-like, as opposed to continuous, decrease of T_{max} with $-\Delta V/V$, addressed already above, is clearly seen. In order to understand this behavior we compare our results to a pressure study [Hashiguchi et al (2000)]. We include the T_{max} vs p data from this investigation on a polycrystalline sample under quasi-hydrostatic pressure in Fig. 9 (crosses and grey fit to these). From a value $T_{max} \approx 20$ K at $p = 0$, similar to the value for our polycrystal, T_{max} increases continuously with increasing pressure. This behavior is typical for Ce-based heavy fermion compounds: with increasing pressure the hybridization between the $4f$ electron of Ce and the conduction electrons increases and thus the Kondo temperature which is proportional to T_{max} increases. Our data follow this trend only at low non-stoichiometry (low $-\Delta V/V$ values) while at higher non-stoichiometry T_{max} decreases quickly. Thus we conclude that the volume effect on T_{max} dominates in our sample series $pc - sc1t - sc1b - sc2 - sc3 - sc4$ up to $sc1b$ while the Kondo disorder effect dominates from $sc2$ on.

Next we analyze the evolution of the ordering temperature T_L along our sample series. For this purpose T_L, extracted from the specific heat data in Fig. 6a, is plotted vs RRR in Fig. 9b and vs $-\Delta V/V$ (bottom axis) or p (top axis) in Fig. 8b. T_L decreases continuously with decreasing RRR (Fig. 9b). This might be taken as indication for the strong influence of disorder on T_L. However, in our sample series, the variation of RRR is intimately linked to a variation of ΔCe (Fig. 8a) and thus to a variation of the lattice parameter a, the relative volume change $-\Delta V/V$, and the corresponding pressure p. Thus, an alternative

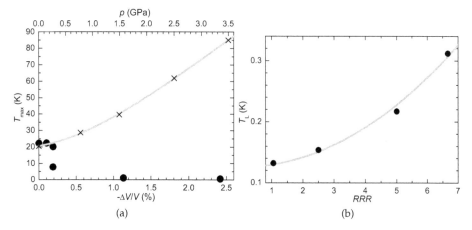

Fig. 9. a) Temperature T_{max} of the maximum in the electrical resistivity vs relative volume shrinkage $-\Delta V/V$ of our $Ce_3Pd_{20}Si_6$ single crystals with respect to the polycrystal (dots) and, for comparison, T_{max} vs pressure p for the polycrystal of Hashiguchi et al (2000) (crosses and grey line, which is fit to the data). $\Delta V/V$ and p are related to each other via the bulk modulus of $Ce_3Pd_{20}Ge_6$. b) Lower transition temperature T_L vs residual resistance ratio RRR for the different $Ce_3Pd_{20}Si_6$ samples. The line represents a quadratic fit to the data. From Ref. Prokofiev et al (2009).

view of the situation is that not increasing disorder but decreasing volume is responsible for suppressing T_L (Fig. 8b). To decide which effect is dominant we come back to our above discussion on Fig. 9a which revealed that the volume effect on T_{max} dominates in the series $pc - sc1t - sc1b - sc2 - sc3 - sc4$ up to $sc1b$ while the disorder effect dominates from $sc2$ on. If this holds true also for the effect on T_L it is Fig. 8b (without $sc2$) that captures the essential physics while Fig. 9b only displays implicit dependencies (except for $sc2$).

In a Kondo lattice picture the physical origin of the suppression of T_L with p is that pressure increases the hybridization between conduction electrons and $4f$ states, thus strengthening Kondo compensation, diminishing Ce magnetic moments, and suppressing the (most probably magnetic) ordering temperature T_L. Extrapolating the T_L vs p dependence of Fig. 8b to higher p suggests that a pressure induced quantum critical point is at reach for $Ce_3Pd_{20}Si_6$. A linear extrapolation of the fit shown in Fig. 8b yields $p_c \approx 0.5$ GPa. Of course, actual low-temperature pressure measurements are needed to verify this possibility.

Finally, we comment on a related study on the germanide compound $Ce_3Pd_{20}Ge_6$ where the influence of the starting composition on the physical properties was investigated [Kitagawa et al (1999)]. A strong composition effect on C_p and ρ was found to occur inside a narrow homogeneity range where a volume contraction of up to 0.7% takes place with increasing Pd content. In that work the concentration of Pd and not of Ce was concluded to be variable and responsible for the changing physical properties. It should be noted that the real composition of the resulting phases was not investigated. Thus, it is plausible that the increasing Pd content is accompanied by a decreasing Ce content. Under this assumption the composition effect on the properties of $Ce_3Pd_{20}Ge_6$ and $Ce_3Pd_{20}Si_6$ are indeed comparable.

3. CeRu$_4$Sn$_6$

In Kondo insulators (frequently also referred to as heavy fermion semiconductors) a narrow gap develops at low temperatures in the electronic density of states at the Fermi level [Aepli & Fisk (1992)]. While most Kondo insulators known to date adopt a cubic crystal structure (e.g., YbB$_{12}$, SmB$_6$, FeSi, Ce$_3$Bi$_4$Pt$_3$) a few compounds (e.g., CeNiSn, CeRhSb) are orthorhombic. These latter show anisotropic properties which suggest that the energy gap vanishes along certain directions in k-space. CeRu$_4$Sn$_6$, first synthesized by Das and Sampathkumaran [Das & Sampathkumaran (1992)], crystallizes in a tetragonal structure of space group $I\bar{4}2m$ (a = 6.8810 Å, c = 9.7520 Å, c/a = 1.4172) [Venturini et al (1990)]. A peculiarity of this compound is that, in addition to the tetragonal (body-centered) cell with lattice parameters a and c there exists a quasi cubic (face-centered) cell with lattice parameters c' and c, where c' is the diagonal of the tetragonal plane which differs by only 0.2% from c. This makes it very difficult to orient single crystals in an unambiguous way. On the other hand it allows us to study a "tetragonal" and a "quasi-cubic" Kondo insulator within the same material which is very appealing.

3.1 Choice of the growth method and the growth procedure

The compound CeRu$_4$Sn$_6$ melts incongruently. Melting followed by rapid quenching yields CeRuSn$_3$, Ru$_3$Sn$_7$ and a tiny amount of the title phase in the solidified material. The incongruent melting was confirmed by a DTA experiment which showed multiple peaks on the heating and on the cooling curves (Fig. 10). Since crystal growth from the melt is impossible for an incongruently melting compound we searched for an appropriate flux. Usually low melting metals are used as high temperature solvents for intermetallic compounds. For CeRu$_4$Sn$_6$ this might be tin. However CeRu$_4$Sn$_6$ was shown to react with tin with the formation of other phases. To avoid possible contaminations by foreign elements we searched thereafter for a flux composition in the ternary Ce-Ru-Sn system (self-flux). Unfortunately, the Ce-Ru-Sn phase diagram which would be helpful for the choice of the best flux composition has not been reported. A Ce-rich flux for the Ce-poor CeRu$_4$Sn$_6$ is expected to lead to the formation of the highly stable phase CeRuSn$_3$. Therefore, a Ce-free binary Ru-Sn mixture is more appropriate. Taking into account the Ru:Sn ratio of 2:3 in

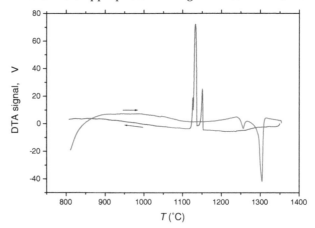

Fig. 10. DTA heating (red) and cooling (blue) curves of CeRu$_4$Sn$_6$.

CeRu$_4$Sn$_6$ the flux composition "Ru$_2$Sn$_3$" is optimal to maintain the Ru:Sn stoichiometry in the crystals. Moreover, the lowest melting (1160°C) composition in the Ru-Sn binary diagram is the eutectic with the composition 42.5 at.% Ru and 57.5 at.% Sn [Massalski (1990)] which is close to the element ratio 2:3. According to Ananthasivan (2002) the system is a (partially immiscible) liquid above 1200°C in the composition range of 37–75 at.% Sn. Finally, the large excess of Ru and Sn in the melt is expected to suppress the formation of the stable Ce-rich CeRuSn$_3$ phase.

The crystals were grown by the floating-solution-zone traveling heater method (THM) in a mirror furnace. The seed and the feed rods had the stoichiometric composition whereas the melt zone was a solution of CeRu$_4$Sn$_6$ in Ru$_2$Sn$_3$. The growth rate was 0.3–1.0 mm/h. The composition of the crystals along their length in the growth direction was investigated by SEM/EDX analysis. No noticeable deviation from the stoichiometric ratio 1:4:6 can be observed (Fig. 11).

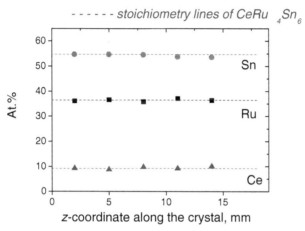

Fig. 11. Element concentration profiles of the CeRu$_4$Sn$_6$ single crystal in the growth direction.

3.2 Physical properties of CeRu$_4$Sn$_6$ single crystals

Figure 12 shows the temperature dependence of the electrical resistivity, $\rho(T)$, of CeRu$_4$Sn$_6$ on a semi-logarithmic scale. With decreasing temperature ρ first increases steeply (range 1), then passes over a maximum at about 10 K, increases again, albeit less steeply (range 2), and finally tends to saturate at the lowest temperatures. A possible explanation of this behaviour is a double-gap structure frequently encountered in simple semiconductors: a larger intrinsic gap visible at high temperatures (range 1) and a smaller extrinsic gap between impurity states and the band edge that dominates the low-temperature behaviour (range 2). Approximating $\rho(T)$ between 120 and 300 K with exponential behaviour ($\rho = \rho_0 \exp(\Delta_1/2k_B T)$, Arrhenius law) yields an energy gap $\Delta_1/k_B = 125$ K, sizeably larger than previously reported for polycrystalline samples [Brünig et al (2006); Das & Sampathkumaran (1992)]. Fitting the data between 0.8 and 1.8 K with the same function yields $\Delta_2/k_B = 0.1$ K (see full red lines in Fig. 12 for both fits). While this gap value may seem incongruous with respect to the fit range, it has to be bourn in mind that the influence of the low-T gap on $\rho(T)$ is expected to diminish by eventual thermal depopulation of the upper states towards $T = 0$ and hence the observed saturation in $\rho(T \to 0)$. Thus, a more complete fitting function is needed to account for these effects.

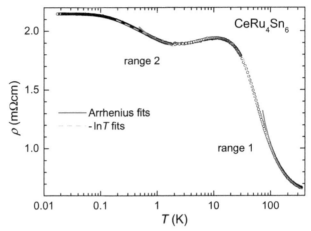

Fig. 12. Temperature dependence of the electrical resistivity, $\rho(T)$, of the CeRu$_4$Sn$_6$ single crystal on a semi-logarithmic scale. From Ref. Paschen et al (2010).

Now, having single crystals an investigation of the anisotropy of CeRu$_4$Sn$_6$ became feasible. The magnetic susceptibility was measured on an oriented single crystal in two mutually perpendicular directions (Fig. 13a). One of these directions is the crystallographic c axis, the other one is situated within the tetragonal plane. A pronounced difference is seen. For both directions Curie Weiss-type behaviour is observed at high temperatures, with an effective magnetic moment that is roughly consistent with the full effective moment of Ce^{3+}, and with the paramagnetic Weiss temperatures $\Theta_{\|c} \approx 395$ K and $\Theta_{\perp c} \approx 155$ K for H$\|c$ and H$\perp c$, respectively.

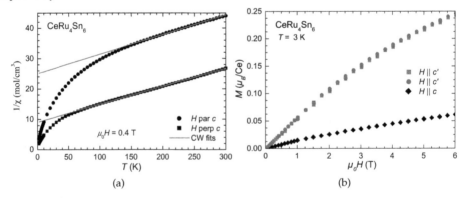

Fig. 13. a) Temperature dependence of the inverse magnetic susceptibility, $\chi^{-1}(T)$, of CeRu$_4$Sn$_6$ for the magnetic field $\mu_0 H = 0.4$ T applied along the crystallographic c axis and within the tetragonal plane. b) Magnetic field dependence of the magnetization, $M(\mu_0 H)$, taken at 3 K for fields applied along the crystallographic c and c' axes. Fig. 13b from Ref. Paschen et al (2010).

In order to test whether anisotropy also exists within the quasi-cubic cell we prepared small single crystalline platelets (with geometries which allowed for specific heat measurements only) cut from one piece in such a way that three mutually perpendicular directions were obtained: for two samples a c' axis is perpendicular to the platelet plane, for one sample it is c. As explained above, our Laue diffractograms cannot identify which sample is which. Our specific heat measurements, however, allow to clearly identify the c- and c'-oriented samples: while the zero field data are very similar for all three samples a magnetic field applied perpendicular to the platelet planes induces sizable anisotropy. The difference is best seen by plotting the relative difference in specific heat induced by a magnetic field, $(c_p(B) - c_p(0))/c_p(0)$, which reaches a maximum of more than 70% at 3.5 K for $B||c'$ but is below 15% for $B||c$ at this temperature (Fig. 14). Two samples ($sc1$ and $sc2$) show very similar behavior and must thus be c'-oriented while sc3 shows distinctly different behavior and is thus identified as the c-oriented sample.

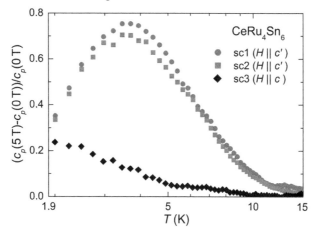

Fig. 14. Temperature dependence of the surplus specific heat induced by a magnetic field of 5 T, $(c_p(5T) - c_p(0))/c_p(0)$, of the three CeRu$_4Sn_6$ single crystals $sc1$, $sc2$, and $sc3$. From Ref. Paschen et al (2010).

A possible interpretation of these data is that a narrow energy gap which is present along c' in zero field is suppressed/diminished by a field of 5 T in this direction. Since the same field applied along c leads to a much weaker increase of c_p we believe that no or a much less field sensitive gap is present along c [Paschen et al (2010)].

4. CeAuGe

A sizable number of Ce- and Yb-based intermetallic compounds demonstrate quantum critical behaviour. Most of the compounds are antiferromagnets, whereas only a handful of Ce and Yb compounds with a ferromagnetic phase transition at low temperatures is known. The occurrence of quantum criticality in a ferromagnetic ground state is a subject of current debate. CeAuGe orders ferromagnetically at a relatively low Curie temperature of $T_C = 10.9$ K [Sondezi-Mhlungu et al (2009), Mhlungu & Strydom (2008)]. Thus it can be expected that the magnetic order can be tuned or fully suppressed by modest variations in magnetic field or pressure.

CeAuGe is a phase with a homogeneity range. Stoichiometric and nearly stoichiometric CeAuGe adopts the NdPdSb structure, an ordered variant of the AlB_2 structure type. The unit cell is hexagonal (space group $P6_3mc$) with the lattice parameters $a=$ 4.4569 Å and $c=$ 7.9105 Å [Pöttgen et al (1998)]. At larger deviations from the elemental 1:1:1 ratio a phase with a slightly different structure forms. The non-stoichiometric $CeAu_{1-x}Ge_{1+x}$ crystallizes in the true AlB_2 structure type and has about twice smaller unit cell. Unlike the ordered 1:1:1 phase, Au and Ge atoms are distributed statistically and the Au/Ge layers are not puckered but planar [Jones et al (1997)].

Since CeAuGe is not cubic it is especially important to investigate its physical properties on oriented single crystalline samples. We have investigated the crystal growth of this phase of various stoichiometries by the floating zone technique. In the course of the growth experiments we encountered a severe non-stoichiometry problem. We report here on our efforts to diminish the deviation from the 1:1:1 stoichiometry and the segregation effects resulting from it.

4.1 Crystal growth using stoichiometric feed rods

As a starting point, growth from the stoichiometric melt was tried. We studied the evolution of the crystal composition during the crystallization by measuring the composition at the starting, middle and final part of the crystallized ingot by the EDX technique. The concentration profiles are represented in Figs. 15 and 16. At the left border of each panel the initial compositions of the respective polycrystalline feed rod is shown. The beginning of the crystallization corresponds thus to the length coordinate $z = 0$. As Fig. 15a shows, the crystals primarily crystallized from the *stoichiometric melt* (i.e. the first portion of the crystalline phase) have a non-stoichiometric composition with a reduced Au content and increased Ce and Ge contents. This leads to a change of the melt composition with an accumulation of Au and a depletion of Ce and Ge. As a consequence, the Au content increases in the crystals and the Ce and Ge contents decrease in the course of further crystallization (Fig. 15a).

Due to the composition change the lattice parameters change too (Fig. 15b). For the first solidified crystals the a-parameter is lower and the c-parameter is higher than the stoichiometric values (Fig. 15b). Further crystallization leads to a decrease of c and to an increase of a.

Thus, the growth using a stoichiometric feed rod results in a non-stoichiometric single crystalline ingot which, along its length, is macro-inhomogeneous with respect to all three constituting elements.

4.2 Crystal growth using off-stoichiometric feed rods

In order to suppress the above discussed starting deviation in the element concentration in the crystals with respect to the melt we used a feed rod enriched in Au and depleted in Ge content, with the off-stoichiometric composition $CeAu_{0.96}Ge_{1.04}$. The *primary* crystals obtained from this run appear to be nearly stoichiometric (Fig. 16a, z=0 mm), but in the course of further crystallization the composition again drifts away from the 1:1:1 stoichiometry: the Au content increases and the Ge content decreases, both summing up to a constant value. However, the Ce content remains constant along the whole solidified ingot (Fig. 16a). This fact is favorable for physical investigations on large crystals because non-stoichiometry of Ce is usually the most disturbing factor in heavy fermion systems. The crucial role of the Ce stoichiometry was demonstrated in Section 2 on $Ce_3Pd_{20}Si_6$. Crystallization from the off-stoichiometric melt $CeAu_{0.96}Ge_{1.04}$ seems to be most promising for the growth of stoichiometric homogeneous

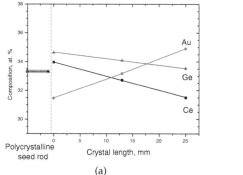
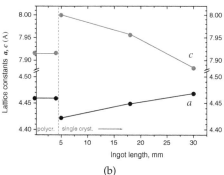

(a) (b)

Fig. 15. a) Element concentration profiles along the growth direction at the crystallization from a stoichiometric feed rod. b) Profiles of the unit cell parameters a and c.

CeAuGe crystals, provided that the growth technique should be modified. In the floating zone method the melt-to-crystal volume ratio is very small. Therefore the segregation phenomena have very strong impact on the composition of the melt zone, and the crystal composition varies strongly during the growth. In other techniques, e.g. the Czochralski method, the melt-to-crystal volume ratio can be rather large, and the growth of homogeneous (also with respect to the Au/Ge ratio) crystals from the melt with a practically constant composition appears feasible.

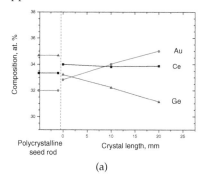
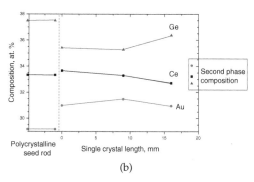

(a) (b)

Fig. 16. Element concentration profiles along the growth direction for crystal growth from off-stoichiometric feed rods with the compositions (a) $CeAu_{0.96}Ge_{1.04}$ and (b) $CeAu_{0.88}Ge_{1.12}$.

Crystallization from a more strongly off-stoichiometric melt $CeAu_{0.88}Ge_{1.12}$ yielded strongly non-stoichiometric crystals with a large excess of Ge and a lack of Au, the Ce content remaining at nearly stoichiometric level (Fig. 16b.)
While the primarily crystallized material is, according to SEM, single phased, the finally solidified ingot consists of two phases (Fig. 17, right). These two phases are the ordered and the disordered variants of the AlB_2 structure (the NdPdSb and the true AlB_2 types). The material from the middle part of the ingot seems to be single-phased at first glance (Fig. 17, upper left), but a closer inspection under higher magnification reveals a two-phase pattern (Fig. 17, lower left). The fine grained microstructure of the material (especially apparent in

comparison with that of the final part (Fig. 17, right)) evidences for crystallisation of a single phase at first which was decaying into two phases later in the course of cooling. This behavior is the sign of a strong temperature dependence of the homogeneity range of the NdPdSb-type phase.

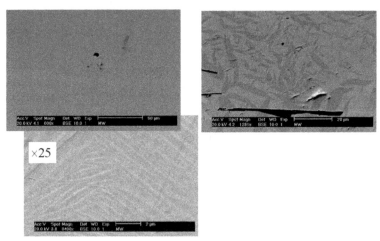

Fig. 17. Microstructure of the ingot crystallized from the feed rod with the composition $CeAu_{0.88}Ge_{1.12}$. Left - middle part, right - upper part. The magnified ($\times 25$) image of the middle part shows a fine inhomogeneity of the material.

5. Conclusions

Single crystals of the $Ce_3Pd_{20}Si_6$ phase were grown from the melt (stoichiometric and slightly off-stoichiometric) under various growth conditions and from high-temperature solutions using Pd_5Si as a flux. $Ce_3Pd_{20}Si_6$ melts quasi-congruently, i.e., the peritectic temperature is very close to the temperature of the complete melting. This fact follows from our DTA experiments and the observation of the melting zone during the growth process. The floating zone with stoichiometric composition was very unstable because of the low surface tension, which made the melt growth problematic. In contrast, the off-stoichiometric flux growth ran stably but resulted in non-stoichiometric single crystals.

While the Si content varies only slightly for different crystals, the Ce and Pd contents do so sizably, the decrease of Ce being partially compensated by an increase in Pd. The existence of a homogeneity range in $Ce_3Pd_{20}Si_6$ is the reason for the strong variation of the properties of single crystals grown by different techniques. The sharpness of the lower (presumably antiferromagnetic) phase transition, its transition temperature T_L, the residual resistivity, and the temperature T_{max} of the (local) maximum in $\rho(T)$ were shown to be measures of the crystal quality. Based on all these properties the upper part of the crystal grown from the stoichiometric melt ($sc1t$) and the whole volume of the crystal $sc5$ grown from a slightly off-stoichiometric melt can be ranked as having the highest perfection among all the grown single crystals. Their lattice parameters together with their compositions indicate that the slight off-stoichiometry is not dominated by Ce on the Pd sites but by Pd vacancies, which do not directly disturb the $4f$ lattice.

$CeRu_4Sn_6$ melts incongruently. Single crystals of $CeRu_4Sn_6$ can be grown from Ru_2Sn_3 flux. The grown single crystals show no marked deviation from the ideal stoichiometry, which indicates a very narrow homogeneity range of the phase. On the grown crystals anisotropies of the magnetic properties of two types were demonstrated: along the tetragonal unit cell axes and along the axes of a quasi-cubic unit cell.

Single crystal growth of CeAuGe is complicated by a wide homogeneity range of the phase. Growth from the stoichiometric melt yields single crystals with an essential deviation from the stoichiometry. The composition of single crystals varies strongly along the growth direction. The crystal composition depends complexly on the melt composition and on the crystallisation temperature (the latter is a function of the former). Growth from the off-stoichiometric melt $CeAu_{0.96}Ge_{1.04}$ is optimal for nearly stoichiometric crystals provided that the melt-to-crystal volume ratio is large enough for keeping the melt composition quasi-constant during the entire growth run.

6. Acknowledgement

We are grateful to A. Strydom and co-workers for feed rod preparation of the CeAuGe phases, to K. Neumaier for C_p measurements on $sc5$, and to H. Ossmer for resistivity measurements on $sc5$. We thank M. Waas and V. Peter for the SEM/EDX measurements and for metallographic sample preparation. The work was financially supported by the Austrian Science Foundation (project P19458-N16) and the European Research Council (Advanced Grant n°227378).

7. References

Aeppli G. and Fisk Z., Comments Condens. Matter Phys. 16, 155 (1992).

Ananthasivan K., Kaliappan I., Vasudeva Rao P.R., Sudha C., Terrance A.L.E.. J. Nucl. Mater., 305, 97-105 (2002)

Brünig E. M., Baenitz M., Gippius A. A., Paschen S., Strydom A. M., and Steglich F., Physica B 378Ű380, 839 (2006).

Chandrasekharaiah M., in *Binary Alloy Phase Diagrams*, 2nd ed., edited by T. Massalski (William W.Scott, Jr., USA, 1990).

Das I. and Sampathkumaran E. V., Phys. Rev. B 46, 4250 (1992).

Dung N., Haga Y., Matsuda T., Yamada T., Thamizhavel A., Okuda Y., Takeuchi T., Sugiyama K., Hagiwara M., Kindo K., Settai R., and Onuki Y., J. Phys. Soc. Jpn. 76, 024702 (2007).

Goto T., Watanabe T., Tsuduku S., Kobayashi H., Nemoto Y., Yanagisawa T., M.Akatsu, Ano G., Suzuki O., Takeda N., Dönni A., and Kitazawa H., J. Phys. Soc. Jpn. 78, 024716 (2009).

Gribanov A. V., Seropegin Y. D., and Bodak O. J., J. Alloys Compd. 204, L9 (1994).

Gribanov A. V., Rogl P., and Seropegin Y. D., in Noble metal systems, in *Landolt-Börnstein, New Series* (Springer, Berlin, 2006), Vol. IV/11B, p. 340.

Hashiguchi T., Takeda N., Kitagawa J., Wada N., Takayanagi S., Ishikawa M., and Mori N., J. Phys. Soc. Jpn. 69, 667 (2000).

Jones C.D.W., Gordon R.A., DiSalvo F.J., Pöttgen R., Kremer R.K. J. Alloys & Comp. 260, 50 (1997)

Kitagawa J., Takeda N., and Ishikawa M., Phys. Rev. B 53, 5101 (1996).

Kitagawa J., Takeda N., Ishikawa M., Yoshida T., Ishiguro A., Kimura N., and T.Komatsubara, Phys. Rev. B 57, 7450 (1998).

Kitagawa J., Takeda N., Sakai F., and Ishikawa M., J. Phys. Soc. Jpn. 68, 3413 (1999).

In: *Binary alloy phase diagrams*, T.B. Massalski, (Ed.), V. 3, 3255, William W. Scott, Jr., ISBN 0-87170-406-4.

Mhlungu B.M., Strydom A.M.. Physica B 403, 862 (2008)

Mitamura H., Tayama T., Sakakibara T., Tsuduku S., Ano G., Ishii I., Akatsu M., Nemoto Y., Goto T., Kikkawa A., Kitazawa H., J. Phys. Soc. Jpn. 79, 074712 (2010).

Nemoto Y., Yamaguchi T., Horino T., Akatsu M., Yanagisawa T., Goto T., O.Suzuki, Dönni A., and Komatsubara T., Phys. Rev. B 68, 184109 (2003).

Paschen S., in *Thermoelectrics Handbook* (ed. D. M. Rowe, CRC Press, Boca Raton, 2006), Chap. 15 (Thermoelectric aspects of strongly correlated electron systems).

Paschen S., Müller M., Custers J., Kriegisch M., Prokofiev A., Hilscher G., Steiner W., Pikul A., Steglich F., and Strydom A. M., J. Magn. Magn. Mater. 316, 90 (2007).

Paschen S., Laumann S., Prokofiev A., Strydom A. M., Deen P. P., Stewart J. R., Neumaier K., Goukassov A., and Mignot J.-M., Physica B 403, 1306 (2008).

Paschen S., Winkler H., Nezu T., Kriegisch M., Hilscher G., Custers J., Prokofiev A., Strydom A. J. Phys.: Conf. Ser. 200, 012156 (2010)

Pöttgen R., Bormann H., Kremer R.K., J. Magn. Magn. Mater. 152, 196 (1996)

Prokofiev A., Custers J., Kriegisch M., Laumann S., Müller M., Sassik H., Svagera R., Waas M., Neumaier K., Strydom A. M., Paschen S., Phys. Rev. B 80, 235107 (2009).

Seropegin Y. D., Gribanov A. V., Kubarev O. L., Tursina A. I., and Bodak O. I., J. Alloys Comp. 317-318, 320 (2001).

Sondezi-Mhlungu B.M., Adroja D.T., Strydom A.M., Paschen S., Goremychkin E.A.. Physica B 404, 3032 (2009)

Strydom A. M., Pikul A., Steglich F., and Paschen S., J. Phys.: Conf. Series 51, 239 (2006).

Takeda N., Kitagawa J., and Ishikawa M., J. Phys. Soc. Jpn. 64, 387 (1995).

Venturini G., Chafik El Idrissi B., Maréché J., and Malaman B., Mater. Res. Bull. 25, 1541 (1990).

Permissions

The contributors of this book come from diverse backgrounds, making this book a truly international effort. This book will bring forth new frontiers with its revolutionizing research information and detailed analysis of the nascent developments around the world.

We would like to thank Nikolai N. Kolesnikov, for lending his expertise to make the book truly unique. He has played a crucial role in the development of this book. Without his invaluable contribution this book wouldn't have been possible. He has made vital efforts to compile up to date information on the varied aspects of this subject to make this book a valuable addition to the collection of many professionals and students.

This book was conceptualized with the vision of imparting up-to-date information and advanced data in this field. To ensure the same, a matchless editorial board was set up. Every individual on the board went through rigorous rounds of assessment to prove their worth. After which they invested a large part of their time researching and compiling the most relevant data for our readers. Conferences and sessions were held from time to time between the editorial board and the contributing authors to present the data in the most comprehensible form. The editorial team has worked tirelessly to provide valuable and valid information to help people across the globe.

Every chapter published in this book has been scrutinized by our experts. Their significance has been extensively debated. The topics covered herein carry significant findings which will fuel the growth of the discipline. They may even be implemented as practical applications or may be referred to as a beginning point for another development. Chapters in this book were first published by InTech; hereby published with permission under the Creative Commons Attribution License or equivalent.

The editorial board has been involved in producing this book since its inception. They have spent rigorous hours researching and exploring the diverse topics which have resulted in the successful publishing of this book. They have passed on their knowledge of decades through this book. To expedite this challenging task, the publisher supported the team at every step. A small team of assistant editors was also appointed to further simplify the editing procedure and attain best results for the readers.

Our editorial team has been hand-picked from every corner of the world. Their multi-ethnicity adds dynamic inputs to the discussions which result in innovative outcomes. These outcomes are then further discussed with the researchers and contributors who give their valuable feedback and opinion regarding the same. The feedback is then collaborated with the researches and they are edited in a comprehensive manner to aid the understanding of the subject.

Apart from the editorial board, the designing team has also invested a significant amount of their time in understanding the subject and creating the most relevant covers. They scrutinized every image to scout for the most suitable representation of the subject and create an appropriate cover for the book.

The publishing team has been involved in this book since its early stages. They were actively engaged in every process, be it collecting the data, connecting with the contributors or procuring relevant information. The team has been an ardent support to the editorial, designing and production team. Their endless efforts to recruit the best for this project, has resulted in the accomplishment of this book. They are a veteran in the field of academics and their pool of knowledge is as vast as their experience in printing. Their expertise and guidance has proved useful at every step. Their uncompromising quality standards have made this book an exceptional effort. Their encouragement from time to time has been an inspiration for everyone.

The publisher and the editorial board hope that this book will prove to be a valuable piece of knowledge for researchers, students, practitioners and scholars across the globe.

List of Contributors

Chaoyang Tu, ZhenYu You, Jianfu Li, Yan Wang and Zhaojie Zhu
Key Laboratory of Photoelectric Materials Chemistry and Physics of CAS, Fujian Institute of Research on the Structure of Matter, Chinese Academy of Sciences, P.R. China

Irina Nicoara and Marius Stef
West University of Timisoara, Timisoara, Romania

Aco Janićijević
Faculty of Technology and Metallurgy, Belgrade, Serbia

Branislav Čabrić
Faculty of Siences, Kragujevac, Serbia

Lukáš Válek
ON Semiconductor Czech Republic, Czech Republic
Institute of Physical Engineering, Brno University of Technology, Czech Republic

Jan Šik
ON Semiconductor Czech Republic, Czech Republic

Lihe Zheng, Liangbi Su and Jun Xu
Shanghai Institute of Ceramics, Chinese Academy of Sciences, P. R. China

Waldemar Wołczyński
Institute of Metallurgy and Materials Science, Polish Academy of Sciences, Poland

Morteza Asadian
Iranian National Center of Laser Science and Technology, Tehran, Iran

Grzegorz Boczkal
AGH-University of Science and Technology, Faculty of Non-Ferrous Metals, Cracow, Poland

Kyoichi Kinoshita and Shinichi Yoda
Japan Aerospace Exploration Agency, Japan

O.V. Avdeev
Nitride Crystals Ltd., Saint-Petersburg,, 194156, Russia

Andrey Prokofiev and Silke Paschen
Institute of Solid State Physics, Vienna University of Technology, Wiedner Hauptstr. 8-10, 1040 Vienna, Austria

Printed in the USA
CPSIA information can be obtained
at www.ICGtesting.com
JSHW011458221024
72173JS00005B/1122